Public Provision and Performance

PUBLIC PROVISION AND PERFORMANCE

CONTRIBUTIONS FROM EFFICIENCY AND PRODUCTIVITY MEASUREMENT

Edited by Jos L.T. Blank

Social and Cultural Planning Office of The Netherlands

With contributions from C.A. Knox Lovell, Shawna Grosskopf e.a.

Emerald

United Kingdom – North America – Japan
India – Malaysia – China

Emerald Group Publishing Limited
Howard House, Wagon Lane, Bingley BD16 1WA, UK

British Library Cataloguing in Publication Data
A catalogue record for this book is available from the British Library

ISBN: 978-0-44-450483-8

INVESTOR IN PEOPLE

Preface

In 1989 the Social and Cultural Planning Office of the Netherlands (SCP) published *Efficient Services* (*Doelmatig dienstverlenen*: Goudriaan et al. 1989). In this publication we presented the results of a research project on the productivity and efficiency of homes for the elderly, police departments, public libraries and secondary schools in the Netherlands. Since then the SCP has published several studies on the productivity and efficiency of other public services, such as the courts, prisons, primary schools, nursing homes and hospitals.

In recent years, efficiency improvement has been an important issue in government policy in The Netherlands as well in many other countries. Although the SCP has published extensively on this matter, it turned out to be difficult for governments to translate the results into concrete policy. From our foreign colleague-researchers we know that we share this experience with them. There is a huge gap between the large pile of scientific research in the field of efficiency measurement and the limited number of cases in which the outcomes of the research has been actually applied to government policy.

One of the major problems seems to be the accessibility of this type of research to policy makers. The SCP is therefore planning to 'popularize' research of this type and to stress its importance. This means that the accessibility of research must improve and that the knowledge of the subject among policymakers and management of public providers must be increased. It was for these reasons that, in 1997, the SCP and the Institute for Research on Public Expenditure (IRPE) organized a seminar for policymakers and others with an interest in the efficiency of public provision. The target group showed a great interest in the seminar and the SCP received many positive reactions to this meeting. Therefore, the SCP decided to continue its efforts to bring the subject under attention of a broader audience. The idea came up to publish a volume of papers on efficiency measurement of public service provision. The SCP invited a

number of well-recognized authorities in the field to contribute to the volume. They were requested to write a state-of-the-art paper in a non-technical way on a specific part of the public sector. All the authors we approached accepted our invitation and the results of their efforts are now before you.

The book contains a set of very interesting papers on a variety of public services and using a variety of techniques. The SCP acknowledges all the contributors for their efforts and their excellent papers.

Prof. Dr. P. Schnabel
Director of the Social and Cultural Planning Office of the Netherlands

Acknowledgements

Editing this book has been one of the most interesting activities in my research career so far. It' s a great challenge to deal with a relatively large number of authors from many different countries. However, the authors made my life easy. They all met the dead lines easily and followed the author's instructions carefully. I am indebted to all of them for their cooperation. Many authors stimulated me in my work with their enthusiastic reactions and comments. Some of them offered me assistance with my editing job. Hal Fried and Kris Kerstens commented some chapters. Kris also provided me with many literature suggestions for the concluding chapter of the book. Besides her own contributions to the book, my colleague Evelien Eggink, assisted me with compiling the tables, figures and all the other odd jobs. I am very grateful to them.

Jos Blank
October, 1999

Abstract of the Book

Chapter 1 discusses the objectives of the book and presents an outline. Governments are searching for possibilities to improve efficiency and reduce cost of the public sector services. Privatizing, deregulation, changing property rights have become popular policy items in recent politics. Although there has been strong progress in methodological issues and the availability of reliable data, there is still a huge gap between efficiency research and the policy implementation of the results. The aim of this book is, then, to try to bridge this gap. The book provides a summary of recent research, an overview of methods, an insight of factors affecting efficiency, a relativisation of the findings and possible policy recommendations. One of the major problems of translating the outcomes of efficiency analysis has probably been the accessibility of the research to non-economists. The book tries to 'popularize' this type of research and to do more to underline its importance. This means improving the accessibility of the research and increasing the familiarity with the subject in policy and management circles.

Chapter 2 begins by providing a menu of eight analytical frameworks which can be used to evaluate the efficiency with which individual public sector agencies provide public services. The choice among the alternative frameworks is determined by the data available to the analyst, and by the behavioral objective deemed most appropriate for the service providers. It also provides a discussion of the characteristics of service delivery technology that influence the overall performance of the service delivery network. It concludes with a discussion of two techniques that are often used to conduct an empirical evaluation of the efficiency of public sector service delivery. Special attention is devoted to the possibility of distinguishing variation in service delivery efficiency from variation in the operating environment in which public services are provided.

Chapter 3 examines how different definitions of inputs and outputs affect measures of efficiency. Using a rich Australian private hospital data set collected by the Australian Bureau of Statistics, the impact of different input and output definitions is analyzed

when using Data Envelopment Analysis (DEA) and Stochastic Frontier Analysis (SFA). Both techniques were quite sensitive to changes in input and output definitions. For example, even a small change in an input definition can produce quite different results. The results suggest that when analyzing the efficiency of service providers it is important to provide sound reasons for choosing one definition of inputs and outputs rather than another, and to examine whether different definitions will lead to different conclusions.

Chapter 4 aims to gain insight into the complexity of the production structure of Dutch hospitals. It provides empirical measures of hospital efficiency and simulations that are useful for formulating public policy. The empirical analyzes were performed on a sample containing 90% of Dutch hospitals in the period from 1985-1995. The main empirical conclusions of the study are that hospitals behave as cost minimizers rather than revenue maximizers, and that the possibilities to improve efficiency are modest. Collectively, efficiency measures such as increasing cooperation between health care institutions and decreasing the sickness rate could at best lead to a savings in variable costs of about 4%. There are small diseconomies of scale and technical changes that have had hardly any effect on costs. Moreover, the production of medical research and the education of physicians grows rapidly and has a large effect on hospital costs. As in Chapter 3, this chapter also stresses the importance of employing and comparing various empirical models, differing in measures of output and in assumptions on economic behavior and empirical methods. Without such a robustness-test, the efficiency results could be misleading and suggest incorrect policy measures.

Chapter 5 explores the possibilities of improving existing low capacity utilization among French surgery units by reallocating over the sector. This is done using an industry production frontier whereby existing capacity limits are imposed at first. In addition, a model variation is considered in which short-run capacities are ignored. Finally, these models are adapted to the regional organization of the French health care system. To illustrate their policy relevance, these models are applied to a sample of 88 French hospitals in the region Nord/Pas-de-Calais. The principal goal is to maximize the savings in all types of surgery rooms. The underlying concern being that a critical number of interventions is a prerequisite to attain a minimal level of quality. Whether such goal is compatible with the current industry allocation of variable resources is verified. In France, it is widely believed that surgeons are a critical resource, especially since the effects of a numerus clausus become apparent. The empirical results show, however, that the current allocation of personnel is more than sufficient at the industry level. Simultaneously, important savings can be achieved by eliminating about 15% to 29% of surgical rooms in the short respectively long run. These are mainly rooms with below

average equipment. Furthermore the industry reallocation allows closing down surgical activities in about 12% of hospitals in the short term.

Chapter 6 evaluates the efficiency of Dutch nursing homes in the period 1984-1993. Special attention is paid to the impact of different resource price measures on DEA-cost efficiency. The mean cost inefficiency varies between 17% to 29%, depending on the year of observation and the way input prices are measured. About 6% inefficiency is related to the size of production, 7-10% to a non-optimal mix of resources and 11-13% to technical issues. The scale inefficiency shows that by enlarging the scale of production of small homes, some savings can be achieved. A second step analysis shows that only one third of the inefficiency can be explained by background characteristics. Only a very small part of the inefficiency can be reduced to managerial or process variables. The main conclusion is that only small efficiency gains can be achieved. The largest part of inefficiency is still unexplained and most of the explanatory variables are out of the control of the government or management. The results further indicate that the efficiency scores for individual homes may change considerably when a different price definition is chosen. In addition, the mean efficiency may change significantly if another set of prices is used. However, the relationship between efficiency and characteristics of the homes is only marginally affected by a different choice of prices.

Chapter 7 explores the possible gains of a budget reform for the Texan school system. It is argued that the public school system of the United States is both regulated and inefficient. If resource decisions are distorted under the current regulatory regime, allowing resource choice should result in improved performance from the same total resources currently used in education. In this Chapter the authors discuss a recently developed model of performance to simulate the potential gains in school efficiency from reducing resource regulations. They determine best-practice frontiers for Texas school districts, first given the observed allocations of labor and nonlabor resources, and then given resource prices and a budget constraint but no other restrictions on their allocation of labor and nonlabor resources. The difference between the two best-practice frontiers represents the potential gains from reallocation of school district resources, and thereby captures allocative inefficiency.

Chapter 8 studies the relationship between school size and costs in Dutch Primary Education. First, the cost structure of the Dutch primary schools is analyzed. A cost function is estimated relating costs to production and prices of labor, material and capital inputs. The estimates indicate that there are considerable economies of scale for small primary schools. However, these economies of scale decrease quickly as school size increases. Costs per pupil are lowest if schools have close to 500 pupils. In 1990

many schools were below that optimum. A policy of school consolidation was implemented to bring costs down. A simulation study was made of the effects of various minimum size standards (closure norms) on the number of schools and costs. It follows that only limited cost reductions of approximately 200 million guilders (or about 3%) can be attained by more stringent closure norms, by closing up to 1300 schools (16%). A further reduction of the number of schools would not yield a higher cost reduction, but would increase transportation costs considerably or diminishes the school choices of parents. The simulation model has been implemented in a policy of school consolidation in the years 1992-1997.

Chapter 9 uses panel data and a *best practice* short run cost function to investigate cost inefficiency in the provision of custodial and security services in Florida prisons. Using different specifications, the findings show that there is a significant cost inefficiency in the provision of custodial and security services in Florida prisons. The estimated cost inefficiency in this chapter is on the same order of magnitude as those estimated by earlier researchers who use data from the public sector or the private sector. Private prisons are no more cost efficient than those operated by the Florida Department of Corrections. The existence of cost inefficiency implies that Florida can provide more prison services with existing resources rather than add more resources. However, privatization of prisons may not be the answer to increasing efficiency in Florida prisons.

Chapter 10 presents a model of the production structure of the Dutch police. It tries to fit the institutional structure and differentiates among to the various outputs of the police. Police departments are regarded as producers which, given their fixed budget and given the fixed quantity of some outputs (e.g. recorded crimes and traffic accidents) maximize the value of their endogenous outputs (the 'revenues') according to a scheme of shadow prices (their value system). Endogenous outputs are crime solutions of different types and the processing of minor offences. Shadow prices of the endogenous outputs are based on the sanction levels, measured by the average severity of punishment in cases in which punishment took place. This makes it possible to formulate and estimate a revenue function of 148 municipal police departments in 1983 and 1986. The estimated revenue function is used to derive various characteristics of the production structure, e.g. the optimal scale of production. The average revenue increases with the size of the department up to a certain level and decreases thereafter. The optimal scale for producing the mean exogenous output mix is around a budget of 18 million guilders or 260 labor years. Some simulations using the estimated revenue function are presented. The estimated effect on productivity of a large restructuring operation of the Dutch police, which took place recently, is negative.

Chapter 11 analyzes the relative performance of local police districts in New South Wales, Australia, using Data Envelopment Analysis (DEA). Unlike other studies, this chapter includes both law enforcement work and crime prevention activities as measures of police services. This chapter develops a suite of sensitivity tests to enhance the usefulness of DEA as a management tool. These sensitivity tests combine statistical methods with qualitative assessments from police experts. The paper demonstrates the sensitivity tests in the context of four DEA model specifications. It also includes an example of how an inefficient police district could use DEA results to improve its performance.

Chapter 12 serves three purposes. First, it evaluates the conceptual, methodological and practical problems to municipal efficiency. Second, the chapter reviews what is known about municipal efficiency in Belgium. Third, the findings are compared to the results available in the international literature to obtain information on the robustness of the determinants of municipal efficiency. Both results on the global performance and on the efficiency of specific services are reviewed. This chapter emphasizes the necessity and the difficulty of global performance evaluation at the level of municipal governments. While the global efficiency measures should not be used for ranking individual municipalities or to formulate policies towards specific local services, the attempts to unveil causal mechanisms is certainly worthwhile going for. However, despite huge problems in defining proper inputs and outputs of the municipal production process some consistent results do follow from the available empirical studies. At the global level the fiscal revenue capacity, the financing mechanisms of local governments, the political characteristics of local governments and the political participation of the local population all are important determinants of efficiency. Specifically, it is consistently found that citizens' average income and the size of per capita grants negatively affect municipal performance. Local tax rates and the educational achievements of the population stimulate efficiency. Moreover, it was shown that political factors are important as well.

Chapter 13 uses data envelopment analysis to evaluate how well the Capital Region of New York State provides government services compared to a sample of 290 metropolitan areas in the U.S. The outcomes show that the Capital Region lies inside the best-practice frontier. By evaluating the Capital Region's efficiency in a succession of models, a long list of dominating regions is identified. These regions are like the Capital Region in that they produce similar levels of government services, but unlike the Capital Region in that they produce these services with lower expenditure. In consultation with New York State and Capital Region local government officials, five regions for intensive case studies were selected in order to identify specific operational prac-

tices that might underlie their superior performance, and which might successfully transfer to the Capital Region.

Chapter 14 concludes the book and provides the reader with some retrospections. Five major issues are being discussed. First, it is stressed that the public sector has a large impact on national economy. An efficient public sector may easily improve GNP by more than 5%. Second, much efficiency research has been conducted, but unfortunately has not been implemented. There is only a small number of examples of successfully implementing research outcomes into concrete policy. Third, theories on efficiency measurement contain advanced mathematics and statistics. The lack of transparency of the methodologies used prevents a widespread use of efficiency measurement amongst policy makers and managers. Nevertheless, various examples in the book show the importance of a valid choice of model specification, a valid choice of empirical technique and a proper definition of service provision and resource usage. The cumbersome theories and techniques are more than just scientific hobbyism. Fourth, efficiency analysis can be beneficiary in various ways. It may be helpful in improving managerial processes, but also in determining financing systems, planning of capacities and choosing the right system of property rights. Fifth, much of the validity of the efficiency measurement depends on the data availability. Much improvement can be accomplished in this area, especially where it comes to the measurement of the quality of service provision and the environmental characteristics.

Contents

Part One: Introductory and Methodological Issues

1 Performance Assessment in the Public Sector

Jos L.T. Blank and C.A. Knox Lovell

1.1 Introduction

Economists, management scientists and public administration scholars have long been interested in the assessment of the performance of the public sector, and in the identification of the likely determinants of performance variation, both across jurisdictions and through time. Their interest is well placed, and motivated by a number of factors.

Their interest is well placed because the sheer size of the public sector in most economies means that its performance has a significant impact on the overall performance of the economy. The impact shows up in two ways: in the final outputs provided to the consumption sector (e.g., health and education), and in the intermediate outputs used as inputs in the private production sector (e.g., energy and infrastructure).

Research interest in public sector performance measurement is motivated by a number of factors:

1. settling on an appropriate definition of performance;
2. revealing the consequences of the economic environment and economic policy on performance;
3. overcoming severe measurement problems.

An initial motive is based on the practical challenge of agreeing on an appropriate *definition* of performance in the public sector, where the bottom line is not as relevant as it is in much of the private sector. In the public sector the profitability yardstick is frequently replaced by the notions of efficiency, effectiveness and productivity change. Efficiency refers to the degree to which service provision is maximized, given the resources at hand. Depending on the circumstances, efficiency can also be measured

from the opposite orientation, as the degree to which resource consumption is minimized in order to satisfy service demand. Effectiveness refers to the concordance of the service mix with the stated objectives of the agency. Productivity change refers to improvement or deterioration through time in the ratio of service provision to resource consumption.

A second set of motives is theory-based, and derives from the proclaimed *consequences* of different competitive conditions facing public sector and private sector managers, of differences in ownership structure in the public and private sectors, and of differences in the objectives and constraints facing public sector and private sector managers.

The first two considerations lead to the expectation that, where both exist as a result of the lack of incentives public service provision will tend to be inferior to private service provision. This expectation has encouraged recent privatization movements in several countries. The third consideration, however, alerts us to the difficulty of developing a framework that enables fair comparisons of public and private service provision to be made.

A third motive arises from the challenges posed by the difficult *measurement* problems in the public sector. There is frequent disagreement about how to define the services being provided. And even when agreement is reached, it is frequently difficult or impossible to obtain reasonably accurate measures of the services being provided. Finally, the services are often unpriced, rendering conventional index number techniques useless as performance indicators. Education, health care and public safety provide good examples of all three problems.

It is easy to draw two conclusions from the foregoing discussion. First, for a number of theoretical and practical reasons, performance assessment within the public sector is a challenging exercise, certainly more so than within much of the private sector. Performance assessment involving both public and private service providers poses an even greater challenge. Second, because the public sector looms so large in most economies, because of the suspicion of greater waste than in the private sector, and consequently because of the growing trend toward privatization in many sectors in many countries, public sector performance assessment is a potentially productive exercise that is likely to reap significant returns.

Prompted by these challenges, many economists have done research in this field in the past three decades. Researchers have published extensively on performance in the provision of a wide range of public sector services. Yet getting these findings converted

into concrete policy measures has proved something different altogether. Probably one of the largest hurdles is to make such research accessible to non-specialists. The complexity of the theories and the advanced statistical methods have frightened off the policymakers. Despite all the efforts to translate research findings into practical policy recommendations, they have been implemented only sporadically. Since substantial savings can be expected from efficiency improving measures, this is very regrettable. In Australia, for example, the Industry Commission estimates that reforms associated with the National Competition Policy will increase GDP by 5.5% in the long run. Various efficiency studies show that in other countries comparable results are probably attainable.

This book contains a representative sampling of some of the most recent evidence on public sector performance assessment. It is intended to 'popularize' this type of research and to underline its importance. This means that this book endeavors to improve the accessibility of the research, and to promote familiarity with the research methodologies in public sector policy and management circles.

The aims of the book are therefore:

1. to summarize recent research;
2. to create an accessible overview of methods;
3. to develop better insight into the underlying factors potentially affecting productivity and efficiency;
4. to put the findings of efficiency analyses into perspective;
5. to demonstrate the potentials of translating these findings into policy.

We hope that the readership will consist mainly of policymakers and public service managers, although we believe that the material is also suitable for students of public administration and economics. However, some knowledge of standard micro-economic theory is required.

This introductory chapter is organized as follows. In Section 1.2 we discuss some of the characteristics that make the public sector special from the private sector. In Section 1.3 we discuss the concept of productive efficiency. In Section 1.4 we examine data issues that plague performance assessment in the public sector and empirical comparison between the public sector and the private sector. In Section 1.5 we offer some suggestions about what can be done with the results of a public sector performance evaluation. We close with a brief outline of the rest of the book in Section 1.6.

1.2 What Is So Special About the Public Sector?

Several distinguishing characteristics make the public sector special. These characteristics combine to make an interjurisdictional and/or intertemporal performance assessment within the public sector a challenging exercise, and they make a comparison between public and private service provision exceedingly difficult.

First and foremost, by almost any definition the public sector is large, arguably too large, and growing. Without getting into polemics, we recommend the historical overview of the debate concerning the appropriate size of the public sector by Stigler (1965), who traces the debate back to Adam Smith.

There are several alternative indicators of the share of the public sector in the aggregate economic activity of a nation, and all would tell roughly the same story. Table 1.1 summarizes OECD calculations of the ratio of government tax revenue to GDP in 1970 and 1995 in a number of member countries. On this definition the share of the public sector in aggregate economic activity is large on average, and growing in most countries; only in the United Kingdom has the indicator declined.[1] It also seems that there has been some convergence in this indicator. Countries with low shares in 1970, like Portugal and Spain, show a large growth in public sector activities. It follows that an improvement in the performance of what has historically been public sector service provision has the potential to bring large benefits. Whether these benefits can best be realized through improved performance within the public sector or by means of privatization is largely a political issue that we intend to ignore in this study.

A *second* characteristic distinguishing the public sector from the private sector is a different ownership form. Private sector firms are owned by their stockholders, whereas public sector entities are "owned" by the public. This has three implications. First, as Alchian (1965) and subsequent writers have argued, the greater attenuation of property rights and the effective nontransferability of ownership among public sector "owners" both suggest that performance monitoring may be more difficult, and performance incentives consequently weaker, in the public sector than in the private sector. In addition, the principal-agent problems that plague the private sector are apt to be even more severe in the public sector, where "owners" tend to be dispersed, unorganized and unable to transfer ownership shares, and where incumbent managers tend to be protected by civil service legislation. Finally, it is argued that relatively modest civil service pay scales tend to deter the best and the brightest from entering public service, and that this tends to dilute the quality of public sector management. To the extent that these three implications are generally valid, they raise two interesting questions.

Table 1.1

Tax / GDP Ratios (%) in selected OECD Countries

	Tax/GDP		
	1970	1995	% change
Australia	24,2	30.9	27.7
Austria	35.7	42.4	18.8
Belgium	35.7	46.5	30.3
Canada	31.3	37.2	18.8
Denmark	40.4	51.3	27.0
Finland	32.5	46.5	43.1
France	35.1	44.5	26.8
Germany	32.9	39.2	19.1
Italy	26.1	41.3	58.2
Japan	20.0	28.5	42.5
Netherlands	37.0	44.0	18.9
New Zealand	27.4	38.2	39.4
Portugal	16.2	33.8	108.6
Spain	16.9	34.0	101.2
Sweden	39.8	49.7	24.9
UK	36.9	35.3	-4.3
USA	27.4	27.9	1.8
OECD mean	29.5	37.4	26.8

First, are public sector "owners" more or less likely to vote under-performing civil servants out of office than private sector owners or corporate raiders are to oust under-performing managers or boards of directors? Second, if a privatization option is followed in place of a reform option, will the potentially most profitable activities be privatized, leading to the creation of a forever unprofitable public sector?

A *third* characteristic distinguishing the public sector from the private sector is the differences in objectives and constraints facing managers in the two sectors. It is easy to think of private sector managers as pursuing conventional textbook objectives subject to conventional textbook constraints. However, it is much more difficult to do so for public sector managers. This is because profit maximization is not an appropriate behavioral objective in the public sector. In the private sector the pursuit of profit is commonly thought to reward good performance and to punish poor performance, the

more so as markets are more competitive. But public sector managers are presumably not motivated by profit, and indeed their objectives and the constraints under which they operate are difficult to identify. Consequently as Fox (1998) has observed, public sector managers do not have comparable control over the mix of services they provide, and so are constrained in their ability to allocate resources and services in an efficient manner. Stated somewhat differently, the service vector is not as endogenous in the public sector as it is in the private sector. Moreover, civil servants do not enjoy the property rights to any residual earnings they may generate.

However, what is an appropriate behavioral objective for public sector managers is an open question. Rees (1984) lists four distinct economic objectives in the public sector: allocative, distributive, financial and macroeconomic. Tulkens (1992) associates the allocative objective with efficiency aspects of public service provision, and assigns priority to the pursuit of efficiency. Although there are bound to be tradeoffs among the different objectives, since our focus is on efficiency and its correlates, we concentrate on Rees' allocative objective. However, the assessment of efficiency requires a framework, a model incorporating objectives and constraints. There are many to choose from.

One possibility would be to assume that public sector managers seek to maximize service provision from given resources. This assumption has the advantage of being quantity-based, and so relatively undemanding of the data, and of being sufficiently general to facilitate comparison between public and private service provision. However in our judgement a more appealing formulation would be to view public sector managers as budget-constrained service maximizers who have (perhaps limited) freedom to substitute among resources. This assumption would also be relatively undemanding of the data, and would also allow for a public-private performance comparison. However such a formulation has been underutilized as a means of assessing public sector performance. One likely reason is that public sector managers frequently operate under additional constraints not typically faced by their private sector counterparts. They face a variety of community service obligations, and occasionally operate under "universal service, uniform price" constraints, for example. A fair assessment of public sector performance requires that these and other constraints be incorporated into the analysis. However it is also possible to analyze and adjust for the impact of these characteristics of the operating environment subsequent to the analysis. This latter approach has the advantage of allowing the use of either the resource-constrained or the budget-constrained service maximization framework, without losing sight of variation in the operating environment across public and private sectors.

Much like the pursuit of profit, the forces of competition are thought to reward good performance and to punish poor performance, and so a *fourth* characteristic distinguishing the public sector from the private sector concerns the consequences of the nature of competition facing public sector managers. Years ago Hicks (1935) warned that managers unconstrained by the rigors of competition "...are likely to exploit their advantage much more by not bothering to get very near the position of maximum profit, than by storming themselves to get very close to it. The best of all monopoly profits is a quiet life." Although Hicks was referring to private sector monopolists, his admonition is equally applicable to public sector managers undisciplined by the forces of competition. It is frequently the case that public sector managers face less competitive service markets than their private sector counterparts do, although limited inter-jurisdictional competition may exist, as may competition from the private sector as well. Although some consumers of some publicly provided services can vote with their feet, Hirschman's (1970) exit option is constrained by the presence of so many public monopolies. Competition issues thus reinforce the expectation that public service provision tends to be less efficient or less effective than private service provision.

The third and fourth characteristics overlap. Just as various constraints specific to the public sector limit the ability of public sector managers to pursue conventional efficiency objectives, the lack of competition in the public sector provides an opportunity to pursue alternative objectives that are inconsistent with efficiency objectives. Limited competition in the public sector allows the pursuit of a variety of self-serving objectives, and has led to the formulation of alternative hypotheses concerning managerial behavior and the consequences of such behavior. One such hypothesis was prefered by Niskanen (1971), who somewhat uncharitably thought of public sector managers as utility-seeking budget maximizers, rather than as service provision maximizers. This view of public sector managers is reminiscent of Williamson's (1964) managerial discretion hypothesis concerning the behavior of private sector managers. Closely related models proposed by de Alessi (1969) and others view public sector managers as adopting inefficient resource mixes, both to exploit the lack of competitive pressure and to satisfy self-serving objectives. Yet another view of public sector managers was put forth by Lindsay (1976), who views them as concentrating on the production of "visible" outputs at the expense of potentially more valuable but less visible outputs. Thus while competition issues lead to the expectation that public sector service provision is less efficient than private sector service provision, Lindsay's thesis questions the effectiveness of public sector decision-making. The behavioral implications of these and other models of bureaucracy are compared by Orzechowski (1977).

A *final* characteristic distinguishing the public sector from the private sector involves the challenges posed by generally more difficult *measurement problems* in the public sector. In many sectors it is difficult to reach agreement (both among analysts and, much more seriously, between analysts and practitioners) concerning appropriate definitions of the services being provided. Moreover, even when agreement can be reached, public services are frequently unpriced, and this severely limits the range of techniques that can be used to evaluate performance. Mintzberg (1996: 79) puts what he calls "the myth of measurement" into perspective, noting that "[m]any activities are in the public sector precisely because of measurement problems; if everything was so crystal clear and every benefit so easily attributable, those activities would have been in the private sector long ago."

Griliches (1994) has lamented the fact that the measurement of productivity has been made increasingly more difficult by the decline in what he calls the "reasonably measurable" sector of advanced economies. Government and other sectors that are frequently publicly or jointly owned are assigned to the "unmeasurable" sector. If public sector measurement issues plague productivity measurement at the macro level, they also plague monitoring and managerial accountability at the micro level. Thus measurement issues provide yet another reason to expect public sector performance to be inferior to private sector performance, although at the same time they bedevil empirical testing of the conjecture. We return to measurement issues in Section 1.4.

In the presence of differing competitive conditions, differing objectives and constraints, and data difficulties, how is public sector performance best evaluated, and perhaps compared with private sector performance? The many bureaucracy models cited above are demanding of data, idiosyncratic in their behavioral objectives and constraints, and consequently difficult to employ in a comparative evaluation. The resource-constrained or budget-constrained service maximization models are less demanding of data, less restrictive in their behavioral objectives and constraints, and consequently relatively easy to employ in a comparative evaluation. In this regard it is worth quoting at length from Pestieau and Tulkens (1993: 297), who advocate the evaluation of public enterprise performance primarily on the basis of efficiency, for two reasons: "...on the one hand, the global performance evaluation problem, i.e. measuring how close a public firm comes to achieving all the objectives just listed is, in our opinion, too ambitious. Both data and techniques of analysis currently available make such an undertaking unachievable. On the other hand, as productive efficiency allows for evaluations that are consistent with the manifold objectives of the firms at stake, it does definitely constitute a step in the right direction." This argument echoes a point raised by Smith and Mayston (1987: 187), who remind us that "[t]he resources released

by any improvement in technical efficiency can then be diverted to satisfy other objectives of the government."

We conclude this section with a dilemma. Each of the characteristics of the public sector leads to the same expectation, that public sector performance is likely to be less efficient than, or less effective than, private sector performance. Yet the empirical evidence is surprisingly mixed. Numerous studies summarized by Bös (1991), and also by Pestieau and Tulkens (1993), report conflicting findings, although the reported performance differentials tend to be relatively small in either direction.

1.3 Performance Measurement

In measuring the performance of public sector enterprises, one should bear in mind that performance is a relative concept. The performance of any service provider is always evaluated relative to that of other providers in the sample. Yet the possibility remains that the latest technology has not yet been fully incorporated by any of the providers, so that all of them could conceivably realize better performance. Although performance measurement is applied mainly to providers at the micro level, it is also possible to make comparisons at higher levels of aggregation, such as a branch of the public sector either through time or across regions.

One of the most popular concepts in performance measurement is productivity. In a single-resource single- service sector, productivity is measured as the ratio of service provided to resource consumed. However most public sector entities provide multiple services and use multiple resources. In such a case services and resources must be aggregated into quantity indexes. Ideally the service and resource quantity indexes would include service and resource prices to act as weights, but these are often missing in the public sector. Unfortunately some researchers are therefore tempted to use partial measures of productivity, such as the quantity of a single service provided per employee, or labor productivity. Although these are easy to compute and to understand, they yield a two-dimensional characterization of an inherently multidimensional problem. Moreover, alternative partial productivity measures can send conflicting signals concerning relative performance, and so they must be interpreted with extreme caution.

Productivity depends on the structure of the service provision technology, the efficiency with which the technology is implemented, and the characteristics of the operating environment in which service provision occurs. Obviously, providing health care in

1950 was very different from providing it in the year 2000. New technologies change the way services are provided. The operating environment describes the circumstances that providers have to reckon with which are beyond their control. Providing highway maintenance is more energy-intensive in cold and mountainous Norway than in mild and relatively flat Portugal, and so the productivity of highway maintenance is likely to be higher in Portugal.

Efficiency measurement involves a comparison between observed and optimal values of services and resources. It can be defined as the ratio of observed to maximum feasible service provision obtained from given resources, or vice versa. If the comparison is made in terms of service and resource quantities, the resulting efficiency measure is expressed in physical or "technical" terms. Another possibility is to define efficiency in terms of the economic goals and constraints of the service providers, such as cost minimization at given service quantities and resource prices. In this case the efficiency measure is expressed in monetary or "economic" terms. Clearly the definitions of services, resources, economic goals and constraints are crucial in this matter. A variety of analytical techniques have been applied to public sector performance assessment. Chapter 2 provides more information on them.

Efficiency measurement is used here as a tool for evaluating public sector performance. It is obvious that these techniques can be applied to other areas as well. Performance evaluations of branches of banks or insurance companies are also popular among researchers. A recent new research area is the efficiency measurement of environmental effects of agricultural production. (See, for example, Reinhard (1999) on Dutch dairy farms).

1.4 Data Issues

Much effort has been devoted to the empirical assessment of public sector performance, and to the empirical comparison of public sector and private sector performance. But the confidence we have in the findings of such assessments depends, or should depend, on the quality of the underlying data. There is ample reason to question the quality of the underlying data, and also to plea for improved quality. Griliches (1994) laments what he calls "the data constraint," and pleads for the allocation of additional resources to data collection. Tulkens (1992: 377) puts the plea in a different perspective, arguing that "...public enterprises can only survive when their citizen-owners are convinced that they actually do serve their purposes. In the absence of evaluative means, public ownership is all too easily put into question and subjected to criticisms

of all kinds.... On the contrary, well-structured information, based on well-designed and faithful indicators, obtained from detailed measurement, is the only way to provide transparency to the democratic debate on the merits and demerits of public undertakings."

Unfortunately the data constraint is a fact of life which is not likely to be relaxed to our satisfaction in the foreseeable future. It is particularly serious in the public sector, in three respects that hamper performance evaluation. First, it is frequently difficult to define the services being provided. We have already mentioned education, health care and public safety, and the list can easily be extended. If public sector output is thought of as including the ultimate public services which it provides (e.g. human capital, good health, police protection), then the measurement problem is insurmountable. On the other hand, if public sector output is thought of as a flow of intermediate public services (e.g. the number of students attending or graduating from school, the number of patients admitted to or discharged from hospitals, the number of arrests made or convictions obtained), the problem is manageable. Nonetheless disagreement concerning the most useful set of intermediate service indicators is bound to persist. Because of this difficulty, government output is typically measured in neither way, but rather is valued in terms of labor compensation (however in the cases of education and health care, excluding the opportunity cost of students' and patients' time). Consequently a bureaucrat can increase measured output merely by employing more labor. Moreover, on this definition productivity change in government is identically zero. As government grows, this accounting practice exerts a dampening impact on aggregate productivity growth (unless, of course, the skeptics are correct and productivity growth in government really is negative).

A second data constraint hampering public sector performance evaluation is that service prices which would serve as weights in the construction of an aggregate service quantity index are frequently missing (e.g. national defense, the "price" of which is buried in taxes), or distorted to serve other objectives (e.g. the post, where uniform prices grossly misrepresent marginal costs). Again, the list of examples can be extended. Thus even if we had service quantity indicators, without service prices to weight these indicators it would still be impossible to obtain a real value of public sector output from which to construct a public sector productivity index. Or, as Searle and Waite (1980: 335) put it, "...no one knows how to obtain an output index for any important government function except by measuring inputs."

Actually the situation is not quite that bleak. In some circumstances we have adequate service quantity data, and either budget data or resource quantity and price data. Given

enough such data (either across service providers or through time or both), it is possible to adopt the following strategy. First, select an analytical framework depending on the management's objectives and the boundaries of the environment, such as budget-constrained service maximization. Second, use the framework to infer service provision efficiency for each provider. Third, derive provider-specific shadow weights for the services whose prices are either missing or distorted. Finally, use these shadow weights to construct a service quantity index and a productivity index. Caves, Christensen and Swanson (1980) have used conventional econometric techniques to estimate a cost function and derive shadow weights for railroads whose service prices were distorted. Grosskopf, Margaritas and Valdmanis (1995) have used unconventional econometric techniques to estimate a distance function and derive shadow weights for hospitals whose service prices were missing. It is also possible to use linear programming techniques to meet the same objective, to derive shadow weights for services whose prices are either missing or distorted. These techniques are discussed in Chapter 2.

A third data constraint hampering public sector performance evaluation is that, particularly in the provision of services, quality matters. Thus it is desirable to include quality indicators together with quantity indicators when constructing a service quantity index. Unfortunately service quality is notoriously difficult to define, much less to measure. Consequently service quality indicators are reported even less frequently than are service quantity indicators. A second problem is that although service quality (e.g. hospital waiting time, electricity supply reliability) is valued by consumers, it is rarely priced. However given sufficient quantity and quality data, it would be possible to use the shadow weight construction techniques alluded to above to generate shadow weights for service quality indicators. This would provide all the information required to construct a productivity index incorporating service quality.

Despite these and other daunting data constraints, we offer the following data wish list. The list is intended to serve as a guide for public sector data collection efforts, with the ultimate objective of fostering a reliable and credible assessment of public sector performance. The wish list contains variables measuring:

- the quantities of the services provided;
- one or more indicators of service provision quality;
- the quantities of resources consumed, or an operating budget, or both;
- service prices if prices are actually charged;
- resource prices; and
- characteristics of the operating environment;

Ideally these data would be available through time, and updated on a regular basis. They would also be comprehensive (including all agencies providing the list of services), inclusive (not omitting any relevant services or resources) and consistent (in the definition of services and resources, both across agencies and through time). Since this is a wish list, we will probably have to accept less, at least in the foreseeable future. How much less will influence the methods we use to evaluate public sector performance, and the confidence we have in our findings. However this wish list also serves a secondary purpose. To the extent that public sector performance evaluation founders on inadequate data, it sends a useful signal to government to reallocate its resources toward improved data collection procedures.

1.5 After the Assessment...

Suppose that, despite the data constraint, an assessment of public service providers has been conducted, and that measured service delivery performance varies across providers. Two additional questions remain to be addressed. First, what is the explanation for the measured performance variation? Second, what is to be done with the findings? The quality and the usefulness of the answers to both questions are enhanced by cooperation between the analysts and representatives of the sector whose service providers have been evaluated.

There are two potential sources of performance variation. One is variation in managerial performance, and the other is variation in the operating environment in which services are provided. The initial performance assessment is not credible unless and until variation in the operating environment is either ruled out or accounted for. It is important to have information on such variables, because if a performance evaluation is conducted in their absence, managements of service providers operating in relatively unfavorable environments will rightly object that the performance evaluation is unfair.

The first objective of the exercise is to generate an assessment of the performance of service providers that distinguishes variation in managerial performance from variation in the environment in which services are provided. A good example of an effort to make this distinction is provided by Gathon and Pestieau (1995), who evaluated the performance of European railways. They were able to decompose initial railway performance evaluations into a component attributable to managerial performance and a component attributable to a feature of the environment in which managements make decisions, namely the degree of hiring, pricing and commercial autonomy granted managements by governments.

The second objective of the exercise is to use the findings to enhance overall performance. Both components – managerial and environmental variations – can be used for this purpose, and distinguishing between the two is essential for governments to identify the source of the problem. The managerial component refers to the way the production process is organized. Insights into the effects of organization may lead to adaptation of the production process. The environmental component consists partly of government regulation and government policy. Measured effects of these environmental variables on performance may result in recommendations to change government policy or regulation. Gathon and Pestieau (1995) concluded that the mediocre performance of the Finnish railways was due not so much to mediocre managerial performance as to the limited decision-making autonomy granted management.

Overall performance is most easily enhanced by improving the performance of the laggards. This requires identification of the laggards, identification of one or more relatively efficient providers who can serve as role models for each laggard, and effective communication between the two. Such communication can lead to the identification of managerial practices and environmental characteristics contributing to the failure of the laggard to approach best practice service delivery standards. Identification of laggards and their role models is an integral part of both empirical techniques discussed in Chapter 2.

1.6 Outline of the Book

As stated in Section 1.1, the aims of this book are:

1. to summarize recent research;
2. to create an accessible overview of methods;
3. to develop better insight into the underlying factors potentially affecting productivity and efficiency;
4. to put the findings of efficiency analyses into perspective;
5. to demonstrate the potentials of translating these findings into policy.

These aims are addressed by various case studies. The book is divided into six parts. Besides the general introductory and concluding chapters (part I and part VI), the book contains four other parts, each referring to a specific segment of the public sector. Part II addresses the performance of health care facilities (Chapters 3-6). Part III analyzes the performance of educational service providers (Chapters 7-8). Part IV contains studies of the performance of law enforcement agencies (Chapters 9-11) and part V provides analyses of the performance of regional service providers (Chapters 12-13). We have

chosen these specific sectors because they probably account for the bulk of previous studies of efficiency and productivity in the public sector. By further including extensive literature overviews in each chapter, we have tried to create a comprehensive picture of the state of the art of current research in this field.

Our selection of chapters has also been motivated by the variety of analytical frameworks and empirical techniques used. Chapter 2 discusses the most important frameworks and techniques. We have endeavored to include a broad range of countries to achieve some variation in economic environments and degrees of regulation. Differential economic environments may prompt different choices of frameworks and techniques. The wide variation among countries analyzed can also enrich the more general debate on policy issues such as privatization, reimbursement schemes, etc.

In most of the case studies presented here, performance analysis is not just a matter of measurement, but of explanation as well. We stress the importance of explaining performance variation, because this can deliver instruments to governments and public sector managers designed to improve performance. Unless performance variation is explained, the analyses are little more than empirical exercises leaving unopened the "black box" within which service delivery decisions take place. Several chapters in the book provide explicit examples of policy recommendations based on such explanations for observed performance variation.

Part II comprises four contributions evaluating the performance of health care facilities. Chapter 3 discusses the efficiency of Australian private hospitals, and Chapter 4 analyzes the efficiency of Dutch general and academic hospitals. Both chapters stress the robustness of the results of the efficiency analysis. Chapter 5 deals with the efficiency of the spatial allocation of hospital capacities in France. Chapter 6 evaluates the efficiency of Dutch nursing homes, likewise devoting special attention to the robustness of the results.

Part III contains two contributions assessing the performance of educational service providers. Chapter 7 shows how efficiency analysis can be used to estimate the potential efficiency gain from a relaxation of resource constraints imposed on school administrators. Chapter 8 presents findings of an efficiency analysis of Dutch elementary schools. It discusses in particular the use of a simulation model designed to predict the consequences of school consolidation.

Part IV includes three contributions evaluating the performance of law enforcement agencies. The effects of the privatization of prisons in South Florida in the U.S. are

reported in Chapter 9. Chapter 10 analyzes the efficiency of police departments in the Netherlands. Here some special economic behavior is assumed: the management of the police departments responds to the probability of conviction and the severity of punishment for different categories of crime. Chapter 11 deals with the efficiency of police patrols in New South Wales, Australia, with special attention devoted to the sensitivity of the findings to alternative specifications of the services provided and the resources used.

Part V provides a pair of analyses of the performance of regional service providers. Chapter 12 discusses the efficiency of Belgian municipalities, and shows that the complexity of the services sometimes inhibits a proper efficiency analysis. The efficiency of the counties in New York State in the U.S. is the subject of Chapter 13, with a special focus on identifying peers or role models for laggard counties.

Note

[1] Table 1.1 may give the wrong impression that there has been a steady growth over the years. However, in most countries the share stabilized or even decreased from the mid-eighties.

REFERENCES

Alchian, A.A. 1965. Some Economics of Property Rights. *Il Politico* 30 (4): 21-37.

Bös, D. 1991 *Privatization: A Theoretical Treatment*. Oxford: Clarendon Press.

Caves, D.W., L.R. Christensen, and J.A. Swanson. 1980. Productivity in U.S. Railroads, 1951-1974. *Bell Journal of Economics* 11 (1): 166-181.

De Alessi, L. 1969. Implications of Property Rights for Government Investment Decisions. *American Economic Review* 59 (1): 13-24.

Fox, K.J. 1998. Efficiency at Different Levels of Aggregation: Public vs. Private Sector Firms. Working Paper. Sydney, NSW 2052, Australia: School of Economics, University of New South Wales.

Gathon, H.J., and P. Pestieau. 1995. Decomposing Efficiency into its Managerial and its Regulatory Components: The Case of European Railways. *European Journal of Operational Research* 80 (3): 500-507.

Griliches, Z. 1994. Productivity, R&D, and the Data Constraint. *American Economic Review* 84 (1): 1-23.

Grosskopf, S., D. Margaritas, and V. Valdmanis. 1995. Estimating Output Substitutability of Hospital Services: A Distance Function Approach. *European Journal of Operational Research* 80 (3): 575-587.

Hicks, J.R. 1935. The Theory of Monopoly: A Survey. *Econometrica* 3 (1): 1-20.

Hirschman, A.O. 1970. *Exit, Voice and Loyalty: Responses to Decline in Firms, Organizations and States*. Cambridge, MA: Harvard University Press.

Lindsay, C.M. 1976. A Theory of Government Enterprise. *Journal of Political Economy* 84 (5): 1061-1077.

Mintzberg, H. 1996. Managing Government, Governing Management. *Harvard Business Review* (May/June): 75-83.

Niskanen, W.A. 1971. *Bureaucracy and Representative Government*. Chicago: Aldine Publishing Co.

Orzechowski, W. 1977. Economic Models of Bureaucracy: Survey. Extensions and Evidence. In: Borcherding, T.E. (ed.), *Budgets and Bureaucrats: The Sources of Government Growth*. Durham, NC: Duke University Press.

Pestieau, P., and H. Tulkens 1993. Assessing and Explaining the Performance of Public Enterprises. *Finanz Archiv* 50 (3): 293-323.

Rees, R. 1984. *Public Enterprise Economics*, second edition. London: Weidenfeld and Nicholson.

Reinhard, S. 1999. *Econometric Analysis of Economic and Environmental Efficiency of Dutch Dairy Farm*. Wageningen: Wageningen Agricultural University.

Searle, A.D., and C.A. Waite 1980. Current Efforts to Measure Productivity in the Public Sector: How Adequate for the National Accounts? In: Kendrick, J.W., and B.N. Vaccara (eds.), *New Developments in Productivity Measurement and Analysis*. Chicago: University of Chicago Press for the National Bureau of Economic Research.

Smith, P., and D. Mayston 1987. Measuring Efficiency in the Public Sector. *Omega* 15 (3): 181-189.

Stigler, G.J. 1965. The Economist and the State. *American Economic Review* 55 (1): 1-18.

Tulkens, H. 1992. Economics and the Performance of the Public Sector. *Annals of Public and Cooperative Economics* 63 (3): 373-385.

Williamson, O.E. 1964. *The Economics of Discretionary Behavior: Managerial Objectives in a Theory of the Firm*. Englewood Cliffs, NJ: Prentice-Hall.

2 Measuring Efficiency in the Public Sector

C.A. Knox Lovell

2.1 Introduction

The objectives of this chapter are twofold: to provide a menu of analytical frameworks within which to evaluate the efficiency of public sector service provision, and to discuss a pair of alternative empirical techniques with which to conduct the empirical evaluation. The selection of an analytical framework leads to the specification of a theoretical "best practice" service delivery standard. The selection of an empirical technique leads to the construction of a data-based best practice service delivery standard against which the performance of providers can be evaluated in practice.

The choice of *analytical framework* depends on two factors: the behavioral objective assigned to service providers, and the data available to the analyst. These two factors influence the choice between a technical or an economic framework, between a service expanding or a resource conserving framework, and between a direct or an indirect framework. Thus there are eight possible analytical frameworks within which to evaluate the efficiency of public sector service provision. It is important to select an analytical framework that is consistent with the behavioral objective assigned to service providers, although inadequacies of the available data frequently constrain the selection to a "second best" framework. Happily, the cost of settling for a second best framework is likely to be a less rigorous performance evaluation than is desired, rather than an inaccurate performance evaluation.

The choice of *empirical technique* depends on the perceived accuracy of the data on which the performance evaluation is based. If the data are thought to be reasonably accurate, then either a deterministic mathematical programming technique known as Data Envelopment Analysis (DEA) or a stochastic econometric technique known

as stochastic frontier analysis (SFA) is appropriate. However, if the data are thought to be sufficiently noisy, then SFA is preferred. The choice is meaningful because it is important not to confuse the effects of inaccurate data with those of variation in managerial performance. However, despite widespread reservations concerning the accuracy of the underlying data, both empirical techniques have been widely applied to the performance measurement problem in the public sector, and both techniques are employed in subsequent chapters in this book.

The chapter is organized as follows. In Section 2.2 I describe the eight alternative analytical frameworks that can be used to evaluate the performance of public sector service providers. I also discuss eight alternative notions of the efficiency of service provision that can be obtained from each of the analytical frameworks. In Section 2.3 I briefly discuss a few important economic characteristics of service delivery technology. These characteristics are largely unrelated to the efficiency of individual service providers, but they are important nonetheless. Their significance arises from the fact that the characteristics of the service delivery network may be as critical to overall performance as is the efficiency of each individual provider within the network. In Section 2.4 I discuss the two alternative empirical techniques available for the construction of best practice service delivery technology and the measurement of service delivery efficiency. Either technique can be employed within each of the eight analytical frameworks. In Section 2.5 I briefly discuss alternative ways of accommodating variation in the operating environment that may influence the performance of service providers. This is important because, within a network of service providers, it is inevitable that some providers enjoy relatively favorable operating environments and others endure relatively unfavorable operating environments. The ability to distinguish between the largely uncontrollable influence of variation in the operating environment and the controllable influence of variation in managerial efficiency is central to the credibility of any service delivery performance analysis. In Section 2.6 I offer some concluding observations.

2.2 The Analytical Frameworks

I begin with some notation. What follows is a list of variables that is potentially available to the analyst. Suppose that we observe the activities of a set of $i = 1,\ldots,I$ public sector service providers. Each provider uses N resources $x = (x_1,\ldots,x_N)$ to provide M services $y = (y_1,\ldots,y_M)$. In addition, each provider may face resource prices $w = (w_1,\ldots,w_N)$ and service prices $p = (p_1,\ldots,p_M)$. Finally, each provider may be assigned an operating budget B or a revenue target R, but presumably not both. It is possible

that we observe the activities of these I providers through a sequence of $t = 1,\ldots,T$ time periods. The availability of such temporal information would enhance the credibility of the performance evaluation itself, since it would enable us to observe the activity of each provider a total of T times rather than just once. However, since the availability of such information would not expand the list of analytical frameworks available to the analyst, I suppress the time dimension in this section.

The first issue to consider is one of *data availability*. In the unlikely event that the entire list of variables (y, x, p, w, B or R) is available, the set of criteria on which to evaluate service delivery performance is large. In the much more likely event that only a subset of variables is available, the set of criteria is reduced accordingly. If, for example, only quantity data (y, x) are available, only a primal ("technical") approach to performance evaluation is feasible, while if price data (p, w) are also available, a dual ("economic") approach is also feasible. Finally, the availability of target data (B or R) makes an indirect approach feasible.

The second issue to consider is the *behavioral objective* assigned to service providers. One pair of possibilities is that providers are expected to maximize either service provision or service revenue (subject either to resource supply constraints or to resource price and operating budget constraints). Another pair of possibilities is that providers are expected to minimize either resource consumption or resource expenditure (subject either to service demand constraints or to service price and revenue target constraints). In the former pair of cases it is appropriate to adopt a service-oriented approach to the evaluation of provider performance, while in the latter pair of cases it is appropriate to adopt a resource-oriented approach to the evaluation of provider performance.

The third issue to consider is whether or not service providers must attempt to satisfy an *operating budget constraint* or must attempt to meet a *service revenue target*. If they are not so constrained, a direct approach is appropriate, while if they are so constrained, an indirect approach is appropriate.

Collecting the three sets of alternatives, a total of eight approaches to service delivery performance evaluation are potentially available. The choice among them depends on data availability (primal or dual), behavioral objective (service expansion or resource conservation orientation) and the relevance of targets (direct or indirect). The eight alternative ways of representing service delivery technology, and of evaluating service provider performance, are listed in Table 2.1. Data requirements are listed in the first column, and associated service enhancing and

resource conserving efficiency measures are listed in the second and third columns respectively. Direct performance measures are listed in the first three rows, and indirect performance measures are listed in the next four rows. In the remainder of this section I discuss each alternative analytical framework. In an effort to facilitate communication and comprehension, I keep the discussion informal by dispensing with mathematical rigor; a formal treatment of the eight analytical frameworks is available in Färe and Primont (1995), which I recommend highly.

Table 2.1
Alternative Analytical Frameworks for Performance Evaluation

	Orientation	
Data	*Service Enhancing*	*Resource Conserving*
Direct Performance Evaluation		
(y,x)	$TE_o(x,y)$	$TE_i(y,x)$
(y,x,p)	$RE(x,y,p) = TE_o(x,y) \cdot AE_o(x,y,p)$	
(y,x,w)		$CE(y,x,w) = TE_i(y,x) \cdot AE_i(y,x,w)$
Indirect Performance Evaluation		
(y,w,B)	$ITE_o(w/B,y)$	
(y,w,B,p)	$IRE(w/B,y,p) = ITE_o(w/B,y) \cdot IAE_o(w/B,y,p)$	
(x,p,R)		$ITE_i(p/R,x)$
(x,p,R,w)		$ICE(p/R,x,w) = ITE_i(p/R,x) \cdot IAE_i(p/R,x,w)$

$TE_o(x,y)$ = service expanding technical efficiency
$ITE_o(x,y)$ = indirect service expanding technical efficiency
$TE_i(y,x)$ = resource conserving technical efficiency
$ITE_i(y,x)$ = indirect resource conserving technical efficiency
$RE(x,y,p)$ = revenue efficiency
$IRE(x,y,p)$ = indirect revenue efficiency
$CE(y,x,w)$ = cost efficiency
$ICE(y,x,w)$ = indirect cost efficiency

A primitive representation of the structure of service delivery technology is provided by the *production set* T, which describes the entire set of technologically feasible combinations of services and resources. The production set is illustrated as the shaded region in Figure 2.1. Corresponding to T are a family of *service sets* P(x), con-

sisting of all service bundles that can be provided with any given resource bundle x. A single service set is illustrated as the shaded region in Figure 2.2. Also corresponding to T are a family of *resource sets* L(y), consisting of all resource bundles that are capable of providing any given service bundle y. A single resource set is illustrated as the shaded region in Figure 2.3. These three sets provide alternative representations of what is technologically feasible, and their boundaries provide alternative representations of "best practice."

Figure 2.1
Production Set

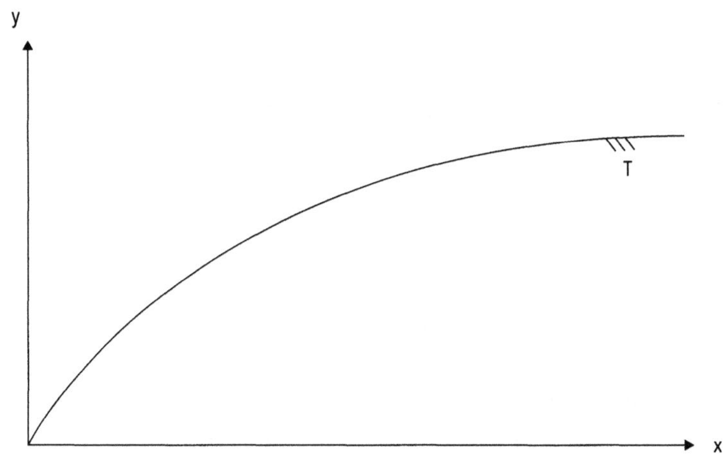

Figure 2.2
Service Set

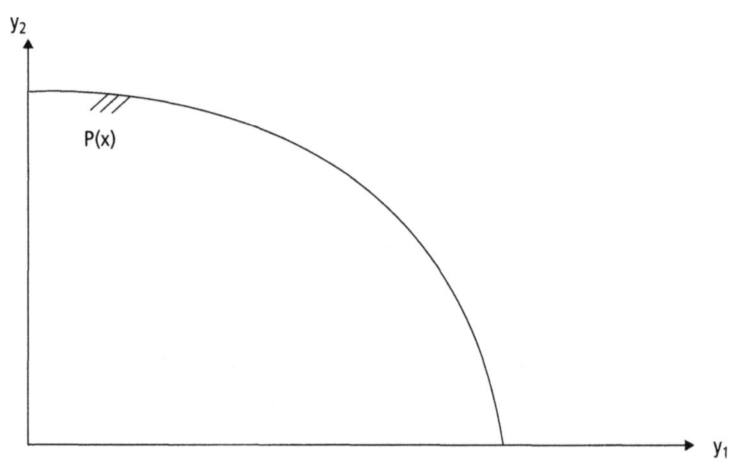

Figure 2.3
RESOURCE SET

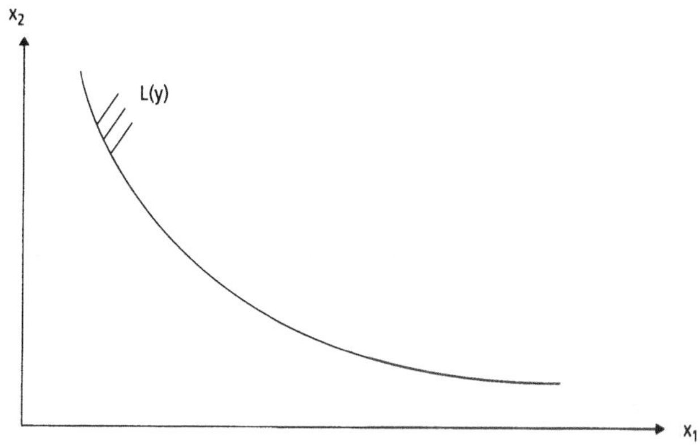

2.2.1 *Direct Performance Evaluation*

I begin by assuming that only quantity data (y, x) are available. In this case only a direct primal framework is available, and attention centers on the distance of providers from the boundaries of T or P(x) or L(y). Distance is determined by orientation. If one adopts a service-expanding orientation, then an *output distance function* $D_o(x,y)$ measures the maximum proportion by which service bundle y can be expanded and still be provided with resource bundle x. If one adopts a resource-conserving orientation, then an *input distance function* $D_i(y,x)$ measures the maximum proportion by which resource bundle x can be contracted and still remain capable of providing service bundle y. The two distance functions are illustrated in Figures 2.4 – 2.6.

Both distance functions are illustrated in Figure 2.4, which is based on the production set T. A provider such as A located on the boundary of T is incapable of expanding its service provision, given its resource usage, and equally incapable of contracting its resource usage, given its service provision, and so $D_o(x^A,y^A) = D_i(y^A,x^A) = 1$. A provider such as B located in the interior of T is capable of expanding its service provision, given its resource usage, or of contracting its resource usage, given its service provision, and so $D_o(x^B,y^B) < 1 < D_i(y^B,x^B)$. The output distance function is also depicted in Figure 2.5, which is based on the service set P(x). A provider such as A located on the upper boundary of P(x) is incapable of expanding its service provision, given its resource usage, and so $D_o(x^A,y^A) = 1$. A provider such as B located beneath the upper boundary of P(x) is capable of expanding its service provision, given its resource usage, and so $D_o(x^B,y^B) < 1$.

Figure 2.4
INPUT AND OUTPUT DISTANCE FUNCTIONS

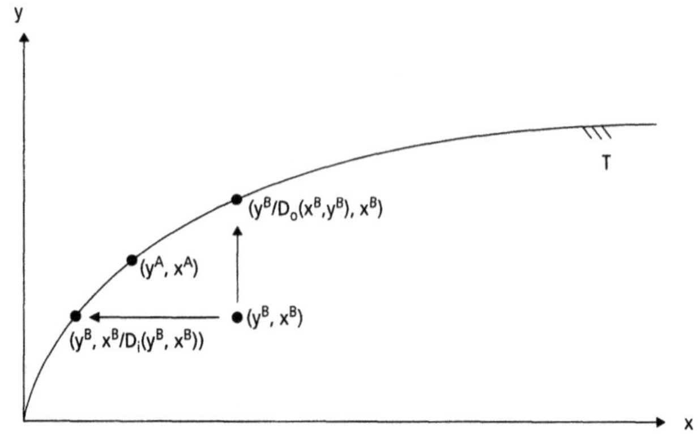

Figure 2.5
OUTPUT DISTANCE FUNCTION

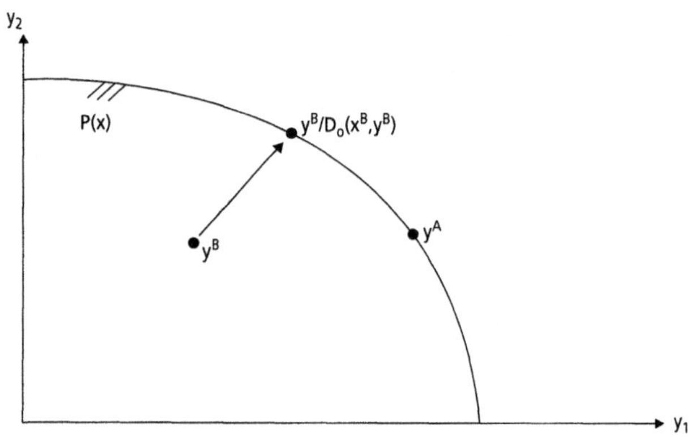

Figure 2.6
INPUT DISTANCE FUNCTION

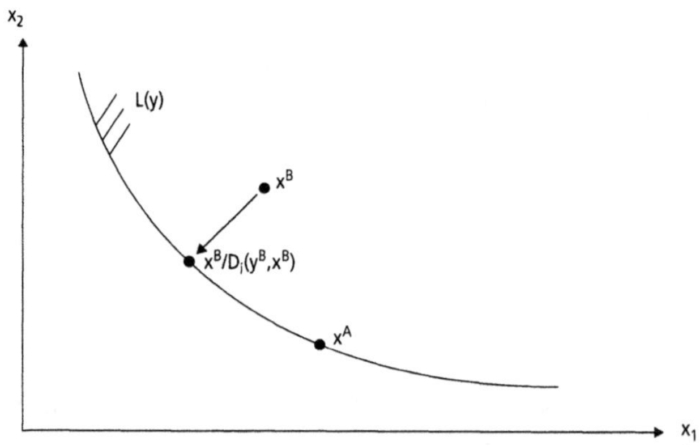

The input distance function is also depicted in Figure 2.6, which is based on the resource set L(y). A provider such as A located on the lower boundary of L(y) is incapable of contracting its resource usage, given its service provision, and so $D_i(y^A,x^A) = 1$. A provider such as B located above the lower boundary of L(y) is capable of contracting its resource usage, given its service provision, and so $D_i(y^B,x^B) > 1$. In all three Figures provider A is a best practice service provider because $D_o(x^A,y^A) = D_i(y^A,x^A) = 1$, while provider B is not up to best practice service delivery standards because $D_o(x^B,y^B) < 1 < D_i(y^B,x^B)$.

This suggests a pair of primal measures of the technical efficiency of service providers. If a service-expanding orientation is appropriate, then an output-oriented measure of service delivery efficiency is provided by the ratio of observed service provision to maximum feasible service provision, given observed resource usage, and so $TE_o(x,y) = D_o(x,y) \leq 1$. If a resource-conserving orientation is appropriate, then an input-oriented measure of service delivery efficiency is provided by the ratio of minimum feasible resource usage to observed resource usage, given observed service provision, and so $TE_i(y,x) = [D_i(y,x)]^{-1} \leq 1$. In common parlance, $TE_o(x,y)$ measures the ability of providers not to squander service delivery opportunities, while $TE_i(y,x)$ measures the ability of providers not to waste scarce resources.

Summarizing, if the available data are limited to (y, x), then all that can be calculated is the technical efficiency of service providers. All that is required is to construct the best practice service delivery technology represented by T, P(x) or L(y), and then to measure the technical efficiency of each provider by calculating the distance of each provider to best practice by means of either $TE_o(x,y)$ or $TE_i(y,x)$. The two technical efficiency measures appear in the first row of Table 2.1. Empirical techniques for constructing the best practice service delivery technology, and for measuring the technical efficiency of service providers, are discussed in Section 2.4.

I continue by assuming that, in addition to quantity data (y, x), service price data p are also available. In this situation I assume that service providers are assigned the behavioral objective of maximizing service revenue generated from given resources. Then the *revenue frontier* R(x,p) is defined as the maximum service revenue that can be generated from resource bundle x when service prices are p. Since observed service revenue may not be maximal, $py \leq R(x,p)$. The revenue frontier is illustrated in Figures 2.7 and 2.8.

In Figure 2.7, which is based on the service set P(x), provider A is maximizing revenue, given (x, p), while provider B is not. Provider B is capable of increasing revenue in either of two ways: by increasing service provision while maintaining

the current service mix, and by adjustments to the service mix. In Figure 2.8, which is based on the revenue frontier R(x,p), provider A is located on R(x,p) because it is incapable of utilizing its resources to generate increased revenue, given the service prices it faces. Provider B is located beneath R(x,p) because it is capable of utilizing its resources to generate increased revenue, given the service prices it faces. In both Figures provider A is revenue efficient and provider B is not.

Figure 2.7
REVENUE EFFICIENCY USING P(X)

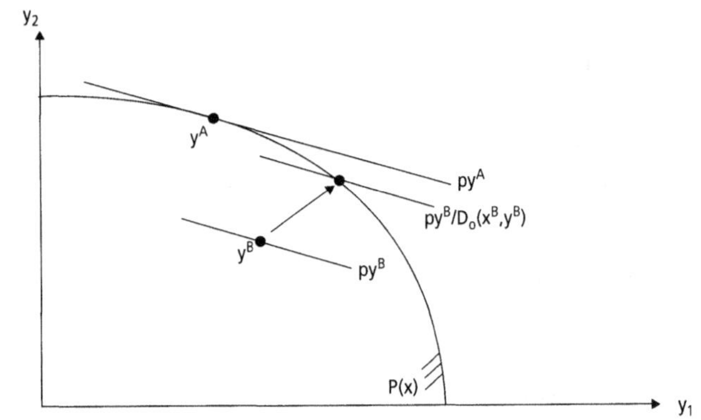

Figure 2.8
REVENUE EFFICIENCY USING R(X,P)

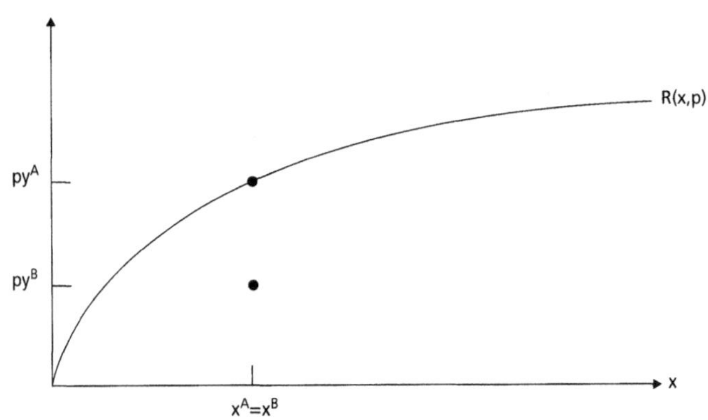

This suggests a measure of the revenue efficiency of service provision, provided by the ratio of observed revenue to maximum feasible revenue, given resource usage and service prices, and so $RE(x,y,p) = py/R(x,p) \leq 1$. Revenue efficiency has two components, one technical and one allocative. In Figure 2.7 provider A is revenue efficient because it is technically efficient and because it provides an optimal mix of

services in light of their relative prices. However, provider B is not revenue efficient, and is capable of improving its revenue efficiency in either of two ways. The first way is by improving its technical efficiency, by expanding service provision from $y^B$to $y^B/D_o(x^B,y^B)$. This would increase revenue from its observed value of py^B to $py^B/D_o(x^B,y^B)$. The second way is through an improvement in its allocative (or service mix) efficiency, by reallocating service provision from $y^B/D_o(x^B,y^B)$ to y^A. This would further increase revenue from $py^B/D_o(x^B,y^B)$ to its maximum feasible value of $py^A = R(x,p)$.

Somewhat more generally, the revenue efficiency of a provider can be expressed as the product of technical efficiency and allocative efficiency components, and so $RE(x,y,p) = TE_o(x,y) \cdot AE_o(x,y,p)$, where $TE_o(x,y) \leq 1$ is a measure of technical efficiency and $AE_o(x,y,p) \leq 1$ is a measure of allocative efficiency. Thus if revenue efficient service delivery is an appropriate economic objective, then all that is required is to construct the revenue frontier $R(x,p)$, measure the revenue efficiency $RE(x,y,p)$ of each provider, and decompose the measured revenue efficiency of each provider into its two components. Thus managerial efforts to improve the revenue efficiency of service providers can focus either on expanding service provision or on improving the allocation of services in light of their relative prices. The revenue efficiency measure and its two components appear in the second line of Table 2.1.

The decomposition of revenue efficiency into its technical and allocative components illustrates an important methodological point I raised in Section 2.1. It should be apparent from the decomposition of $RE(x,y,p)$ that the attainment of service-oriented technical efficiency is necessary, but not sufficient, for the attainment of revenue efficiency. Suppose that the behavioral objective assigned providers is one of revenue maximization. But suppose that providers are evaluated on the basis of their service-oriented technical efficiency. Then the evaluation holds them to a less rigorous standard than is desired. The performance evaluation is informative, but it is incomplete because it addresses only one of two components of revenue efficiency.

I now reverse the orientation. Suppose that, in addition to quantity data (y, x), resource price data w are also available. I assume that service providers are assigned the behavioral objective of minimizing expenditure on resources used to satisfy given service demand. Then the *cost frontier* $C(y,w)$ is defined as the minimum expenditure on resources required to provide service bundle y at resource prices w. Since observed expenditure may not be minimal, $wx \geq C(y,w)$. The cost frontier is illustrated in Figures 2.9 and 2.10.

Figure 2.9
COST EFFICIENCY USING L(Y)

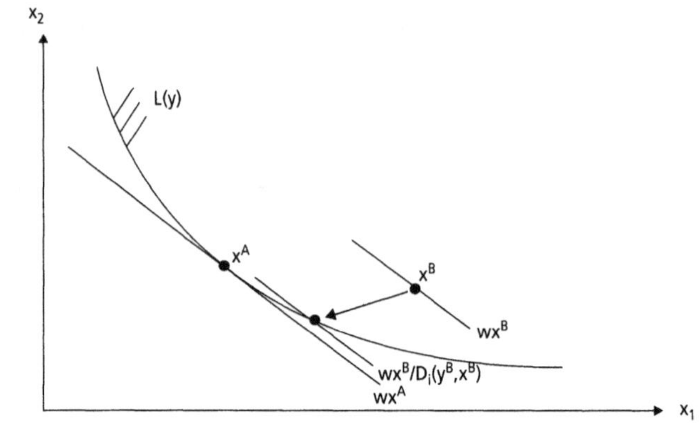

Figure 2.10
COST EFFICIENCY USING C(Y,W)

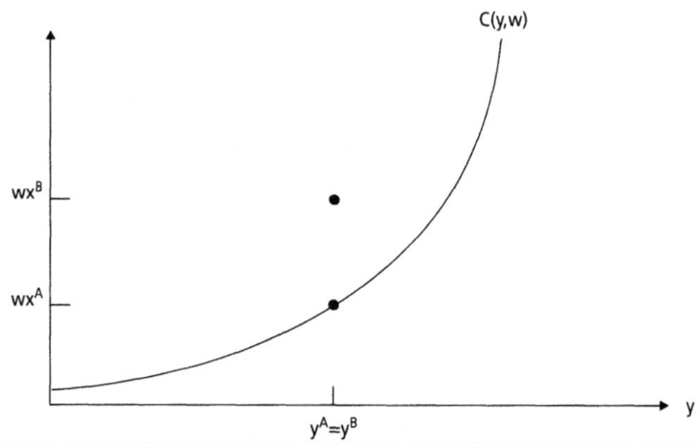

In Figure 2.9, which is based on the resource set L(y), provider A is minimizing expenditure, given (y, w), while provider B is not. Provider B is capable of reducing expenditure in either of two ways: by reducing resource usage while maintaining the current resource mix, and by adjustments to the resource mix. In Figure 2.10, which is based on the cost frontier C(y,w), provider A is located on C(y,w) because it is incapable of providing its services at lower cost, given the resource prices it faces. Provider B is located above C(y,w) because it is capable of providing its services at lower cost, given the resource prices it faces. In both Figures provider A is cost efficient and provider B is not.

This suggests a measure of the *cost efficiency* of service provision, provided by the ratio of minimum feasible expenditure to observed expenditure, given service pro-

vision and resource prices, and so $CE(y,x,w) = C(y,w)/wx \leq 1$. Cost efficiency has two components, one technical and one allocative. In Figure 2.9 provider A is cost efficient because it is technically efficient and because it uses an optimal mix of resources in light of their relative prices. However, provider B is not cost efficient, and is capable of improving its cost efficiency in either of two ways. The first way is by improving its technical efficiency, by contracting resource usage from x^B to $x^B/D_i(y^B,x^B)$. This would reduce resource expenditure from its observed value of wx^B to $wx^B/D_i(y^B,x^B)$. The second way is through an improvement in its allocative (or resource mix) efficiency, by reallocating resource usage from $x^B/D_i(y^B,x^B)$ to x^A. This would further reduce resource expenditure from $wx^B/D_i(y^B,x^B)$ to its minimum feasible value of $wx^A = C(y,w)$.

Somewhat more generally, the cost efficiency of a provider can be expressed as the product of technical efficiency and allocative efficiency components, and so $CE(y,x,w) = TE_i(y,x) \cdot AE_i(y,x,w)$, where $TE_i(y,x) \leq 1$ is a measure of technical efficiency and $AE_i(y,x,w) \leq 1$ is a measure of allocative efficiency. Thus if cost efficient service delivery is an appropriate economic objective, then all that is required is to construct the cost frontier $C(y,w)$, measure the cost efficiency $CE(y,x,w)$ of each provider, and decompose the measured cost efficiency of each provider into its technical and allocative components. Thus managerial efforts to improve the cost efficiency of service providers can focus either on the reduction of waste or on an improvement in the allocation of resources in light of their relative prices. The cost efficiency measure and its two components appear in the third line of Table 2.1.

The methodological point I raised concerning revenue efficiency also applies to cost efficiency. If cost efficiency is the appropriate behavioral objective, but providers are evaluated on the basis of their resource-oriented technical efficiency, then the evaluation is useful but incomplete.

Summarizing the direct approach to performance evaluation, if the available data include either w or p in addition to (y, x), then the economic efficiency of service providers is measured in either of two ways. If (x, y, p) data are available and a service expansion orientation is appropriate, their economic efficiency is measured by their revenue efficiency $RE(x,y,p)$, which can then be decomposed into its technical and allocative components. If (y, x, w) data are available and a resource conservation orientation is appropriate, their economic efficiency is measured by their cost efficiency $CE(y,x,w)$, which can then be decomposed into its technical and allocative components. The empirical construction of the relevant frontiers is discussed in Section 2.4.

2.2.2 *Indirect Performance Evaluation*

I now introduce targets, and move from direct efficiency measurement (both technical and economic) to indirect efficiency measurement (also both technical and economic). Indirect efficiency measurement is appropriate when service providers are constrained by operating budgets rather than by resource supplies, or by revenue targets rather than by service demands.

Suppose first that providers face an operating budget B, but that they are (relatively) free to select any technologically feasible resource bundle that satisfies the operating budget constraint $wx \leq B$. Since providers are free to choose among feasible resource bundles satisfying the operating budget, their service provision opportunities expand from the service set P(x) to the cost indirect service set IP(w/B). This is illustrated in Figure 2.11, where the cost indirect service set IP(w/B) is the outer envelope of all service sets P(x) for which $wx \leq B$.

Figure 2.11
INDIRECT SERVICE SET IP(W/B) AND DIRECT SERVICE SET P(X) FOR WHICH $wx \leq B$

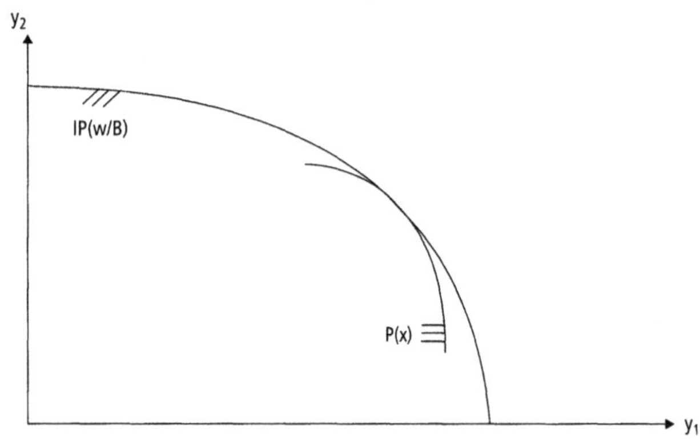

I begin by assuming that providers are assigned the behavioral objective of maximizing service provision, subject to the constraints imposed by the nature of service delivery technology and the operating budget constraint $wx \leq B$. In this exercise the available data consist of (y, w, B); resource quantity data x are not required. Then a *cost indirect output distance function* $ID_o(w/B,y)$ measures the maximum proportion by which service provision can be expanded, given the operating budget. Just as in the direct case, a cost indirect service-oriented efficiency measure is provided by the ratio of observed service provision to maximum feasible service provision, in

light of the nature of service delivery technology and the operating budget constraint, and so $ITE_o(w/B,y) = ID_o(w/B,y) \leq 1$. This is illustrated in Figure 2.12, where provider A is efficient in this sense since $ITE_o((w/B)^A,y^A) = 1$. The cost indirect measure of service delivery efficiency appears in the fourth line of Table 2.1.

Figure 2.11 illustrates another important point. A resource-constrained provider has service set P(x), while a budget-constrained provider has larger service set IP(w,B). This expanded service provision opportunity is a reflection of the value of being able to substitute among resource bundles satisfying an operating budget constraint. A measure of the potential gain from having the freedom to choose from among feasible resource bundles is provided by the ratio $ITE_o(w/B,y)/TE_o(x,y) \leq 1$.

Figure 2.12
INDIRECT REVENUE EFFICIENCY USING IP(W/B)

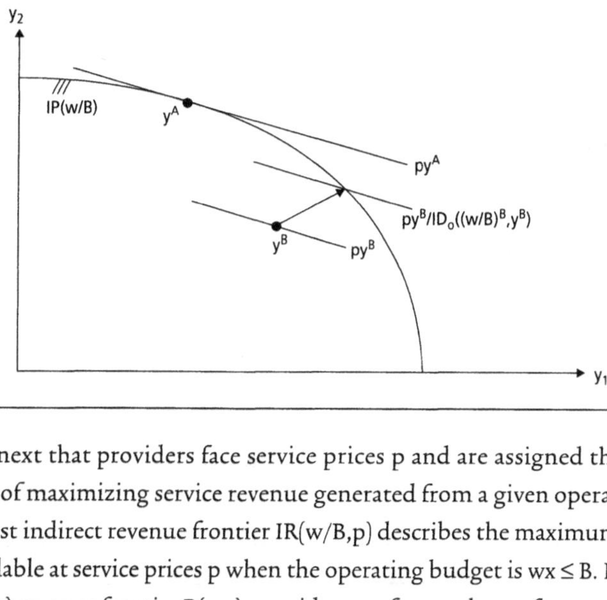

Suppose next that providers face service prices p and are assigned the behavioral objective of maximizing service revenue generated from a given operating budget. Then a cost indirect revenue frontier IR(w/B,p) describes the maximum service revenue available at service prices p when the operating budget is $wx \leq B$. In contrast to the (direct) revenue frontier R(x,p), providers are free to choose from among feasible resource vectors satisfying the operating budget constraint. Since observed service revenue may not be maximal, $py \leq IR(w/B,p)$. The cost indirect revenue frontier is illustrated in Figure 2.12, which is based on the cost indirect service set IP(w/B). A provider such as A, located on the boundary of IP(w/B), is maximizing revenue, given (w/B, p), while a provider such as B, located in the interior of IP(w/B), is not. Provider B is capable of increasing revenue in either of two ways: by increasing service provision while maintaining the current service mix, or by adjustments to the service mix. Thus provider A is cost indirect revenue efficient, and provider B is not.

This suggests that a measure of the cost indirect revenue efficiency of service provision is provided by the ratio of observed revenue to maximum feasible revenue, given resource prices, operating budgets and service prices, and so IRE(w/B,y,p) = py/IR(w/B,p) ≤ 1. Cost indirect revenue efficiency has two components, one technical and one allocative. Consequently IRE(w/B,y,p) = ITE_o(w/B,y)·IAE_o(w/B,y,p), where ITE_o(w/B,Y) ≤ 1 measures cost indirect technical efficiency and IAE_o(w/B,y,p) ≤ 1 measures cost indirect allocative efficiency. The measure of cost indirect revenue efficiency and its decomposition appear in the fifth line of Table 2.1.

Summarizing, if the available data include service prices p and the operating budget B in addition to (y, w), then the efficiency with which service providers utilize their operating budgets is measured in either of two ways. If only (y, w, B) data are available and a service expansion orientation is appropriate, it is measured by their cost indirect technical efficiency ITE_o(w/B,y). If service price data p are also available, it is measured by their cost indirect revenue efficiency IRE(w/B,y,p), which can be decomposed into its technical and allocative components ITE_o(w/B,y) and IAE_o(w/B,y).

Suppose finally that providers face a revenue target py ≥ R, but that they are (relatively) free to select any technologically feasible service bundle that satisfies this constraint. Since providers are free to choose among feasible service bundles satisfying the revenue target, their resource usage opportunities expand from the resource set L(y) to the revenue indirect resource set IL(p/R). This is illustrated in Figure 2.13, where the revenue indirect resource set IL(p/R) is the inner envelope of all resource sets L(y) for which py ≥ R.

Figure 2.13
INDIRECT RESOURCE SET IL(P/R) AND DIRECT RESOURCES SET L(Y) FOR WHICH PY ≥ R

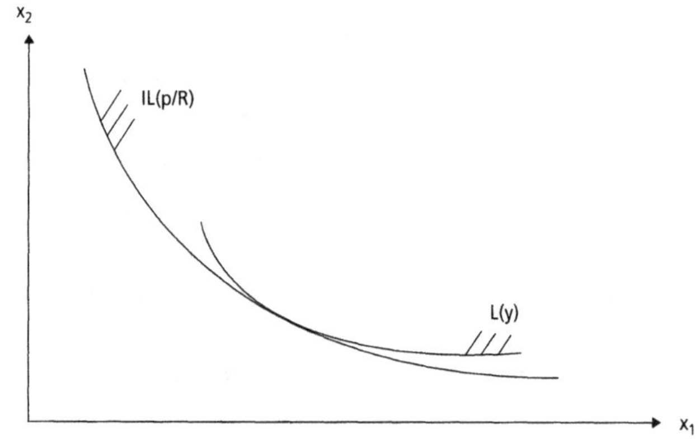

I begin by assuming that providers are assigned the behavioral objective of minimizing resource usage, subject to the constraints imposed by the nature of service delivery technology and the revenue target $py \geq R$. In this exercise the available data consist of (x, p, R); service quantity data y are not required. Then a *revenue indirect input distance function* $ID_i(p/R,x)$ measures the maximum proportionate reduction in resource usage consistent with meeting the revenue target $py \geq R$. In contrast to the (direct) input distance function, providers are free to choose among feasible service bundles satisfying the revenue target $py \geq R$.

Just as in the direct case, a revenue indirect resource-oriented measure of service delivery efficiency is provided by the ratio of observed resource usage to minimum feasible resource usage, in light of the nature of service delivery technology and the service revenue target, and so $ITE_i(p/R,x) = [ID_i(p/R,x)]^{-1} \leq 1$. This is illustrated in Figure 2.14, where provider A is efficient in this sense since $ITE_i((p/R)^A,x^A) = 1$. The revenue indirect measure of service delivery efficiency appears in the sixth line of Table 2.1.

Figure 2.13 illustrates another important point. A service-constrained provider has resource set $L(y)$, while a revenue target-constrained provider has larger resource set $IL(p/R)$. This expanded resource conservation opportunity is a reflection of the value of being able to choose from among service bundles satisfying a revenue target. A measure of the potential gain from having the freedom to choose from among feasible service bundles is provided by the ratio $ITE_i(p/R,x)/TE_i(y,x) \leq 1$.

Suppose next that providers face resource prices w and are assigned the behavioral objective of minimizing resource expenditure required to meet a revenue target R. Then a *revenue indirect cost frontier* $IC(p/R,w)$ describes the minimum expenditure on resources required to satisfy the revenue target $py \geq R$. In contrast to the (direct) cost frontier, providers are free to choose from among feasible service vectors satisfying the revenue target. Since observed resource expenditure may not be minimal, $wx \geq IC(p/R,w)$. The revenue indirect cost frontier is illustrated in Figure 2.14, which is based on the revenue indirect resource requirement set $IL(p/R)$. A provider such as A, located on the boundary of $IL(p/R)$, is minimizing expenditure, given $(p/R, w)$, while a provider such as B, located in the interior of $IL(p/R)$, is not. Provider B is capable of reducing expenditure in either of two ways: by reducing resource usage while maintaining the current resource mix, and by adjustments to the resource mix. Thus provider A is revenue indirect cost efficient, and provider B is not.

Figure 2.14
INDIRECT COST EFFICICENCY USING IL(P/R)

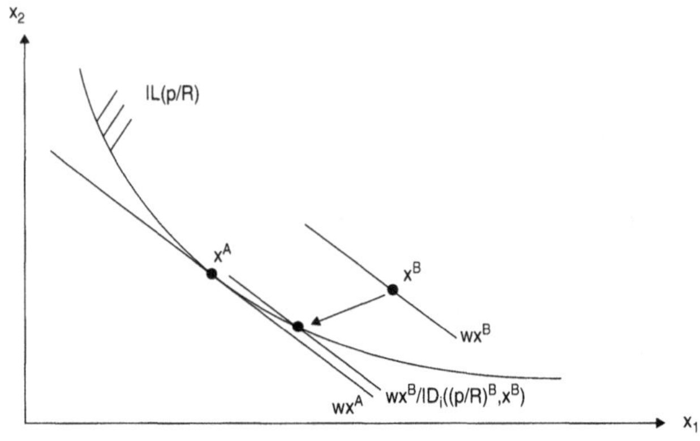

This suggests that a measure of the revenue indirect cost efficiency of service provision is provided by the ratio of minimum feasible expenditure to observed expenditure, given resource prices, service prices and revenue targets, and so $ICE(p/R,x,w) = IC(p/R,w)/wx \leq 1$. Revenue indirect cost efficiency has two components, one technical and one allocative. Consequently $ICE(p/R,x,w) = ITE_i(p/R,x) \cdot IAE_i(p/R,x,w)$, where $ITE_i(p/R,x) \leq 1$ is revenue indirect technical efficiency and $IAE_i(p/R,x,w) \leq 1$ is revenue indirect allocative efficiency. The measure of revenue indirect cost efficiency and its decomposition appear in the final line of Table 2.1.

Summarizing, if the available data include resource prices w and a revenue target R in addition to (x, p), then the efficiency with which service providers pursue revenue targets is measured in either of two ways. If only (x, p, R) data are available and a resource conservation orientation is appropriate, it is measured by their revenue indirect technical efficiency $ITE_i(p/R,x)$. If resource price data w are also available, it is measured by their revenue indirect cost efficiency ICE(p/R,x,w), which can be decomposed into its technical and allocative components $ITE_i(p/R,x)$ and $IAE_i(p/R,x,w)$.

This concludes the presentation of the eight alternative analytical frameworks within which a comparative performance evaluation can be conducted. The choice among the frameworks depends on orientation, behavioral objective, and data availability. At this point it is worth noting that a ninth analytical framework exists. If providers are free to choose both y and x, and face prices p and w, then it is possible to formulate a profit maximization framework, and to measure and decompose profit efficiency. However, while such a behavioral objective may be appropriate in

the private sector, it is unlikely to be appropriate in the public sector, and consequently I have not developed this framework. Färe and Primont (1995) do.

2.3 Characteristics of Best Practice Service Delivery Technology

In Section 2.2 I discussed eight analytical frameworks within which to evaluate the efficiency of individual service providers. Each framework is associated with a particular behavioral objective. Each framework also leads to a particular specification of best practice service delivery technology. I did not discuss the economic characteristics of best practice technology. This omission is unimportant if interest centers exclusively on evaluating and enhancing the performance of individual providers. However, it is very important if interest also focuses on evaluating and enhancing the overall performance of the service delivery network. In this section I briefly redress this omission.

To keep the exposition simple at the outset, suppose that we have service data y and resource data x for each provider in the network, and suppose that we are interested in the service-oriented technical efficiency of each provider. Then the production set T establishes best practice standards and the service-oriented technical efficiency measure $TE_O(x,y)$ provides the requisite performance evaluation for each provider. Now suppose that every provider is technically efficient, unable to increase service provision with available resources and existing technology. Is it possible to squeeze additional services out of the network without increasing its aggregate resource base? Perhaps, and in at least three different ways.

Suppose that service delivery technology is characterized by *scale economies*, or by *increasing returns to scale*. In this event each provider is undesirably small, in the sense that each is capable of expanding service delivery by proportionately more than it expands its resource base. But with a fixed network resource endowment, this is not possible for every provider. The solution may be to amalgamate providers. Fewer providers, each with an expanded resource base, can provide a more than proportionate increase in services, even though each was operating efficiently to begin with. Exploitation of scale economies enables an increase in network service provision, without an increase in the network resource base, and without increases in individual provider efficiency. Thus restructuring the service delivery network serves as a substitute for improving the efficiency of individual providers. Of course, some of the potential gains may be partially offset by efficiency declines

associated with the difficulties of managing larger agencies, but this is not guaranteed. In addition, a network comprised of fewer larger agencies may lead to a reduction in accessibility and customer convenience, although this is not certain either. It is the responsibility of central management to weigh the costs and benefits of amalgamation in the presence of scale economies.

A similar scenario can be constructed for the opposite case in which service delivery technology is characterized by *diseconomies of scale*, or by *decreasing returns to scale*. In this event each provider is undesirably large, in the sense that each is capable of contracting its resource base by proportionately more than it contracts its service delivery. The solution now is to restructure the network into a larger number of smaller providers. This avoidance of diseconomies of scale also enables an increase in network service provision, without an increase in the network resource base, and without increases in individual provider efficiency. Although some of the potential gains may be partially offset by the need to hire and train new managers, they may be enhanced by increases in accessibility and customer convenience associated with an increase in the number of providers.

A second characteristic of the structure of service delivery technology that may influence the overall performance of the network is the nature of *economies or diseconomies of diversification*. Loosely speaking, economies of diversification are present if the cost of providing a range of services collectively is less than the collective cost of providing each service (or a subset of services) individually. The presence of economies of diversification makes it desirable to operate a network of diversified agencies providing "one-stop shopping." Conversely, diseconomies of diversification are present if the cost of providing a range of services collectively exceeds the collective cost of providing each service (or a subset of services) individually. The presence of diseconomies of diversification makes it desirable to unbundle the services by operating a network of relatively specialized agencies. Just as in the case of economies or diseconomies of scale, the exploitation of economies or diseconomies of diversification enables an increase in network service provision, without an increase in the network resource base, and without increases in individual provider efficiency. Consequently restructuring the service delivery network serves as a substitute for improving the efficiency of individual providers. Once again, it is the responsibility of central management to evaluate the impact on customer convenience of any decision to reconfigure the service delivery network to exploit economies or diseconomies of diversification.

A third characteristic of the structure of the service delivery technology is the way it improves through time. *Technical progress* is a fact of life, as is its adaptation and diffusion. Improvements in service delivery technology, and its diffusion through the service delivery network, both allow for increases in service provision at proportionately smaller increases in operating cost. It is the responsibility of central management to weigh the cost of adopting and installing new technology against the benefits of increased service provision and increased customer convenience.

Each of these three characteristics of service delivery technology, and others as well, can be incorporated into the eight analytical frameworks discussed in Section 2.2. Since all frameworks are linked, each characteristic of technology influences all eight notions of best practice standards. Thus the nature of economies of scale or economies of diversification influence the shape of the production set T, which in turn influences the shape of the cost frontier C(y,w), and so on. Technical progress enlarges the production set T, which in turn leads to a downward shift in the cost frontier C(y,w), and so on. Although it is useful from an analytical perspective to know how the structure of service delivery technology influences the eight notions of best practice standards, the primary value of this knowledge is empirical. Only by quantifying the influence of these characteristics can management adapt the service delivery network in an optimal fashion.

2.4 The Empirical Techniques

DEA and SFA are empirical techniques employed to construct best practice technology from a sample of service providers, and to provide performance indicators for each service provider in the sample. In this Section I provide a brief non-technical exposition of how DEA and SFA work. A brief, moderately technical, introduction to the two empirical techniques is provided in the Appendix to this chapter. I refer the reader to the references cited in the next paragraph for advanced treatments of the mathematical and statistical details of the two empirical techniques.

The two techniques were developed independently at about the same time, and by now both are widely used to evaluate performance in both the public and the private sectors. The original DEA contributions were made by Charnes, Cooper and Rhodes (1978) and Banker, Charnes and Cooper (1984), and the original SFA contributions were made by Aigner, Lovell and Schmidt (1977), Meeusen and Van den Broeck (1977) and Jondrow, Lovell, Materov and Schmidt (1982). Recent compre-

hensive overviews are available in Charnes, Cooper, Lewin and Seiford (1994) for DEA, Kumbhakar and Lovell (2000) for SFA, and Fried, Lovell, Schmidt and Yaisawarng (1993) and Coelli, Prasada Rao and Battese (1998) for both. A thoughtful collection of applications of DEA to public sector performance evaluation is provided by Ganley and Cubbin (1992). A variety of user-friendly software packages has been developed to implement both techniques.

The basic idea behind both techniques is the same. The first objective is to construct a best practice envelopment of data describing the economic activities of a sample of service providers. Such an envelopment provides an empirical approximation to the best practice standards established in the sample. Depending on behavioral objective and data availability, these standards can be defined by the production set T and its associated service sets P(x) and resource sets L(y); by the revenue frontier R(x,p) or the cost frontier C(y,w); by the indirect service sets IP(w/B) or the indirect resource sets IL(p/R); and by the indirect revenue frontier IR(w/B,p) or the indirect cost frontier IC(p/R,w).

The second objective is to generate performance evaluations for each provider in the sample. Again depending on behavioral objective and data availability, these evaluations consist of two direct measures of technical efficiency, $TE_o(x,y)$ and $TE_i(y,x)$; two direct measures of economic efficiency, RE(x,y,p) and CE(y,x,w); two direct measures of allocative efficiency, $AE_o(x,y,p)$ and $AE_i(y,x,w)$; two indirect measures of technical efficiency, $ITE_o(w/B,y)$ and $ITE_i(p/R,x)$; two indirect measures of economic efficiency, IRE(w/B,y,p) and ICE(p/R,x,w); and two indirect measures of allocative efficiency, $IAE_o(w/B),y,p)$ and $IAE_i(p/R,x,w)$.

Both empirical techniques readily accommodate multiple resources and multiple services, and both can incorporate resource prices and service prices, as well as operating budgets or revenue targets, if they are available and deemed to be both reliable and relevant. Thus both empirical techniques can be used within any of the eight analytical frameworks to construct best practice standards and to evaluate provider performance. Rather than demonstrate the workings of DEA and SFA within each of the eight analytical frameworks, I focus on the simplest framework and refer the reader to the references cited above for extensions to the remaining frameworks.

To keep the exposition simple, and to facilitate a graphical approach, suppose that we observe the annual performance of a sample of providers, and that each uses a single resource to provide a single service. Thus we want to construct an empirical

approximation to the production set T, and we want to generate measures of the technical efficiency (either $TE_o(x,y)$ or $TE_i(y,x)$) of each producer in the sample. Hypothetical data for a sample of providers, together with the DEA and SFA envelopments of the data, are illustrated in Figure 2.15. The dots indicate resource use and service provision for each provider in the sample. Note that providers with relatively large resource use generally provide relatively large service quantities, although there are exceptions to this generalization.

Figure 2.15
DEA and SFA Best Practice Frontiers

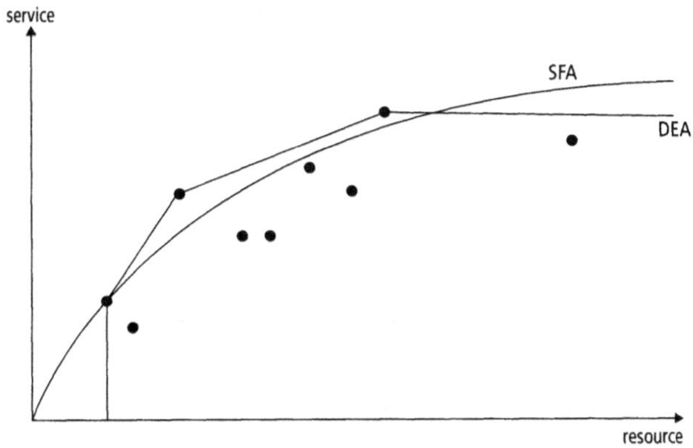

The difference between the two empirical techniques lies in the way envelopment is accomplished. DEA achieves envelopment by solving a sequence of linear programs, one for each provider. Since the programs are linear, the DEA best practice service delivery frontier is piecewise linear. Since the programs are deterministic, *all* providers are located on or beneath the DEA frontier. SFA achieves envelopment by means of statistical estimation of a single regression model. Since the regression model is parametric, the SFA best practice service delivery frontier is smooth. Since SFA is regression-based, it is stochastic. However unlike conventional regression models, the SFA error term is skewed, having both a conventional symmetric component capturing the effects of measurement error and other sources of random noise, and a one-sided component capturing the effects of inefficiency in service delivery. Consequently providers are more likely to end up beneath the SFA frontier than above it. Nonetheless it is possible that for some providers the influence of favorable noise dominates the influence of inefficiency, permitting them to end up above the SFA frontier. However, *most* sample providers are located on or beneath the SFA frontier. Thus the two frontiers provide alternative empirical approximations to the boundary of the true but unknown production set T.

DEA and SFA have offsetting advantages and disadvantages. The main advantage of SFA is the main disadvantage of DEA. As its name implies, SFA is stochastic, and so in principle is capable of distinguishing performance variation across providers from the statistical noise (due perhaps to measurement error in the included variables) that pervades all economic data. Thus a small proportion of providers may end up above the estimated SFA frontier, not because they are better than best practice but due to the influence of measurement error and other sources of statistical noise. DEA is deterministic, and so no providers end up above the calculated DEA frontier because the effects of statistical noise are ignored. Consequently the presence of measurement error and other sources of statistical noise may cause DEA to mis-specify the structure of the best practice service delivery frontier. This in turn would have an adverse impact on inferences concerning provider performance relative to DEA best practice standards.

Conversely, the main advantage of DEA is the main disadvantage of SFA. DEA is nonparametric, and lets the data create the structure of the best practice service delivery frontier. SFA is parametric, and imposes a pre-specified (albeit generally flexible) structure on the frontier. Consequently, SFA is also capable of mis-specifying the structure of the best practice service delivery frontier, although for a different reason. This in turn would have an adverse impact on inferences concerning provider performance relative to SFA best practice standards.

Once the best practice frontier has been constructed, either by DEA or by SFA, the next step is to calculate the service delivery performance of each provider. This can be accomplished by taking either a service enhancing (vertical) orientation or a resource conserving (horizontal) orientation to performance evaluation. If we adopt the former orientation, we simply compute for each provider the ratio of observed to maximum feasible service provision, given the resource base of each provider. This is the service-oriented technical efficiency measure $TE_o(x,y)$ illustrated in Figures 2.4 and 2.5. If we adopt the latter orientation, we compute for each provider the ratio of minimum feasible resource use to observed resource use, given the service provision of each producer. This is the resource-oriented technical efficiency measure $TE_i(y,x)$ illustrated in Figures 2.4 and 2.6. Regardless of the orientation selected, providers located on the best practice service delivery frontier receive an efficiency score of unity, and are labeled best practice providers. Providers located beneath the frontier receive an efficiency score of less than unity, and have room for varying degrees of improvement in their service provision.

Since the best practice standards established by the DEA and SFA frontiers differ, efficiency scores are not generally invariant to the choice of technique. However in a sufficiently large sample with sufficiently small measurement error, the empirical divergence between the two frontiers is likely to be minimal. In such favorable circumstances, DEA and SFA can be expected to define similar best practice service delivery standards, and to generate similar conclusions concerning the relative performance of providers. This is important, because it would be uncomfortable to discover that one's conclusions concerning the structure of best practice service delivery standards, and the relative performance of each provider in the sample, were to depend on the choice of empirical technique. Although this possibility cannot be eliminated, it can be minimized by the generation and use of large data bases with small measurement error. Moreover, even in less favorable circumstances the identification of best practice and worst practice providers is unlikely to be particularly sensitive to the choice of empirical technique. This is important, because interest frequently centers not on a ranking of all providers, but on the identification of those under-performing providers most in need of managerial attention, as well as on the identification of those exemplary providers most likely to serve as role models for the laggards.

I conclude this section with an observation. The vast majority of public sector applications I have seen have used DEA, despite its obvious drawback of being deterministic. I have heard three explanations, the evaluation of which I leave to the reader. (i) DEA more easily accommodates multiple services in the performance evaluation, and public sector agencies invariably provide multiple services that are difficult, if not impossible, to aggregate. (ii) Civil servants and public sector managers are more receptive to performance evaluations based on DEA, perhaps because the analysis is more transparent. (iii) Those who employ DEA are more gifted marketers than those who employ SFA. However, the Netherlands, with its long and distinguished tradition in econometrics, is likely to provide an exception to this general rule.

2.5 The Operating Environment

In developing the alternative analytical frameworks in Section 2.2, and in discussing the alternative empirical techniques in Section 2.4, I incorporated two sets of variables: (i) resources and services over which management is assumed to have control; and (ii) prices and budgets or targets over which management is assumed to have no control. I abstracted from a third set of variables: (iii) uncontrollable features of the operating environment that influence the ability of man-

agement to convert resources into services. To the extent that these environmental features vary across service providers and influence service delivery performance, they cannot be ignored. To omit them from a comparative performance evaluation is to invite managerial complaints of unfairness and to endanger their cooperation. In this section, I suggest two general ways in which features of the operating environment can be incorporated into either a DEA or an SFA performance evaluation.

Suppose we observe a set $z = (z_1,...,z_K)$ of characteristics of the operating environment that exert a systematic, non-random influence on the performance of providers. Several examples come quickly to mind: (i) socio-economic status (education, welfare); (ii) age distribution (health care, public safety); and (iii) population density (transport, energy). These and other environmental features are known to influence service delivery performance, but they are not under the control of service providers.

It is possible to incorporate these features into a DEA performance evaluation in either of two ways. The first way is to add a set of K additional constraints, one for each environmental variable, to the linear program. Adding these constraints ensures that each provider is evaluated relative to other providers in equally or less favorable operating environments. It is also possible to evaluate each provider relative to other providers in equally or more favorable operating environments. This approach is feasible only if the direction, although not the magnitude, of the effect of each environmental variable is known in advance. Assuming this to be the case, a comparison of efficiency scores with and without the environmental constraints imposed provides an indication of the magnitudes of each individual environmental effect on the performance of each provider.

A second approach does not require that the directions of the environmental effects be known in advance. In this approach the initial DEA efficiency scores are regressed against the environmental variables in a second stage regression model. This regression seeks to explain variation in initial efficiency scores in terms of three sources: (i) variation in the operating environment; (ii) the impact of statistical noise (which was ignored in the deterministic DEA model); and (iii) managerial inefficiency remaining after the impacts of the operating environment and statistical noise have been accounted for. Details and extensions of both DEA approaches are available in Fried, Lovell, Schmidt and Yaisawarng (1999).

Environmental variables can be incorporated into an SFA performance evaluation in two very similar ways. The first approach parallels the first DEA approach. The stochastic frontier regression is modified by including the environmental variables as

additional explanatory variables, so that service provision depends not only on resource use, but also on characteristics of the operating environment. The advantage of this approach in the SFA context is that the direction of the impact of each environmental variable does not need to be known in advance. The direction and magnitude of each impact is determined by the signs and magnitudes of the relevant regression coefficients. This procedure provides a convenient way for distinguishing the impacts of resource use from those of variation in the operating environment.

The second approach to incorporating environmental variables into an SFA performance evaluation simply relocates the environmental variables from the stochastic frontier to the one-sided error component. The only difference between the two SFA approaches is the interpretation of how the environment influences performance. In the first approach the environment influences the technology with which resources are converted into services, and so indirectly influences performance. In the second approach the environment influences the efficiency with which providers utilize the technology at their disposal, and so directly influences performance. Details and extensions of both SFA approaches are available in Kumbhakar and Lovell (2000; Ch. 7).

2.6 Concluding Observations

A comparative performance evaluation of public sector service providers can be conducted within any of a number of analytical frameworks, using either of two empirical techniques. The primary objective of this chapter has been to outline the frameworks and introduce the techniques. A secondary objective has been to stress the desirability and the feasibility of incorporating features of the service delivery technology and characteristics of the operating environment into the performance evaluation.

In my judgement the two essential ingredients of a comparative performance evaluation are (i) the selection of an appropriate analytical framework; and (ii) the incorporation of features of the service delivery technology and characteristics of the operating environment into the empirical analysis. Both require close cooperation between the analyst and representatives of the public sector, both those responsible for the evaluation and those being evaluated. However, both are essential if the evaluation is to generate credible results that can inspire confidence and cooperation, and for the evaluation to provide the basis for a continuing program of overall performance enhancement.

APPENDIX

A2.1 DATA ENVELOPMENT ANALYSIS

The basic idea behind DEA can be described as follows. Suppose that a group of providers, indexed $i = 1,...,o,...,I$, uses a nonnegative vector of N resources $x_i = (x_{1i},...,x_{Ni})$ to provide a nonnegative vector of M services $y_i = (y_{1i},...,y_{Mi})$, and consider the linear programming problem

$$\max_{\phi,\lambda} \phi \quad \textit{subject to} \quad \begin{array}{ll} \phi y_{mo} \leq \sum_i \lambda_i y_{mi} & m = 1,...,M \\ \sum_i \lambda_i x_{ni} \leq x_{no} & n = 1,...,N \\ \lambda_i \geq 0 & i = 1,...,I \\ \sum_i \lambda_i = 1. & \end{array} \tag{A2.1}$$

This problem seeks the maximum equiproportionate expansion in the service vector of provider "o" which is attainable without exceeding best practice standards observed in the sample. Solving this problem I times, once for each provider in the sample, generates an M+N-dimensional DEA best practice service delivery frontier analogous to that illustrated in Figure 2.15. The frontier provides the tightest possible envelopment of the data, consistent with the maintained hypotheses of convexity and monotonicity (free disposability of resources and services).

In addition to constructing a best practice service delivery frontier, DEA also generates efficiency scores for each provider. Best practice providers are located on the service delivery frontier and receive a service-oriented technical efficiency score $TE_o(x,y) = \phi^{-1} = 1$, because it is impossible for them to expand their service provision without expanding their resource bases. Inefficient providers are located beneath the service delivery frontier and receive service-oriented technical efficiency scores $TE_o(x,y) = \phi^{-1} < 1$, because they are capable of expanding their service provision by $100\cdot(\phi^{-1})$% without violating best practice standards observed in the sample. The linear programs also identify, by means of the nonzero values of λ, the best practice role models for each relatively inefficient provider.

The DEA program can be altered in a number of ways, depending on data availability and behavioral objective. For example, given the same data, a resource conserving performance evaluation can be conducted by reversing the orientation to

$$\max_{\theta,\lambda} \theta \quad \textit{subject to} \quad \begin{array}{ll} y_{mo} \leq \sum_i \lambda_i y_{mi} & m = 1,...,M \\ \sum_i \lambda_i x_{ni} \leq \theta x_{no} & n = 1,...,N \\ \lambda_i \geq 0 & i = 1,...,I \\ \sum_i \lambda_i = 1. & \end{array} \tag{A2.2}$$

which provides resource-oriented measures of technical efficiency $TE_i(y,x) \leq \theta \leq 1$ relative to the same best practice standards. The best practice role models identified by the nonzero values of— are generally different, because the orientation is different.

Both DEA models allow for variable returns to scale. The local nature of scale economies can be inferred by comparing the results of either model with the results of a similarly oriented model in which constant returns to scale is imposed by deleting the constraint $\sum_i \lambda_i = 1$. Details are available in Färe,

Grosskopf and Lovell (1983). In addition, both DEA models impose convexity on the production set, implying that if a pair of production plans is feasible, so are all convex combinations of them, regardless of what the engineers say. It is possible to maintain free disposability and relax convexity to generate a production set that is bounded by a monotonic step function. Such an approach has been proposed by Deprins, Simar and Tulkens (1984), who call their approach FDH ("free disposal hull").

In a similar fashion, indirect measures of technical efficiency can be derived by replacing x with (w/B) to obtain $ITE_o(w/B,y) = \phi^{-1} \leq 1$, or by replacing y with (p/R) to obtain $ITE_i(p/R,x) = \theta \leq 1$. Alternatively, direct and indirect measures of economic efficiency can be obtained from alternative modifications to the above programs.

Characteristics of the operating environment can be incorporated into a DEA performance evaluation in either of two ways. The first way is to add a set of K additional constraints of the form

$$\sum_i \lambda_i z_{ki} \geq z_{ko} \qquad k = 1,\ldots,K \tag{A2.3}$$

to DEA problems such as those given above. Adding these constraints ensures that each provider is evaluated relative to other providers in equally or less favorable operating environments. It is also possible to reverse the inequality and evaluate each provider relative to other providers in equally or more favorable operating environments. However, Kerstens and Vanden Eeckaut (1999) express concerns about this approach. A second approach does not require that the directions of the environmental effects be known in advance and is not subject to the concerns raised by Kerstens and Vanden Eeckaut (1999). In this approach the initial DEA efficiency scores are regressed against the environmental variables in a second stage model of the form

$$\ln \phi_i = \ln g(z_i;\gamma) + v_i - u_i, \quad i = 1,\ldots,I. \tag{A2.4}$$

This regression seeks to explain variation in initial efficiency scores in terms of three sources: (i) variation in the operating environment, captured by the term $g(z_i;\gamma)$; (ii) the impact of statistical noise (which was ignored in the deterministic DEA model), captured by the symmetric error component $v_i \sim N(0,\sigma_u^2)$; and (iii) managerial inefficiency remaining after the impacts of the operating environment and statistical noise have been accounted for, captured by the nonpositive error component $-u_i \leq 0$. In this approach $[g(z_i;\gamma) + v_i]$ is interpreted as a stochastic best practice managerial performance frontier, and the performance of individual providers is evaluated by means of $\phi_i \,/\, [g(z_i;\gamma) + v_i] = \exp\{-u_i\} \leq 1$, $i = 1,\ldots,I$.

A 2.2 Stochastic Frontier Analysis

The basic idea behind SFA can be described as follows. Suppose that a group of providers, indexed $i = 1,\ldots,I$, uses a vector of N resources $x_i = (x_{1i},\ldots,x_{Ni})$ to provide a single service y_i, and consider the regression model

$$\ln y_i = \ln f(x_i;\beta) + v_i - u_i, \quad i = 1,\ldots,I. \tag{A2.5}$$

where $f(x_i;\beta)$ is the deterministic kernel of the stochastic service delivery frontier $[f(x_i;\beta)\cdot\exp\{v_i\}]$, and b is a service delivery technology parameter vector to be estimated. The deterministic kernel of the SFA stochastic service delivery frontier is illustrated in Figure 2.15. The parameter vector β characterizes the structure of the service delivery frontier, and among other things enables one to draw inferences concerning the nature of economies of scale in service delivery.

The composed error term $(v_i - u_i)$ is what distinguishes this regression model from old-fashioned production function regression models. The first error component $v_i \sim N(0,\sigma_v^2)$ is traditional, and captures the effects of statistical noise. The second error component $u_i \geq 0$ is relatively new, and is intended to capture the effects of inefficiency in service delivery. Since $u_i \geq 0$, actual service provision is less than or equal to maximum possible service provision attainable in an operating environment characterized by v_i. Thus the stochastic service delivery frontier $[f(x_i;\beta)\cdot\exp\{v_i\}]$ envelops the data from above, but in a stochastic fashion allowing for the effects of statistical noise captured by the symmetric error component v_i. Departures from the frontier are captured by the one-sided error component u_i, and are attributed to inefficiency in service delivery.

The regression is estimated by maximum likelihood techniques because the composed error term has nonzero mean, since $E(v_i - u_i) = E(-u_i) \leq 0$. All that is required is to impose a tractable distributional assumption on $u_i \geq 0$, such as half normal or exponential. After estimation, the output-oriented technical efficiency of each provider is estimated as the ratio of observed output to maximum output attainable in a noisy environment, and so $TE_o(y_i,x_i) = \exp\{-u_i\} = y_i / [f(x_i;\beta)\cdot\exp\{v_i\}] \leq 1$, $i = 1,\ldots,I$; details are available in Jondrow, Lovell, Materov and Schmidt (1982).

The SFA approach can be modified to generate several alternative measures of efficiency. For example, a service-oriented indirect measure of technical efficiency $ITE_o(w/B,y)$ can be obtained by replacing x_i with $(w/B)_i$, a direct measure of revenue efficiency $RE(x,y,p)$ can be obtained by replacing y_i with $(py)_i$, and an indirect measure of revenue efficiency $IRE(w/B,y,p)$ can be obtained by replacing x_i with $(w/B)_i$ and y_i with $(py)_i$. Finally, if the error structure is modified, a measure of cost efficiency $CE(y,x,w)$ can be derived from

$$\ln (wx)_i = \ln C(y_i,w_i;\beta) + v_i - u_i, \tag{A2.6}$$

where $C(y_i,w_i;\beta)$ is the deterministic kernel of the stochastic cost frontier $[C(y_i,w_i;\beta)\cdot\exp\{v_i\}]$, v_i represents the effects of statistical noise, and $u_i \geq 0$ represents cost inefficiency. After estimation, the cost efficiency of each provider is measured by $CE(y_i,x_i,w_i) = [C(y_i,w_i;\beta)\cdot\exp\{v_i\}] / (wx)_i \leq 1$, $i = 1,\ldots,I$. Alternative direct and indirect technical and economic efficiency measures can be generated within the SFA framework with suitable modifications to either the data or the error structure.

Environmental variables can be incorporated into an SFA performance evaluation in two ways. The first approach parallels the first DEA approach. An SFA model is modified by including the environmental variables in the deterministic kernel of the stochastic service delivery frontier to obtain

$$\ln y_i = \ln f(x_i,z_i;\beta) + v_i - u_i, \qquad i = 1,\ldots,I. \tag{A2.7}$$

The resulting output-oriented measure of technical efficiency is given by $TE_o(y_i,x_i;z_i) = y_i / [f(x_i,z_i;\beta)\cdot\exp\{v_i\}] = \exp\{-u_i\} \leq 1$, $i = 1,\ldots,I$. This measure incorporates the environmental effects into the performance evaluation. This approach can be applied to any of the SFA models discussed in Section 2.4.

The second approach to incorporating environmental variables into an SFA performance evaluation simply relocates the z vector from the deterministic kernel of the stochastic frontier to the one-sided error component. Again using the service-oriented model given above, we have

$$\ln y_i = \ln f(x_i,z_i;\beta) + v_i - u_i(z_i;\gamma), \quad i = 1,\ldots,I. \tag{A2.8}$$

In this formulation the directions and magnitudes of the environmental impacts on performance are determined by the estimated environmental effects parameter vector γ. The only difference between the two SFA approaches is the interpretation of how the environment influences performance. In the first

approach the environment influences the technology with which resources are converted into services, and so indirectly influences performance. In the second approach the environment influences the efficiency with which providers utilize the technology at their disposal, and so directly influences performance.

References

Aigner, D.J., C.A.K. Lovell, and P. Schmidt. 1977. Formulation and Estimation of Stochastic Frontier Production Function Models. *Journal of Econometrics* 6 (1): 21-37.

Banker, R.D., A. Charnes, and W.W. Cooper. 1984. Some Models for Estimating Technical and Scale Inefficiencies in Data Envelopment Analysis. *Management Science* 30 (9): 1078-1092.

Charnes, A., W.W. Cooper, A.Y. Lewin, and L.M. Seiford (eds.). 1994. *Data Envelopment Analysis: Theory, Methodology and Applications*. Dordrecht: Kluwer Academic Publishers.

Charnes, A., W.W. Cooper, and E. Rhodes. 1978. Measuring the Efficiency of Decision-Making Units. *European Journal of Operational Research* 2 (6): 429-444.

Coelli, T., D.S. Prasada Rao, and G.E. Battese. 1998. *An Introduction to Efficiency and Productivity Analysis*. Dordrecht: Kluwer Academic Publishers.

Deprins, D., L. Simar, and H. Tulkens. 1984. Measuring Labor-Efficiency in Post Offices. in Marchand, M., P. Pestieau, and H. Tulkens (eds.), *The Performance of Public Enterprises: Concepts and Measurement*. Amsterdam: North-Holland.

Färe, R., S. Grosskopf, and C.A.K. Lovell. 1983. The Structure of Technical Efficiency. *Scandinavian Journal of Economics* 85 (2): 181-190.

Färe, R., and D. Primont. 1995. *Multi-Output Production and Duality: Theory and Applications*. Dordrecht: Kluwer Academic Publishers.

Fried, H.O., C.A.K. Lovell, and S.S. Schmidt (eds.). 1993. *The Measurement of Productive Efficiency: Techniques and Applications*. New York: Oxford University Press.

Fried, H.O., C.A.K. Lovell, S.S. Schmidt, and S. Yaisawarng. 1999. Accounting for Environmental Effects and Statistical Noise in Data Envelopment Analysis. Working Paper. New York 12308, USA: Department of Economics, Union College, Schenectady.

Ganley, J.A., and J.S. Cubbin. 1992. *Public Sector Efficiency Measurement*. Amsterdam: North-Holland.

Jondrow. J., C.A.K. Lovell, I.S. Materov, and P. Schmidt. 1982. On the Estimation of Technical Inefficiency in the Stochastic Frontier Production Function Model. *Journal of Econometrics* 19 (2/3): 233-38.

Kerstens, K., and P. Vanden Eeckaut. 1999. A New Criterion for Technical Efficiency Measures: Non-monotonicity Across Dimensions Axioms. *Managerial and Decision Economics* 20 (1): 45-59.

Kumbhakar, S.C., and C.A.K. Lovell. 2000. *Stochastic Frontier Analysis*. New York: Cambridge University Press.

Meeusen, W., and J. van den Broeck. 1977. Efficiency Estimation from Cobb-Douglas Production Functions with Composed Error. *International Economic Review* 18 (2): 435-44.

Part Two:
Health Services

3 Comparing Techniques for Measuring the Efficiency of Australian Private Hospitals

Richard Webster, Steven Kennedy and Leanne Johnson[1]

3.1 Introduction

The impetus for this paper was an Australian Bureau of Statistics (ABS) project that aimed to improve the measurement of outputs and inputs for the non-market sector, and more generally for the services sector, of the Australian economy. In the first phase of the project, the Methodology Division of the ABS constructed experimental estimates of outputs and inputs for the health industry. The primary measure of hospital output was constructed using Diagnostic Related Group (DRG) cost weights to aggregate the treatments provided by hospitals – though it was noted that other measures such as occupied bed days and deflated patient revenue had also been used by other investigators to measure output. An ABS dataset on Australian private hospitals provided a unique opportunity to compare direct or volume-based measures of output, such as DRG based measures or occupied bed days, with measures that arise out of the market context in which private hospitals operate, that is, patient revenue-based measures. This enabled comparison of (and, to a degree validation of) our use of certain output indicators for the non-market sector and more broadly for the whole service sector of the economy.

The recent emergence of unit record or firm-based frontier techniques (which not only provide important firm-level information to managers or policy makers, but also provide aggregate information on the change in efficiency and productivity over time) has given further impetus for the ABS to undertake this style of research. The ABS is in a unique position to do such analyzed, given its access to unit record files like the Australian private hospitals dataset.

Section 3.2 of this chapter describes the institutional and financial structure of the Australian hospital sector. Section 3.3 briefly introduces the techniques used in the

analysis. Section 3.4 discusses the measurement of variables representing input quantities and prices, output quantities and total operating costs. Section 3.5 presents the results of applying the Data Envelopment Analysis (DEA) and Stochastic Frontier Analysis (SFA) techniques and compares selected results.[2] Section 3.6 concludes this chapter and contains some suggestions for further work. The paper also contains an appendix which presents correlation coefficients for selected variables.

3.2 Australian Hospitals

The Australian health services sector consists of two major components, the public health system and the private health system. The funding for these systems is complex, and is sourced from all three levels of Australian government[3] and the private sector. A major proportion of the funding for each of these systems is consumed by hospitals, particularly acute care hospitals. In 1994-95, 36.6% of recurrent health services expenditure was accounted for by public acute care and private hospitals. Within these expenditures, public acute care hospitals accounted for 28.7% and private hospitals for 7.9% (Australian Institute of Health and Welfare 1998)

The unequal expenditure shares of public and private hospitals are indicative of their unequal market share. In 1994-95, there were 672 officially recorded public acute care hospitals, which accounted for 3.37 million patient separations and 13.9 million bed days. In the same period, there were 328 private hospitals, accounting for 1.4 million separations and 5.3 million bed days. Further evidence of the public hospital systems' dominance of the market is that private hospitals tend to be smaller than public acute care hospitals. In 1994-95, on average, public acute care hospitals had approximately 80 available beds whereas on average private hospitals had approximately 68 available beds. Before the 1980s, the private hospitals tended to be discrete and segmented. However, since the 1980s, the sector has moved more towards chains of private hospitals and in some cases, doctors and private health insurance funds have formed partnerships. There has also been an increase in communication and co-operation between private and public hospitals although as recently as 1992, funding barriers prevented patients from easily switching between the two sectors (Grant and Lapsley 1993). These changes reflect the desire of Australian governments to consolidate the various small, scattered hospitals into larger facilities offering a greater variety of services. The private sector is beginning to reflect this change with various examples of partnerships between and co-location of private and public hospitals to allow cost sharing and an extension of available facilities (Dusevic 1998). Largely this is driven by the increasing costs of health care and a desire to minimize these costs to the public.

Despite an increase in private hospital activity, membership in private health insurance funds has been declining since 1984 (Fine 1995), coinciding with the introduction of Medicare, a Commonwealth and State Government system of universal health insurance available to all persons normally resident in Australia. The proportion of the population holding private health cover fell from 50% in June 1984 to 38.4% in December 1993 (Australian Institute of Health and Welfare 1998) . On 1 July 1997, the Commonwealth Government introduced the Private Health Insurance Incentives Scheme (PHIIS) designed to provide payments to low and middle income earners with private health insurance (Commonwealth Department of Health and Family Services 1998). The Commonwealth Government also introduced new tax measures to encourage high income earners to take out private health insurance. Both policies are attempts to encourage a greater proportion of the population to take up private health insurance.

Table 3.1 Summarizes the descriptive statistics for the private acute hospitals in Australia in 1994-1995.

Table 3.1

Australian Private Acute Care Hospitals, 1994-1995

	1994-1995
Number of Private Hospitals	
Acute	301
Psychiatric	27
Total	328
Size of Private Hospitals	
0-25 beds	69
26-50 beds	99
51-100 beds	101
101-200 beds	45
over 200 beds	14
Number of available beds	
Acute	20,900
Psychiatric	1,430
Total	22,400
Total number of staff	36,600
Staff per occupied bed	2.5
Total number of separations (x 1,000)	1,350
Number of hospitals with specialized units	237
Total revenue (x AUD mln.)	2,760
Percentage of total revenue which is patient revenue	94.7

3.3 The Techniques

3.3.1 *Data Envelopment Analysis*

The use of DEA in the study of hospital efficiency, both public and private, is relatively common (for example, Banker, Conrad and Strauss 1986; Grosskopf and Valdmanis 1987; Register and Bruning 1987; Färe, Grosskopf and Valdmanis 1989 and Valdmanis 1992). Most authors cite the inherent flexibility of the DEA technique as a major attraction for its use in such studies. Another reason for the use of the DEA technique arises when there is a lack of realistic price data for hospital inputs and outputs.

While there is general agreement about the applicability of DEA to evaluate hospital efficiency, a number of features of the technique may worry many researchers in the field. Two important problem areas are: the assumption that there is no 'noise' (or error) in the data being studied; and the lack of a definite functional form encapsulating the production technology. The latter, whilst a strong argument for the technique in many studies, raises the problem of what method should be used to evaluate the results of a DEA study, mainly due to the inability to perform the usual diagnostic tests associated with regression estimation (see also Section 2.4).

Valdmanis (1992) (based on Nunamaker 1985) suggests, as a possible answer to these problems, that DEA researchers run a number of different models from each dataset and evaluate the sensitivity of the results to changes in model specification. These changes may take the form of alternative input and output definitions, or even different populations within each dataset. The purpose of this sensitivity analysis is to assess whether the ranking and efficiency of an individual firm is variable-specific (or model-specific) or whether the results are robust to changes in dataset specification. Valdmanis (1992) cautions that '... for a model to be considered robust, it must be shown that minor changes in the list of variables cannot alter fundamentally the conclusions of the DEA model.'

Another method of evaluation is to compare the results of a DEA study with results from other efficiency evaluation methods applied to datasets comprising similar observations and variables. One alternative efficiency evaluation method is SFA.

In this paper, we assess the DEA technique by comparing results obtained using a selected input-output specification with results from different dataset specifications and SFA.

3.3.2 *Stochastic Frontier Analysis*

A number of recent studies have applied SFA to hospital datasets, usually to measure relative efficiencies (for example, Newhouse 1994; Vitaliano and Toren 1994 and Zuckerman, Hadley and Iezzoni 1994). In these studies, cost functions were estimated rather than production functions. Cost functions are often estimated to control for the biases that arise if the production function is estimated directly,[4] to represent a multiproduct firm, or because analysts are interested in measuring allocative efficiency as well as technical efficiency or a combination of these two efficiencies, cost efficiency. This paper estimates both production and cost functions for a number of reasons: to facilitate comparisons between other techniques; because of the likely fallibility of the price data used in this study; and because of uncertainty in determining what environment private acute hospitals operate in and how their economic behavior is affected.[5]

SFA is often used to compare firms' relative efficiencies, though it can also be used to derive estimates of productivity change over time.[6] The technique has a number of benefits when compared to standard econometric estimation of production functions.[7] It estimates a 'true' production frontier rather than an average frontier, thus it fully represents the maximal properties of the production function. One important implication of estimating the frontier is that measured productivity change will represent pure technological change rather than a combination of efficiency change and technological change, which is the case when using non-frontier techniques.

SFA has some advantages over non-parametric techniques, such as DEA, for estimating frontiers, efficiency and productivity; in particular, it is able to account for measurement error. However, in this study, the stochastic frontier modeling had some constraints which DEA did not including: only one output could be accommodated when modeling production functions, and the need to select functional forms for both the production structure and error components. Thus, the SFA technique for measuring efficiency and productivity provided an alternative approach for dealing with errors, but at the cost of using a more restrictive model specification than DEA.

3.4 Data

Data were obtained from the ABS Private Health Establishments Collection (PHEC), an annual census of private hospitals covering all private acute care and psychiatric hospitals operating in Australia. We chose to exclude private psychiatric hospitals due to their substantially different characteristics and mode of operation. Various char-

acteristics of private acute hospitals can be identified in the dataset including: whether a hospital is characterized as (1) For-profit (FP), (2) Not-for-profit, religious or charitable (NFPr) or (3) Not-for-profit, other (NFPo); size proxied by the number of beds. The population of acute care private hospitals in 1994-95 comprised 301 observations, composed of 155 FP, 70 NFPr and 76 NFPo. When estimating cost functions only 255 hospitals were used due to missing data or zeros problems.[8]

3.4.1 *Input Variables*

Input variables were constructed in order to measure the quantity of labor, capital and intermediate inputs. Table 3.2 shows the definition of the various input variables used in this study. Labor inputs were measured by total full-time equivalent (FTE) staff. This measure included salaried medical officers (SMOs) but did not include visiting medical officers (VMOs), as data on the hours worked or days worked by VMOs were not available. To capture the input of VMOs a labor measure, based on wages paid to all staff plus the cost (contract value) of VMO services, was derived.[9] Neither the FTE or dollar based measure of labor input is ideal, the FTE measure because of the high level of aggregation of employment groups, and the dollar-based measure because movements represent changes in not only volumes but also prices of labor (wages).

In the DEA study, models including lower levels of aggregation in measuring labor inputs were able to be developed. Variables representing the inputs of SMOs, nursing staff and other staff are used in a number of DEA models.

In all models, the number of available beds is used as a proxy for the level of capital stock. This is a common feature of studies of hospital efficiency, since reliable measures of the value of assets are often unavailable or difficult to obtain, as was the case for this study.[10]

Intermediate inputs were measured as the total dollar value of all non-labor, non-capital expenditure, covering expenditure on drugs and medical supplies, food, repairs and maintenance, and patient transport services. When volume-based measures of labor inputs were used the contract costs of VMOs were also included in the measure of intermediate inputs. This was not entirely unsatisfactory as VMO services were purchased similarly to other intermediate services and unlike salaried labor. However, the characteristics of VMO services indicate that they are more like a labor service than a capital or non-labor service.

A fourth input class representing the total patient input was considered in the light of the argument stated by Valdmanis (1992), '... just as raw materials such as iron are inputs to a steel mill, so "sick" people are inputs to the hospital production process'. The actual variable used was the total number of inpatient separations. This approach to explaining hospitals' production processes tends to view the output of hospitals as health outcomes rather than a process type output. The output measures in our analysis are focused on the process type or production volume style estimates of output and we are only interested in health outcomes in the sense that outcomes reflect the quality of outputs. In any case, the dataset does not contain measures which would enable us to measure the improvement in the health of sick persons over the course of their hospital stay.
Costs were calculated as the total expenditure of hospitals minus expenditure on new capital goods.

3.4.2 *Output Variables*

This study examined a number of measures of private acute care hospital output. A frequently used measure of hospital output is case-weighted separations where the case weights reflect the severity of the different cases treated by the hospital. This type of output measure was constructed using disease-costing weights provided by the Australian Institute of Health and Welfare. However, it was found that the number of disease classifications was too low to reweight effectively the raw separations. This led to movements in weighted separations being almost the same as movements in unweighted separations.[11]

To avoid this problem, a composite estimate of hospital output was constructed based on occupied bed days. Although this measure of output is less than ideal, it at least incorporates some element of adjustment for disease severity. The occupied bed day measure of output is focused on inpatient care provided by hospitals. In order to capture non-inpatient care and reflect this in a single measure of output, non-inpatients were weighted according to their relative costs to form an overall measure of occupied bed days.

When undertaking DEA, output was disaggregated, for example, into types of occupied bed days and non-inpatient activities. Output disaggregation was not possible in SFA, due to the problem of logarithms of zero values.

Patient revenue was also used to measure hospital output. This was done for two reasons; firstly as an alternative to the occupied bed day measure of outputs; and secondly

because deflated revenues are often used in National Accounting exercises to measure output and it would be interesting to see how this measure of output compared with direct volume measures using occupied bed days. When conducting analysis on cross-section data there is no need to deflate revenue, however it still needs to be assumed that prices are the same across hospitals. The patient revenue measure of output may be less appropriate for NFP hospitals than FP hospitals.

Despite having alternative measures of output, neither measure is ideal and some aspects of the outputs of hospitals have failed to be accounted for, for example, research and development.[12] Another aspect which has not been adequately accounted for in measuring outputs has been quality dimensions of outputs. The occupied bed day measure does not adjust at all for changes in quality between hospitals. The patient revenue measure may adjust for quality between hospitals if difference in price between hospitals reflects differences in quality. The authors plan to extend this analysis by either augmenting the output measure or directly including variables which measure differences in the quality of outputs between firms .[13]

A future option for developing an improved output measure is to make use of the increasing availability of information on separations and occupied bed days classified by detailed DRG. Each DRG represents a class of patients with similar clinical conditions requiring similar hospital services. A cost-weighted separations measure, based on DRG data, would adjust for changes in quality which are due to changes in the mix of cases across hospitals or over time, for example, hospitals which focus on providing complex, high-technology surgical services will record higher output than hospitals which focus on nursing home type care but have the same number of separations. However, changes in quality of output within DRGs will still not be captured.

3.4.3 *Input Prices for Cost Functions*

An average price of labor was calculated by dividing wages by FTE for all employees. It was possible to calculate prices for labor at a more disaggregated level, for example, nursing and non-nursing prices for the DEA analysis.[14]

We were unable to calculate directly a price for intermediate inputs as volume measures of intermediate inputs were unavailable. As a result, the dollar-based measure or value of intermediate inputs was divided by the composite occupied bed day measure of output to produce a price per unit of output for intermediate inputs.[15]

The price of capital or the price of the flow of capital services was also a difficult concept to measure adequately. The standard way of measuring the price of capital would be to divide the value of capital inputs (i.e. the user cost of capital) by the quantity of capital inputs (e.g. the real value of the capital stock). The user cost of capital consists of depreciation, the nominal opportunity cost of holding capital and changes in the nominal price of capital.[16] However, due to data limitations the measure of the price of capital used in this paper can only be considered a rough approximation to this concept. The measure of the price of capital was based on interest payments plus reported depreciation divided by the number of beds. An additional issue is the potential inappropriateness of this standard formulation of the price of capital to the non-market sector (e.g. NFP hospitals).

3.4.4 *Descriptive Statistics*

Correlation coefficients for a range of input and output measures are given in Appendix A3.1. Table 3.2 presents a consolidated list of the definitions of input and output quantity variables and some descriptive statistics for these variables. An interesting feature of the descriptive statistics is the relatively small variation in input variables, between hospitals, but the larger variation in disaggregated output variables such as nursing home bed days and different types of surgery. This is an indication that hospitals may use similar quantities of inputs but produce quite different outputs.

Table 3.2

VARIABLE DEFINITIONS AND DESCRIPTIVE STATISTICS

	Definition	*Mean*	*Standard Deviation*
SMO	Total FTE professional medical officers	1.14	2.70
VMO	Total contract value (AUD) allied and medical health services	79,800	288,000
Nursing staff	Total FTE nursing staff (registered, enrolled and student/trainee and other nurses)	65	87
Other staff	Total FTE of staff other than medical professionals and nursing staff	51	72
Beds	Average number of total available beds (calculated on monthly figures)	70	161
Materials	Total value of recurrent expenditure on non-labor items (total recurrent expenditure less wages and salaries, superannuation, payroll tax, depreciation and VMO contract services) (x 1,000)	2,730	3,950

Table 3.2 (continued)

Variable Definitions and Descriptive Statistics

	Definition	Mean	Standard Deviation
Admissions	Weighted sum of inpatient, same day and accident and emergency separations, and non-inpatient occasions of service	5,780	8,720
Total staff I	Total FTE all staff employed	120	160
Total staff II	Total value (AUD) of wages, salaries and contracts for all staff and professionals employed (x 1,000)	4,780	6,550
Acute care inpatient days	Total of inpatient and same day patient occupied beds days less nursing home type occupied bed days and surgery bed days	6,680	6,760
Accident and emergency treatments	Total number of accident and emergency (or casualty) treatments less patients admitted by presentation at accident/emergency department	450	1,970
Non-inpatient occasions	Total non-inpatient occasions of service (excluding accident/ of service emergency and admitted patients)	2,970	14,400
Nursing home ID	Total nursing home type occupied bed days	650	3,970
Surgical procedures	Total number of surgical procedures performed (including advanced and minor surgery and obstetric procedures)	3,560	4,410
Advanced surgery	Total number of advanced surgery occupied bed days	2,610	4,770
Surgery	Total number of surgery occupied bed days	4,130	5,240
Minor surgery	Total number of minor surgery occupied bed days	1,430	2,300
Obstetrics	Total number of obstetrics occupied bed days	1,300	2,840
Total Surgery	Sum of Advanced, Surgery, Minor and Obstetrics	9,480	13,100
Psychiatric care	Total number of psychiatric occupied bed days	140	750
Rehabilitation	Total number of rehabilitation occupied bed days	780	2,950
Medical	Total number of medical occupied bed days	5,750	6,250
Composite inpatient separations	Cost weighted sum of separations, with separations classified by principal diagnosis into 18 categories (in accordance with ICD9-CM Vol 1)[a]	4,340	5,010
Composite output I	Weighted sum of inpatient occupied bed days and non-inpatient occasions of service	17,600	19,600
Revenue	Total patient revenue from admitted and non-admitted patients (x 1,000)	8,300	11,300

a ICD9-CM : *International Classification of Diseases Volume 9 - Clinical Modification.*

3.5 Results

3.5.1 DEA Results

The first part of this section discusses results obtained from applying the DEA technique to a number of dataset model specifications. These specifications, based on the variable definitions in Section 3.4, are presented below in Table 3.3, which shows the input and output combinations used in 12 model specifications. Models 2 to 8 are based on slight modifications of model 1, which is used as the preferred model for studying hospital efficiency. Models 9 to 12 are based on models used later in SFA.

Model 1 was chosen as the preferred model because it was decided that occupied bed days was conceptually a better measure of output than separations. The model also gives a sensible spread of efficiency scores for the whole sample and contains a plausible number of variables (input plus outputs) when compared with the size of the overall sample. Models 9 to 12 are derived directly from models used in the SFA section of this paper, and are used as comparison tools between DEA and SFA.

The DEA technique provides relative efficiency scores for a particular sample. One important consideration in this analysis is whether any patterns observed are common to a number of models when variable definitions and numbers change between the models.

Sensitivity tests were conducted by varying the initial model, through the use of different input and output definitions and by dividing the sample into appropriate sub-samples, and testing for similarities between efficiency score distributions. We would conclude that the results were robust if similar patterns were observed across a range of model specifications.[17]

We examined the reliability of the DEA results by comparing the efficiency rank or ordering of hospitals between models. If the DEA results are to be considered reliable the efficiency ranking of hospitals should be similar between different models.[18]

The DEA technique is very susceptible to outliers in the data because of its non-stochastic nature. This is particularly the case where an observation contains inputs which are significantly smaller, or outputs which are significantly larger, than other observations employing a similar input mix or producing a similar output level. An outlier analysis of the current dataset[19] indicates that most hospitals use inputs and produce outputs commensurate with size, so that no significant outliers were discovered.

Table 3.3

MODEL SPECIFICATIONS[a]

	1	2	3	4	5	6	7	8	9	10	11	12
Inputs												
SMO (FTE)	•	•	•	•	•	•	•	•				
VMO (AUD contract value)	•	•	•	•	•	•	•	•				
Nursing staff (FTE)	•	•	•	•	•		•	•				
Other staff (FTE)	•	•	•	•	•	•[b]	•	•				
Beds	•	•	•	•	•	•	•	•[c]	•	•	•	•
Materials (non-labor costs)	•	•	•	•	•	•	•[d]	•	•[e]	•	•[e]	•
Admissions			•									
Total staff I (FTE)									•		•	
Total staff II (labor costs, AUD)										•		•
Outputs												
Acute care Inpatient days	•	•		•		•	•	•				
Psychiatric care Inpatient days			•									
Rehabilitation days			•									
Medical care days			•									
Surgery inpatient days	•			•		•	•	•				
Advanced surgery days			•									
Surgery days			•									
Minor surgery days			•									
Obstetrics days			•									
Non-inpatient occasions of service	•	•	•	•	•	•	•	•				
Nursing home type inpatient days	•	•	•	•	•[f]	•	•	•				
Surgical procedures		•			•							
Acute care inpatient separations					•							
Accident/emergency	•	•	•	•	•	•	•	•				
Composite output I									•	•		
Total inpatient revenue											•	•

a *An X in the table indicates that the variables are included in the model.*
b *Other staff in this case includes nursing FTE.*
c *Non-discretionary input.*
d *Material costs (drugs and medical supplies, food and other domestic services).*
e *Non-labor costs plus VMO contract valuation.*
f *Nursing home type separations.*

One interesting feature of the dataset related to the input-output mix of large hospitals. A significant number of the largest hospitals in the sample (those with more than 225 beds) showed significantly higher than average use of most (or all) inputs, but in most cases showed above average production in only a small number of the outputs. As a result, larger hospitals may appear at the lower end of the efficiency score range in the DEA analysis, in part because the data fails to adequately capture the complexity and technological advancement of procedures performed.

The results of applying DEA to each model outlined in Table 3.3 for the entire sample of private hospitals (301 hospitals) are presented in Table 3.4,[20] showing the mean efficiency score and standard deviation for each model for various measures of efficiency.[21]

Table 3.4

MEAN EFFICIENCY SCORES, BY MODEL SPECIFICATIONS

	Technical efficiency		*Pure technical efficiency*		*Scale efficiency*	
All hospitals	*Mean efficiency score*	*standard error*	*Mean efficiency score*	*standard error*	*Mean efficiency score*	*standard error*
1	0.73	0.17	0.82	0.18	0.91	0.11
2	0.54	0.22	0.68	0.25	0.82	0.20
3	0.86	0.16	0.90	0.16	0.96	0.07
4	0.85	0.14	0.88	0.14	0.97	0.05
5	0.44	0.22	0.59	0.28	0.81	0.23
6	0.74	0.19	0.79	0.20	0.94	0.09
7	0.77	0.20	0.81	0.20	0.94	0.10
8	0.70	0.18	0.80	0.19	0.88	0.14
9	0.29	0.11	0.41	0.18	0.76	0.20
10	0.28	0.11	0.39	0.18	0.77	0.19
11	0.67	0.16	0.69	0.17	0.96	0.06
12	0.73	0.15	0.75	0.15	0.97	0.06

The results in Table 3.4 indicate that:

1. As expected, the inclusion of additional variables or the disaggregation of existing variables (while holding the number of observations constant) has the effect of increasing efficiency scores for observations which were not previously fully efficient. This effect can be seen by the difference in average efficiency between models 3 and 4 and model 1. This effect is discussed in detail in Nunamaker (1985).[22]

2. In all models scale efficiency is greater than 75% (and greater than 90% in over half of the models) indicating that scale inefficiency is less important than pure technical inefficiency as a source of private hospital inefficiency.
3. In some cases, a change in the definition of a variable (such as the definition of surgery output from inpatient days to number of procedures (model 1 to 2)) had a large effect on the average level of efficiency of the sample.
4. Changing the output measure from inpatient days to separations (models 1 to 5), changed the average technical efficiency score for the sample from 73.4% in model 1 to 44.1% in model 5.
5. Changing the definition of output from composite bed days to total patient revenue had the effect of significantly increasing the average efficiency score (comparing model 9 with 11 and model 10 with 12).

We identified two explanations for average technical efficiency falling when separations are used instead of occupied bed days as a measure of output. Firstly, the occupied bed days measure captures the output mix of different types of hospitals better than separations (which may discriminate against hospitals which specialize in treatments requiring longer hospital residence). An alternative explanation is that by using occupied bed days we are "rewarding" hospitals which have a relatively slow patient turnaround over hospitals which treat patients 'efficiently' and quickly. Possibly the reason for the result is a combination of the two explanations, an observation which led the authors to look at models including average inpatient days per separation or including both inpatient days and separations output variables. The results of these models (not reported here) were almost identical to those from model 1, prompting the authors to adopt inpatient days as the output measure in the preferred model. The diversified mix of hospital types, and the 'output' of each, seems to be captured best by inpatient days in this instance.[23]

In the style of Valdmanis (1992), the results from each model were analyzed by considering a number of sub-sample structures within the sample of efficiency scores.[24] In most model specifications, the differences in technical and pure technical efficiency between NFP and FP hospitals were insignificant, and therefore robust to model specification changes. For NFPo and NFPr hospitals the mean technical efficiency and technical efficiency distributions were not robust to even minor changes in model specification. However, the results from pure technical and scale efficiency did appear to be more consistent over each model. In all cases, NFPr hospitals had a significantly higher level of pure technical efficiency and NFPo hospitals a significantly higher level of scale efficiency. Similar to the analysis of NFPo and NFPr models, using revenue as an output measure gave very different results to models using occupied bed days.

Another approach to examining the sensitivity of DEA results to model specification is to examine the breakdown of hospital efficiency score distributions by size. Hospital size was defined by the number of available beds, with the following definitions for three size categories: small (fewer than 25 beds), medium (25 to 100 beds), and large (more than 100 beds). Statistical tests suggested that the distribution of efficiency scores was different for different size categories.[25]

In summary, it appears as if the main cause of inefficiency in small and medium size hospitals is pure technical inefficiency, whereas in large hospitals the major cause is scale inefficiency. The results, in terms of average technical and pure technical efficiency, appear to be quite sensitive to the choice of model. This is particularly so when comparing the average efficiency scores for various sub-samples of hospitals (by ownership type and size).

While general patterns can be identified from the results not all models follow the identified patterns. The sensitivity analysis indicates that the technique is not robust to model specification changes, even when they are quite minor such as a definitional change to a single variable, implying that caution must be used when choosing a model for this type of analysis. When results are sensitive to variable set changes, a policy analyst must be able to determine what effect the particular variable set has on the results obtained and how results may differ when a different set is used in the analysis.

The results indicate that a majority of hospitals appear to operate in areas of decreasing returns to scale, and this is particularly true for hospitals in the medium and large size groups.[26] We note that the number of hospitals operating with constant returns may also depend on the chosen model specification, since hospitals which are categorized as operating under constant returns to scale (CRS) are those which appear fully efficient using the CRS model formulation. Models with fewer variables (inputs and outputs) will tend to show fewer hospitals operating under constant returns, due to the nature of the DEA technique (see Nunamaker 1985 for a discussion).

Congestion occurs in DEA when a firm is unable to dispose of unwanted inputs costlessly, creating a situation of negative marginal returns to inputs. The results indicate that congestion plays the smallest role in determining overall technical efficiency. On average, hospitals which have congestion inefficiency could reduce their inputs by a further 8.0%. Overall, hospitals which are not fully efficient could reduce their use of inputs by 26.6% compared with the most efficient hospitals in the sample. Scale inefficiency accounts for 9.4% of the inefficiency, with the remainder accounted for by 'pure' technical inefficiency.[27]

Input slacks occur in DEA when after reducing inputs proportionally it may still be possible that one or more inputs can be further reduced. For model 1 the results indicate that hospitals with a positive slack could use, on average, 2.77 fewer SMO FTE, AUD 193,000 fewer in VMO contracts, 16.38 fewer nursing FTEs, 19.78 fewer other staff FTEs, 3.71 fewer available beds and AUD 405,000 fewer in non-labor costs. Input slacks range from 12.2% to 65.2% of total input usage. [28]

3.5.2 *SFA Results*

A number of models were estimated where inputs and outputs were represented by a variety of variables, (see Table 3.5).

Table 3.5

Model Specifications[a]

	Model 1	Model 2	Model 3	Model 4
Inputs				
Beds	•	•	•	•
Materials I (non-labor costs)		•		•
Materials II (including VMOs)	•		•	
Total staff I (total FTE)	•		•	
Total staff II (labor costs)		•		•
Outputs				
Revenue			•	•
Composite output I (occupied bed days)	•	•		

a X indicates that the variable is present in the model. Models 1 through 4 correspond with models 9 through 12 in Section 3.5.

The results of estimating models 2 and 4 rarely differed from models 1 and 3 respectively and therefore are only reported where they did differ or to indicate the effect of variations in defining the input set. All inputs and outputs were transformed into logarithmic form for estimation. The results of estimating models 1 and 3 using a translog production function are presented in Table 3.6.[29]

An interesting feature of model 1 was that material input was negatively related to output, when material input usage rose output fell. This result may indicate the inability of the occupied bed day measure of output to capture appropriate differences in the severity of cases. Hospitals that treat high severity cases and use high levels of material

inputs will appear to have a similar level of output to hospitals that do not treat these cases and use far fewer material inputs.

Table 3.6

TRANSLOG PRODUCTION FUNCTION ESTIMATION, MODELS 1 AND 3 (N = 300)

	Model 1		Model 3	
	Coefficient	*Standard error*	*Coefficient*	*Standard error*
Constant	32.42	4.45	11.41	4.20
Beds	0.49	1.29	2.10	0.90
Materials	-5.31	0.91	-0.82	0.85
Labor	5.83	1.34	0.89	1.25
Beds squared	-0.20	0.26	0.08	0.19
Labor squared	0.20	0.18	-0.04	0.17
Materials squared	0.54	0.09	0.15	0.08
Beds x labor[a]	0.31	0.17	0.13	0.10
Beds x materials[a]	-0.05	0.13	-0.20	0.10
Labor x materials[a]	-0.51	0.14	-0.04	0.13
Diagnostics				
σ_u/σ_v	2.73	0.44	5.06	1.01
$\sqrt{(\sigma^2_u+\sigma^2_v)}$	0.53	0.01	0.42	0.01
Log likelihood	-99.00		4.50	

a *These terms represent the combinations of interactions between each of the three inputs in the translog functional form.*

Mean efficiency scores were calculated for each model and for selected hospital sub-populations within models. Hospital sub-populations were defined by size and profit status as for the DEA. Mean efficiency scores did not vary greatly between models using the same output measure. However, models using revenue typically had higher mean efficiency scores than models using occupied bed days. The pattern of efficiency scores across models was consistent between groups, that is, models which typically have higher efficiency scores do so for all sub-populations.

Comparing the ranks of hospital efficiency scores between models indicates that changing the output measure produces some change in the ranks of hospital efficiency scores. Hospitals which appear highly efficient in a model using occupied bed days as a measure of output won't necessarily appear as highly efficient in a model using revenue as a measure of output.

A preliminary conclusion drawn from estimating production functions was that differences in the inputs set were not impacting greatly on the results; however, differences in the output measure adopted produce different results both in terms of the underlying production technology and for the relative efficiency scores of hospitals.
One aspect of estimation which was further explored was whether the population of hospitals was suitably homogenous, that is, whether the inputs and outputs used were able to capture fully variation in individual production behavior. To understand better the production behavior of hospitals' models were estimated on sub-populations classified by hospital type, bed size and level of technology.

We did not find differences in the production structure of hospitals when we disaggregated by size and type for model 1 however, differences were evident for model 3. Size disaggregation produced similar results to hospital type. The differences identified in these results might not lie in production technologies but in the assumptions about output measurement. Therefore perhaps a better indicator to test differences in subsets might be one which relates more directly to technology.

We examined how similar hospitals were in terms of the level of technology inherent in their operation using a 'high tech' indicator. The 'high tech' indicator was calculated as the percentage of occupied bed days that were classified as surgical bed days (SBD). Four groups were derived representing hospitals with less than 25% of SBDs, 25 to 50%, 50 to 75% and greater than 75%. Differences between hospitals were evident when grouped this way indicating that significantly different technologies were likely to be present in the population of hospitals.

An interesting result that arose out of regressions using the 'high tech' indicator as an explanatory variable was the differences in the sign on the 'high tech' dummies between models with different output types. In models which use occupied bed days as measure of output the sign on the dummy variables was positive suggesting that output increases with an increase in technology. However, the model using revenue as an output measure suggested that output decreases with an increase in technology; as the signs of 'high tech' dummy regression coefficients were negative (although insignificant).

The question of which output proxy, revenue or occupied bed days, is the better measure has not been answered by this analysis. The positive relationship between all inputs and output suggested by models using revenue as a measure of output and the negative relationship between material inputs and output in models using occupied bed days as a measure of output may suggest that revenue is a more appropriate measure. However, there are alternative explanations for why we observed a negative relationship between

material inputs and occupied bed days which relate to the measurement of the material inputs, in particular, the fact that it is measured as a value and not a volume, in contrast to output. It can be shown that if volume discounts are available in the purchase of inputs and a simple markup model for pricing is assumed we can produce similar results to those obtained when using the two different output measures. This suggests that the assumption that prices are fixed across hospitals is violated and that the occupied bed day measure of output is more appropriate. This analysis has as not yet resolved these issues.[30]

Given the similarities in results between production function estimation of models using different input sets and because of the limitations imposed by the data in obtaining true volume measures of inputs from which average input prices could be calculated, two cost function models with two different output proxies were estimated (see Table 3.7).

Table 3.7

Cost Function Model Specifications

	Model 1	Model 3
Input prices		
Beds	•	•
Materials II (including VMOs)	•	•
Total staff II (labor costs)	•	•
Outputs		
Revenue		•
Composite output I (occupied bed days)	•	

As for the production function estimation, models using different measures of output produce different results in terms of efficiency ranks of firms and estimated coefficients on price variables. One overriding feature of all models was their relatively poor performance in terms of supporting the standard assumptions underlying well-behaved cost functions. [31]

Comparing Production and Cost Function Estimation

The performance of both cost and production functions was poor in terms of estimating theoretically defensible models. In production function estimation we found that in models using occupied bed days as a measure of output, an increase in material inputs leads to a decrease in output.[32] In estimating cost functions, models which use

revenue as a measure of output suggest that an increase in the price of labor leads to a decrease in costs.[33] These results clearly indicate that the models are not effectively capturing the economic behavior of these hospitals. It may be the case that standard economic explanations of behavior upon which our models are based are inadequate representations of these hospitals' behavior.

Comparing Production and Cost Function Efficiency Scores

The production and cost function efficiency scores were well correlated between like models pointing at least to consistent estimation of efficiency rankings between the two approaches.

3.5.3 Comparison of DEA and SFA Results

One important extension of the analysis is to make informative comparisons between the DEA and SFA techniques. Both of these techniques produce technical efficiency scores and, although the scores cannot be directly compared in terms of their levels,[34] judgements can be made about the rankings of individual hospitals. Information that the two techniques provide on scale and the estimates of aggregate productivity growth can also be examined.

Two methods of comparing the results of efficiency scores are rank correlation analysis and frequency tables:

- Rank correlation analysis involves ranking firms by DEA and SFA technical efficiency and testing the correlation between the two sets of ranks. This will give a measure of how closely matched two sets of results are, with a high correlation indicating that the two techniques tend to rank firms in a similar order.
- A frequency table provides information about the frequency in which high, medium or low efficiency firms in one results set are ranked as high, medium or low efficiency in the other results set. In fact, it is possible to derive a correlation coefficient based on the frequencies in each cell of the frequency table.

The results of this analysis indicate that efficiency scores generated from the DEA and SFA, using the same input-output set for each model, are very comparable in terms of the tendency for similar hospitals to be ranked highly or lowly in each model. The results also indicate that changing input and/or output definitions (or even including more variables in the model) very quickly reduces the correlations for individual rankings between the techniques, to the point where there appears to be no correlation between the efficiency rankings in some cases.

3.6 Conclusions and Possible Extensions

The purpose of this study was to examine the performance of different techniques for measuring efficiency and productivity and to gain a better understanding of the Australian health sector through the operations of private hospitals. Both the SFA and DEA techniques provided insights into this sector but perhaps more importantly for this study, demonstrated their own strengths and weaknesses in measuring efficiency.

An important conclusion from the non-parametric DEA analysis was that hospital efficiency scores were not robust to changes in the sets of inputs and outputs used to define activity. While this was expected, we were surprised to find that sometimes even small changes in input sets can produce very different results, particularly when outputs are disaggregated. Given that DEA results can be sensitive to the choice of inputs and outputs, it is necessary to provide sound reasons for nominating one model rather than another and (as far as possible) to explain just how different model specifications can lead to different conclusions. Overall, technical efficiency appeared to be only marginally influenced by factors such as hospital type (profit-making status) or scale, even though the majority of hospitals appear to operate under decreasing returns to scale.

Despite its immaturity, the parametric SFA also produced a number of interesting results. Perhaps the most emphatic is that modeling with two different measures of output (occupied bed days and revenue) will produce very different results in terms of both the structure of production and the relative efficiencies of hospitals. Preliminary SFA on sub-samples (characterized by size, profit-making status and degree of high technology) points to the likelihood that individual hospitals' activities vary substantially; it is unclear whether the full population of hospitals are so disparate that effective modeling may not be possible, and the population may have to be dissected into sub-industries. However, one of the purposes of this exercise was to obtain aggregate results for the private hospitals segment of the health industry and it is difficult to serve this purpose by breaking it into sub-populations.

It is also clear that the dataset on which these models have been estimated (rich though it is) is not rich enough to effectively characterize the industry using standard economic models. As well, the difficulties in incorporating all the types of variables in the dataset into the various techniques have also contributed to the ambiguity of some findings. In particular, analysis of the production structure and efficiency of Australian private hospitals would benefit from improved capital stock estimates (e.g. through collection of asset data) and a more detailed disaggregation of hospital activity (e.g. linking of DRG data to the PHEC dataset).

APPENDIX

A3.1 VARIABLE DESCRIPTIVE STATISTICS

Table A3.1 presents correlation coefficients for some of the key variables used in parametric analysis.

Table A3.1

CORRELATION COEFFICIENTS, KEY VARIABLES: SFA MODELS

	CAP	LC	LQ	M1	M2	OUT	REV
Capital (CAP)	1.00						
Labor costs (LC)	0.93	1.00					
Labor quantity (LQ)	0.93	0.99	1.00				
Intermediate inputs 1 (M1)	0.89	0.97	0.96	1.00			
Intermediate inputs 2 (M2)	0.90	0.97	0.96	0.99	1.00		
Output (OUT)	0.96	0.95	0.95	0.91	0.92	1.00	
Patient Revenue (REV)	0.93	0.99	0.98	0.98	0.98	0.95	1.00

Notes

[1] This paper is a shortened version of Webster R., S. Kennedy, and L. Johnson. 1998. Comparing Techniques for Measuring the Efficiency and Productivity of Australian Private Hospitals, Australian Bureau of Statistics, Working Papers in Econometrics and Applied Statistics, ABS Catalogue Number 1351.0. It was written whilst the authors were working at the Australian Bureau of Statistics. Views expressed in this paper are those of the authors and do not necessarily represent those of the Australian Bureau of Statistics. Where quoted or used, they should be attributed clearly to the authors. The authors would particularly like to thank Ken Tallis for his assistance with the research project. They would also like to thank Tim Coelli, Annette Dobson, Kathy Kang, Marelle Rawson, John Goss, Helen Stockdale, Jos Blank and Keith Carter for their helpful comments. Responsibility for any mistakes or omissions is entirely our own.

[2] The analyses were applied to both a cross-section of the private hospitals data (for 1994–95) and a panel (for 1991–92 to 1994–95). This paper only discusses results obtained from analyzing the cross-section data.

[3] Australian Government operates at three levels: Commonwealth; State; and Local.

[4] Under the assumptions of profit maximization or cost minimization, parameter estimates will be biased and inconsistent when the production function is directly estimated (Thomas 1985: 224). Direct estimation is only valid when input levels are assumed fixed or expected profit is maximized, that is, when there is uncertainty about output prices or quantities. See also Coelli (1995: 226).

[5] This paper assumes that private hospitals behave in a manner that can be predicted by neoclassical economics; however, it is quite possible that this model of behavior is inadequate, particularly when explaining how private non-profit hospitals operate.

[6] See Lovell (1996) for a review of the ability of SFA and DEA to measure efficiency and productivity change.

[7] This discussion will be mostly couched in terms of production functions though much of the discussion also applies to cost functions.

[8] A problem we encountered in estimating production and cost functions was the existence of zeros for some inputs for some hospitals therefore not enabling us to log this value. This problem was largely solved by moving to a higher degree of aggregation in inputs which resulted in all hospitals having some positive value. Battese (1996) uses a dummy variable technique to overcome the problems of some firms not using certain inputs.

[9] Although it was possible to disaggregate the FTE staff measure, for example, into nursing and non-nursing personnel, this was often not done when estimating parametric models due to the problem of zero values when logarithms of values are required.

[10] Beds are often used to proxy for capital stock in hospital studies usually because a reliable measure of the value of assets is not available. However, the PHEC dataset contains data on depreciation and gross capital expenditure, and an alternative measure of capital stock was derived through inverting the Perpetual Inventory Model and by incorporating a number of assumptions such as the investment his-

tory of firms. These estimates were trialed in production function analysis; however, they were not persisted with due to the substantial inconsistencies in the data at the unit record level. Interestingly an aggregate (industry wide) measure of capital stock derived from this technique proved quite robust to changes in underlying assumptions.

[11] As a measure of output, unweighted separations tend to favor hospitals which treat simpler diseases and provide quicker treatments.

[12] Newhouse (1994) discusses output measurement issues in respect of hospitals, such as the difficulties in measuring outputs and adjusting for quality when the product is heterogeneous and multi-dimensional, and the existence of omitted outputs. He also notes how existing studies have drastically aggregated both inputs and outputs since it is the only way to make headway when estimating production or cost functions.

[13] The ABS is currently undertaking a project investigating how to assess quality change within the health services industry. There are a number of industry initiatives to monitor changes in the quality of service provision, including development of Quality of Care and Patient Satisfaction Indicators, while outcome concepts such as Quality Adjusted Life Years are also gaining increasing attention. In practice, combination and weighting of these indicators to adjust activity-based measures of output is a complex issue and the ABS is unlikely to be able to directly adjust output estimates for quality change for quite some time. However, quality indicators can readily be incorporated in DEA, as illustrated by the use of unplanned readmission rates as an additional (negative) output variable for Victorian hospitals in Gregan and Bruce (1997).

[14] Zuckerman, Hadley and Iezzoni (1994: 260) used an instrument for capturing the price of labor because '... average annual salary per full-time equivalent employee is used as the price of labor and this variable reflects hospitals' choices regarding the number and skill-mix of employees. Therefore, it is endogenous'. This appears to be a sound argument and will be addressed in future analysis.

[15] This approach follows the methodology adopted in Ferrier and Lovell (1990).

[16] A more detailed discussion, formula and example are provided in Steering Committee on National Performance Monitoring of Government Trading Enterprises (1992: 17–18).

[17] This testing involved the use of non-parametric Mann-Whitney and Kruskall-Wallis tests, with the sample disaggregated by both hospital type (profit status) and size (number of beds).

[18] The Spearman rank correlation test was used for this purpose, to analyze whether the ordering of firms between models (specifically between model 1 and each of the other models) was significantly different.

[19] This test identifies observations with inputs or outputs lying more than 25 standard deviations on either side of the sample mean.

[20] All DEA LPs are solved using routines written for the Interactive Matrix Language module of SAS V612.

[21] The authors undertook some two-stage type analysis, using a Tobit regression approach, where it was found that input-output mix captured most of the variation in efficiency score for a particular sample.

[22] Nunamaker (1985) shows that no firm can become 'less' efficient by the additional of a variable, so that firms which were previously fully efficient will remain fully efficient with the addition of extra variables.

[23] As has been pointed out by a referee, this is despite the high correlation between the occupied bed day and separations measures. An explanation is that while the two measures are highly correlated, there is much greater diversity in the separations measure than the bed days measure.

[24] Copies of the tables relating to the sub-sample analyses can be obtained from the authors.

[25] A possible reason for the ordering of mean efficiencies for most models could be that the input-output set does not adequately allow for the more complicated treatments that may occur in larger hospitals. For one of the models, this theory was tested by analyzing efficiency scores in a two-stage, regression-based approach including a measure of treatment technology. The results of this analysis were inconclusive.

[26] A comparison of variable returns to scale (VRS) and non-increasing returns to scale (NIRS) efficiency scores allows information to be inferred about the nature of economies of scale for each hospital, the overall sample and the various sub-samples of hospitals. Comparing the results of constant returns to scale (CRS), VRS and NIRS linear programs allows an observation to be characterized as operating with either constant, increasing or decreasing returns to scale. Another method of analyzing scale and returns to scale, due to Banker, Conrad and Strauss (1986), is the concept of most productive scale size (MPSS).

[27] In this context, pure technical inefficiency accounts for inefficiency due to any factors other than input disposability and scale.

[28] Our examination of slacks should be treated with some caution, particularly since the methods discussed here are not invariant to the unit of measurement for each input, and, in the case of the two-stage method, the approach of maximization may be incompatible with the minimization approach of the first stage. Recently, a multistage method for the treatment of slacks has been developed (Ali and Seiford 1993), and it is planned to trial this method and contrast the results with those presented above in future analysis.

[29] Care should be taken in interpreting these results due to the high degree of multicollinearity between the input variables (see Appendix A3.1).

[30] The choice between revenue and occupied bed days as a proxy for output, where output is assumed to be truly measured by case weighted separations, will also depend on the inter-hospital variation in costs associated with the non-bed day component on clinical costs.

[31] For example, linear homogeneity in prices was not accepted as a restriction in any of the models.

[32] That is, the models produced negative elasticities on the materials variables.

[33] That is, the models produced a negative coefficient on the price labor when linear homogeneity was not imposed.

[34] DEA efficiency scores are calculated relative to the sample and have little meaning when comparing results between DEA studies. On the other hand, the efficiency rankings of observations in DEA studies contain useful information which can be used for comparison purposes.

REFERENCES

Ali, A.I., and L.M. Seiford. 1993. The Mathematical Programming Approach to Efficiency Analysis. In: Fried, H.O., C.A.K. Lovell, and S.S. Schmidt (eds.), *The Measurement of Productive Efficiency: Techniques and Applications*. New York: Oxford University Press.

Australian Institute of Health and Welfare. 1998. *Health Expenditure Bulletin*, No. 14.

Australian Institute of Health and Welfare 1995. In: John Donovan (ed.), *Health in Australia: What You Should Know*. Canberra: AGPS.

Banker, R.D., R. Conrad, and R. Strauss. 1986. A Comparative Application of Data Envelopment Analysis and Translog Methods: an Illustrative Study of Hospital Production. *Management Science* 32: 30-44.

Battese, G.E. 1996. On the Estimation of Production Functions Involving Explanatory Variables Which Have Zero Variables. *Working Papers in Econometrics and Applied Statistics 86*.

Coelli, T. 1995. Recent Developments in Frontier Modeling and Efficiency Measurement. *Australian Journal of Agricultural Economics* 39: 219-245.

Commonwealth Department of Health and Family Services 1998. *Annual Report 1997-98*. Canberra: AGPS.

Dusevic, T. 1998. Private Versus Public; the Great Hospital Debate. *The Australian Financial Review*. March 7th: 30-31.

Färe, R., S. Grosskopf., and V. Valdmanis. 1989. Capacity, Competition and Efficiency in Hospitals: A Non-parametric Approach. *The Journal of Productivity Analysis* 1: 123-139.

Ferrier, G.D., and C.A.K. Lovell. 1990. Measuring Cost Efficiency in Banking: Econometric and Linear Programming Evidence. *Journal of Econometrics* 46: 229-245.

Fine, M. 1995. *The Changing Mix of Welfare in Health Care and Community Support Services*. Social Policy Research Centre. Discussion Papers, August 1995, Number 61, UNSW.

Grant, C., and H.M. Lapsley. 1993. *The Australian Health Care System 1992*. Australian Studies in Health Service Administration, University of NSW.

Gregan, T., and R. Bruce. 1997. Technical Efficiency in the Hospitals of Victoria. In: *Data Envelopment Analysis: A Technique for Measuring the Efficiency of Government Service Delivery*. Steering Committee for the Review of Commonwealth/State Service Provision. Chapter 5.1. Canberra: AGPS.

Grosskopf, S., and V. Valdmanis. 1987. Measuring Hospital Performance: A Non-parametric Approach. *Journal of Health Economics* 6: 89-107.

Lovell, C.A.K. 1996. Applying Efficiency Measurement Techniques to the Measurement of Productivity Change. *Journal of Productivity Analyses* 7 (23): 329-340.

Newhouse, J. 1994. Frontier Estimation: How Useful a Tool for Health Economics? *Journal of Health Economics* 13: 317-322.

Nunamaker, T.R. 1985. Using Data Envelopment Analysis to Measure the Efficiency of Non-profit Organisations: A Critical Evaluation. *Journal of Managerial and Decision Economics* 6: 50-58.

Register, C.A., and E.R. Bruning. 1987. Profit Incentives and Technical Efficiency in the Production of Hospital Care. *Southern Economic Journal* 53: 899-914.

Steering Committee on National Performance Monitoring of Government Trading Enterprises 1992. *Measuring the Total Factor Productivity of Government Trading Enterprises.* Canberra: AGPS.

Thomas, R. 1985. Introductory Econometrics; *Theory and Applications.* London: Longman Group.

Valdmanis, V. 1992. Sensitivity Analysis for Dea Models: an Empirical Example Using Public Vs. Nfp Hospitals. *Journal of Public Economics* 48: 185-205.

Vitaliano, D.F., and M. Toren. 1994. Cost and Efficiency in Nursing Homes: a Stochastic Frontier Approach. *Journal of Health Economics* 13: 281-300.

Zuckerman, S., J. Hadley, and L. Iezzoni. 1994. Measuring Hospital Efficiency with Frontier Cost Functions. *Journal of Health Economics* 13: 255-280.

4 Between Bed and Budget: The Efficiency of Dutch Hospitals

Jos L.T. Blank, Evelien Eggink and Arnold H.Q.M. Merkies [1]

4.1 Introduction

In the Netherlands, the hospital sector accounts for 30% of the cost of health care, or approximately 2.5% of GNP (Statistics Netherlands 1999: 380, 503). Direct hospital costs were over 18 billion guilders (USD 9 billion) in 1997; care by physicians accounted for another 3 billion guilders (USD 1.5 billion). Health insurance funds hospital expenses and the premiums have an effect on the income distribution. Moreover, health care seizes a considerable part of productive capacity. Therefore the efficiency of the hospital sector is important for both policy makers and managers.

The hospital sector is not easy to study. Hospitals are complex organizations, which provide a wide range of services. Quality of care and severity of cases are important but difficult to measure. The resources and service deliveries are not obvious, nor are the associated prices (see e.g. Ganley and Cubbin 1992). Moreover, the objectives of hospital managers are unclear: do managers attempt to minimize cost, maximize revenues, or even maximize profit?

The aim of this chapter is to gain insight into the complexity of the production structure of hospitals. In addition it provides empirical measures of hospital efficiency and simulations that are useful for formulating public policy. To achieve these aims this chapter takes an eclectic approach and considers numerous alternative behavioral models and empirical specifications.

The empirical analyses were performed on a sample of Dutch hospitals in the period from 1985-1995. Over 100 hospitals, i.e. almost 90% of Dutch hospitals, were observed in

each of these years. This makes it possible to analyze production at a point in time, and how production changes over time. The results in this chapter are based on an extensive productivity study of Dutch hospitals by Blank, Eggink and Merkies (1998b).[2]

There is a large volume of literature on the efficiency of hospitals, including both parametric and non-parametric approaches. The parametric literature estimates cost and production functions and identifies aspects like marginal costs, economies of scope and scale effects. Some examples are Cowing and Holtman (1983), Jensen and Morrisey (1986), Granneman, Brown and Pauly (1986) and Gaynor and Anderson (1993). Frontier cost functions generate measures of the efficiency of hospitals, see, for example, Zuckerman, Hadley and Iezzoni (1994), Vitaliano and Toren (1996). The non-parametric literature focuses on the efficiency of individual hospitals, changes in efficiency over time and scale effects. The individual hospital scores can be used to explore the determinants of managerial efficiency. Managerial variables often include dummies for organizational form (for-profit or not-for-profit), market concentration, intensity of care and case-mix. Some examples of non-parametric studies are Grosskopf and Valdmanis (1987, 1993), Register and Bruning (1987), Morey, Fine and Loree (1990),Valdmanis (1992) and Byrnes and Valdmanis (1996), Ferrier and Valdmanis (1996), and Parkin and Hollingsworth (1997) and Ball, Roberts and Thomas (1999). A few studies, including this one, take both parametric and non-parametric approaches (Banker, Conrad and Strauss (1986), Webster, Kennedy and Johnson. (1998) and Chapter 3 of this volume).

This chapter is organized as follows. Section 4.2 describes the Dutch hospital sector. This includes descriptive statistics on general and teaching hospitals and a discussion on the regulatory environment. The incentives established by the regulatory environment may lead to different assumptions on economic behavior. Section 4.3 discusses the methodology used in the analyses. Section 4.4 describes the dataset and the construction of resources and service deliveries. Section 4.5 contains the empirical results. This includes the appropriate assumptions on economic behavior, estimates of the efficiency levels and some simulations that are important from a policy perspective. Section 4.6 is a summary and conclusion.

4.2 THE DUTCH HOSPITAL SECTOR

4.2.1 *Quantitative Overview of Dutch Hospitals*

This study focuses on general and teaching hospitals. Special hospitals, such as eye clinics and rehabilitation clinics are excluded. These specialist hospitals account for only 6% of hospital capacity and costs (Institute for Health Care Management 1996) and their production structure is different from that of the general and teaching hospitals.[3]

Table 4.1 provides some quantitative characteristics of general and teaching hospitals in 1985 and 1995.

Table 4.1

SOME CHARACTERISTICS OF THE DUTCH HOSPITAL INDUSTRY, 1985 AND 1995

	1985	1995
number of general and teaching hospitals	161	118
outpatient visits (x mln.)	19.9	22.3
discharges (x mln.)	1.5	1.6
inpatient days (x mln.)	18.1	14.7
day-care days (x mln.)	0.2	0.6
variable costs (in NLG mln)a	10,800	13,800
capital costs (in NLG mln)a	1,520	2,460
personnel (in ftes x 1,000)b	121	131

a *1 NLG ≈ USD 0.5 (exchange rate 1998), prices 1995.*

b *Including physicians on the hospital payroll.*

Source: Institute for Health Care Management (1986, 1996)

There were 161 hospitals in the Netherlands in 1985 (154 general hospitals and 7 teaching hospitals). As a result of closures and mergers of general hospitals, the number of hospitals had decreased to 118 in 1995. At the same time, hospitals became larger. In 1985 the hospitals "produced" nearly 20 million outpatient visits, and over 1.5 million discharges; 18 million inpatient days and 0.2 million day-care days. Table 4.1 indicates that in 1995 a shift in services towards outpatient treatment took place. The number of outpatient visits and day-care days increased substantially between 1985 and 1995, while the number of discharges remained constant and the number of inpatient days even decreased. The total cost of general and teaching hospitals in 1985 was over NLG 12 billion (about USD 6 billion). Around 86% of hospital costs is accounted for by the variable resources of personnel and material supplies (NLG 10.8 billion). The remaining 14% is spent on capital, which is seen as a fixed resource. The hospital workforce in 1985 consisted of about 120,000 full-time equivalent employees (ftes's). The number of ftes's increased by about 8% between 1985 and 1995.

4.2.2 Regulatory Environment

Dutch hospitals are constrained by some distinctive regulations. The central government controls capacity. Insurance companies reimburse hospitals on a prospective basis. Budgets have a fixed component (related to capacity) and a variable component (related to services). The services -related component is based on local agreements between the hospitals and the insurance companies on the numbers of certain service deliveries such as first-time visits and patient days.[4] Larger hospitals receive higher budgets per case on the assumption that they typically treat more severely ill patients. Moreover, some budget rates depend on the medical specialties supplied by a hospital, also indicating differences in case-mix. Thus, the budgets are based on case-mix only to a certain extent.

Three concepts are important in the budgeting-system for Dutch hospitals: budget, revenues and expenditures. The hospitals are allotted a virtual budget calculated as described above. The actual revenues are based upon the medical procedures actually performed. A price is fixed for each medical procedure by the Central Council of Health Care, and this price is paid by insurance companies. If the total revenues of a hospital are larger (or smaller) than its virtual budget, the difference is subtracted from (or added to) the budget for the subsequent year. Thus, in the long run, revenues are always equal to budgets; hospitals are not allowed to make a profit. However, a surplus can be generated by keeping expenditures (costs of production) lower than the allotted budget. A surplus is added to the financial assets. Wages are regulated as well. Therefore, manage-

ment has little ability to increase their own salaries. Hence, not-for-profit can not be 'for-profit in disguise'. In spite of these regulations, hospitals do compete with each other. Hospitals with deficits and negative financial assets are vulnerable to budgetary cutbacks and closure. Growth is an important goal.

Another important feature of the Dutch hospital sector is that hospitals may not select patients. Patients are referred to a hospital by general practitioners. Hospitals are required to treat any patient, provided that they have the medical resources required for the treatment. In practice, hospitals can attract patients by offering particular specialized procedures or a high quality of care. This implies that expanding high-tech medical treatments may be another important goal.

4.2.3 *Economic Behavior*

An empirical analysis of efficiency is based upon behavioral assumptions and economic constraints (see e.g. Färe and Primont 1995, Färe, Grosskopf and Lovell 1994, or Section 2.2). In the context of the operating environment of Dutch hospitals various behavioral assumptions may be valid. Four commonly used behavioral assumptions are considered: cost minimization given service deliveries or given revenues and revenue maximization given resources or given costs. The assumption of cost minimization may be the most plausible. Both service deliveries (patient mix) and revenues (virtual budget) may then be the appropriate economic constraint. However, the assumption of revenue maximization is appropriate to the extent that growth is a goal of the hospitals. The choice of a model based upon behavioral assumptions and economic constraints has been extensively discussed by Lovell in Sections 2.2 and 2.3 of this volume.

4.3 METHODOLOGY

4.3.1 *Introduction*

The empirical models in this study have three distinct aspects, which are summarized in Figure 4.1, and explained in the following subsections.

Figure 4.1

Schematic Representation of Methodology

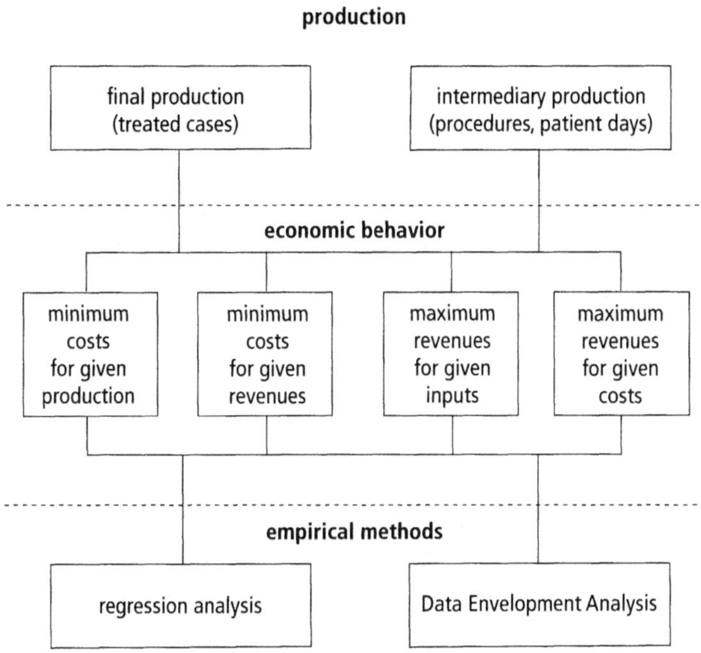

Figure 4.1 has three layers, each standing for one of the aspects, or in other words, a choice to be made by the researcher. The top layer lists two approaches to production: the final production and the intermediary production approach. This is explained in Section 4.3.2. The middle layer concerns the four different types of economic behavior described in Section 4.3.3. The bottom layer shows the two empirical methods discussed in Section 4.3.4. Combining the two approaches to production with four economic models and two empirical methods yields 16 different empirical models. Each model was analyzed separately in Blank, Eggink and Merkies (1998a, 1998b), and the results were compared. The remainder of this section discusses the three aspects of the empirical models in Figure 4.1.

4.3.2 *Approaches to Production*

Hospitals produce a wide range of medical services. There are a number of different approaches to modeling this multi-product firm (see e.g. Butler 1995: 45-83 or Breyer and Zweifel 1992: 299-340). One approach to hospital services is to interpret the

hospital as a medical facility that treats patients. Then the indicator of service deliveries is 'cases treated'. In this study this viewpoint is referred to as the final production approach.[5] Another approach is to envision the hospital as a form of cooperation between two firms (Harris 1977). One firm consists of the physicians who treat patients; the other firm is the hospital that provides services to the physicians. Examples of such services are the care for patients (patient days) and the provision of operating theaters. In this setting, the procedures and the patient days are seen as the service deliveries of the hospital. This approach to hospital service deliveries is called the intermediary production approach.[6]

The two approaches to hospital service deliveries can be combined into a model for the production structure. This is illustrated in Figure 4.2.

Figure 4.2

Simple Representation of the Production Structure

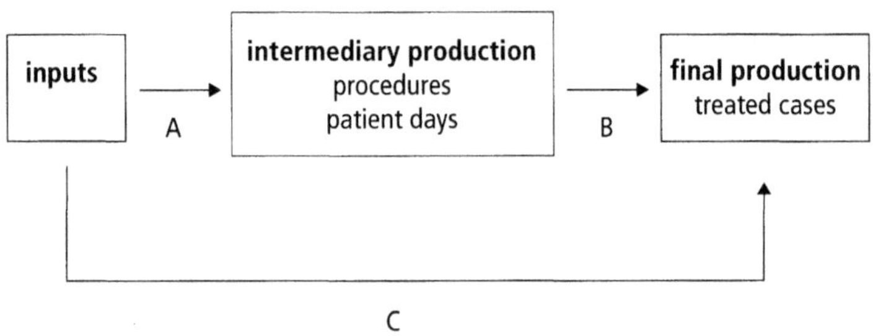

In this study separate models are analyzed which relate the resources to the intermediary and the final production respectively. Arrows A and C in Figure 4.2 are thus examined. By comparing the results of these two models, the relationship between procedures and treated cases is analyzed implicitly (arrow B). This means that the behavior of the physicians who produce treated cases using the procedures as resources is also analyzed implicitly. Since physicians decide which procedures are performed for each patient, they form the link between the procedures and cases treated.

4.3.3 Modeling Economic Behavior

The behavior of Dutch hospitals can be described by four different assumptions on economic behavior. Rather than making an arbitrary choice, models are examined for all four assumptions, i.e. cost minimization given service deliveries or given revenues and revenue maximization given resources or given costs. Blank, Eggink and Merkies (1998a, 1998b) use this approach and show that the model of cost-minimization is the most appropriate one.[7]

4.3.4 Empirical Methods

Section 2.4 present two methods to estimate efficiency. In this study we use Data Envelopment Analysis (DEA) (see Section 2.4) and an alternative for the Stochastic Frontier Analysis. This alternative consists of a set of equations to be estimated simultaneously by regression analysis. Appendix A4.2 provides some technical details of the methods as they are employed here. Blank, Eggink and Merkies (1998a) extensively explores both the regression and DEA settings.

4.4 DATA

4.4.1 Data Sources

Data for this study covering the years 1985-1995 were obtained from the Ministry of Health, Welfare and Sport. The data were collected by the Institute for Health Care Management and are derived from a number of surveys, such as financial, patient and personnel surveys. The surveys contain information on all general and teaching hospitals yielding 120 to 160 observations each year, situated in 27 health care regions. For the purposes of this study, observations on hospitals with missing or unreliable data were excluded from the dataset. Various consistency checks were run on the data to ensure that changes in average values and the distribution of values across time were not excessive. After eliminating observations containing inaccurate or missing values, 1,299 observations over the 11 years of study remained. These include 1,215 observations on general hospitals and 84 on teaching hospitals. Thus, 100 to 112 hospitals were observed in each year.

4.4.2 *Measurement of Service Deliveries*

Indicators of Final Production

There are two primary service delivery measures: the number of discharges and medical research. The discharges have been separated into medical specialties in order to measure case-mix (see also Feldstein 1968: 15 a.f.). The dataset distinguishes over 30 specialties. Since it is not possible to use such a large number of categories, these were aggregated into five categories on the basis of cost homogeneity (see Appendix A4.2). Each category thus contains patients with about the same costs. The number of day-care patients is used as an additional category. Outpatients are ignored since they contribute very little to costs.

A case-mix index and a medical technology index have been included as measures of severity of cases. The case-mix index reflects the hospitals' degree of specialization and is derived from information theory indices (see e.g. Butler 1995). The idea is that patients who are severely ill tend to be treated in hospitals with a high degree of specialization. The construction of the case-mix index is described in Appendix A4.2. The medical technology index measures to what extent hospitals perform special procedures such as open heart surgeries, kidney transplants and neurological surgeries. This is measured by the number of special surgeries per 1,000 discharges.

Hospitals also train nurses and physicians and perform medical research. Even though these functions may require a lot of funding, they are omitted in most productivity studies for lack of information. In this study, the time of physicians not devoted to patient care is used as a proxy for medical research and education. See Appendix A4.2 for the construction of this variable, which is referred to as medical research. It should be kept in mind that this may include other factors such as training physicians or the intensity of the care provided by the physicians.

Indicators of Intermediary Production

In the setting of intermediary production, the quantitative measures of service deliveries include patient days and several medical procedures. The patient days are distinguished into day-care patient days, and four types of inpatient days, each with a different severity (see Appendix A4.2). Here patient days cover only the nursing care given to patients; but not the medical procedures performed. The medical procedures form separate service delivery indicators. The procedures considered here comprise eight types of procedures, such as X-ray procedures, lab procedures, and surgeries. The types of procedures included in the analyses are listed in Table A4.2. The medical research variable is also included as a product in the setting of intermediary production.

Service Delivery Prices

Some of the models analyzed also require service delivery prices. However, the focus of this chapter lies on the cost minimization model, which is the most appropriate one. Since such a model does not require service delivery prices the reader is referred to Blank, Eggink and Merkies (1998b) for the details of the construction of the service delivery prices. Here it must only be noted that the service delivery prices are based on the budget parameters in the financing system.

4.4.3 Measurement of Resources and Resource Prices

Indicators of Resources

Resources include nursing personnel, paramedical personnel (such as physiotherapists and speech therapists), other personnel (including administrative, maintenance and management staff), material supplies and capital. Physicians are not included in these personnel variables, to ensure that hospitals with physicians on their payroll and hospitals with physicians who are self-employed are treated equally.[8] The fourth resource is material supplies, including such aspects as medical supplies, food and heating. Personnel and material supplies are treated as variable resources since the hospital can change these in the short run. Capital is included as a fixed resource, since the capital assets such as buildings and medical equipment can only be changed in the long run.

Resource Quantities and Resource Prices

Volume and price have to be constructed for each of the (variable) resources. This is important since the results of the analyses tend to be sensitive to the price measures (see Chapter 6). There is data on the costs and the quantity for each resource. For each region and time period a resource price is defined as the average cost per resource unit. The resulting resource price is subsequently attached to each hospital in that region and that time period (see Appendix A4.2).

Since there is no natural unit of measurement for material supplies, a circumventing construction was used. The price of material supplies is a weighted index based on components of the consumer index calculated for the Netherlands by Statistics Netherlands. The weights are derived from cost shares.

The quantity of the fixed resource (capital) also had to be constructed. The quantity of capital is distinguished into two categories: buildings and medical inventory. An indirect quantitative measure needed to be constructed for both categories. Capital was measured by an index consisting of the number of beds, number of operating theaters,

number of X-ray rooms and number of delivery rooms. Appendix A4.2 gives a more detailed explanation of the construction of capital. In the analyses, the two types of capital have been aggregated into a single variable. Since capital is only used as a fixed resource a price for capital is not required.

4.4.4 *Background Characteristics*

In addition to the resources, service deliveries and price variables a number of characteristics of the hospitals play a role in the analyses. They are used as explanatory variables to explain for the differences in efficiency amongst hospitals. These characteristics are divided into three groups: characteristics concerning service delivery, institutional and market conditions and process. The *service delivery characteristics* are features of service delivery that are not captured by the service delivery indicators. These are mostly quality measures, such as the average experience of staff measuring the quality of the delivered care. The *institutional and market characteristics* are features that cannot be affected by the hospital. These features include religious affiliation of the hospital, level of competition and budget system. Finally, the *process characteristics* are features of the production process that can be affected by the management of the hospitals. Some of these factors are the average length of the working week of personnel, occupancy rate of beds, operating theaters etc., and the rate of absence due to illness. See Appendix A4.2 for the specific definitions.

4.4.5 *Quantitative Information*

Table 4.2 gives descriptive statistics of the data set. The variable service deliveries and resources are represented by their respective quantities and prices; the fixed service deliveries and resources are represented by their quantities only. The prices of service deliveries (or resources) are given in Dutch guilders per unit of measurement. Table A4.3 contains descriptive statistics of the background variables.

Service Delivery

The *final production* block of Table 4.2 indicates that, on average, the hospitals in the sample have over 10,000 discharges per year, divided into six groups ranging from 600 to 4,300 discharges per year each. Both the case-mix index and the technology index show a large variation, indicating substantial differences in the severity of cases amongst the hospitals. In addition the medical technology index increases in time indicating an increase in the use of complex medical procedures.

Table 4.2

Descriptive statistics service delivery variables general hospitals, 1985-1995

	mean	*minimum*	*maximum*
final production			
discharges group 1 (day-care days)	3,200	0	13,800
discharges group 2 (cardiology, ENT, etc.)	4,300	200	12,700
discharges group 3 (pediatrics, psychiatry etc.)	2,100	0	6,800
discharges group 4 (surgery, orthopedics etc.)	3,000	0	8,600
discharges group 5 (internal)	1,500	0	4,400
discharges group 6 (lung spec, etc.)	600	0	26,900
medical research[a]	100	0	1,490
case-mix index	0.87	0.38	3.37
medical technology	10	0	103
intermediary production			
day-care days	3,200	0	13,800
psychiatric inpatient days	5,000	0	39,600
intensive care patient days	4,100	0	31,200
patient days, children	7,100	0	38,500
other patient days	104,900	14,100	280,800
X-rays	64,400	7,500	215,000
reactivation sessions	43,700	0	203,800
inpatient surgeries	8,300	900	61,000
lab procedures[a]	100	0	610
outpatient surgeries	12,700	100	77,100
outpatient deliveries	290	0	1,500
hemodialysis/heart cath.	3,200	0	40,000
special surgery	180	0	2,150
medical research[a]	100	0	1,490
resources			
nursing personnel	410	40	1,660
paramedical personnel	180	10	980
other personnel	330	40	1,700
material supplies[a]	100	4	740
capital[a]	100	10	440

Table 4.2 (continued)

Descriptive statistics service delivery variables general hospitals, 1985-1995

	mean	minimum	maximum
resource prices			
nursing personnel	56,300	41,800	72,800
paramedical personnel	57,700	41,500	79,500
other personnel	59,600	47,000	76,700
material supplies[a]	100	90	110
cost			
variable cost (NLG. mln.)[b]	80	10	500

a This variable does not have a natural unit of measurement. It is therefore given as an index number with the mean value as base.
b NLG 1 ≈ USD 0.5 (exchange rate 1998).

Source: Social and Cultural Planning Office

The quantity of *medical research*, which is included both in final and intermediary production, is measured as an index number, with the mean value in the sample set at 100. This quantity shows a very large variation. The amount of medical research in large hospitals is substantial.

Table 4.2 also shows that on average, the *intermediary production* of patient days per hospital is about 125,000 patients days per year (including day-care days). Most of these patient days are "other patient days"; the day-care, psychiatric, intensive care, and children's patient days account for only 3,000 to 7,000 per year.
The numbers of other intermediary products all show considerable variation, corresponding to the variation in size of the hospitals. It must be noted that the haemodialysis, heart catheterizations and special surgeries are only performed in larger hospitals. So the mean number of these procedures is relatively low.

Resources
Table 4.2 indicates that almost half of the personnel is nursing personnel, and one third is other personnel. The variation in the personnel variables here is also substantial, due to differences in the size of the hospitals. The use of material supplies and capital was measured by index numbers. These show that the quantities of material supplies vary from 4% of the mean value to 7 times the mean value. The variation in capital is somewhat smaller (from 10% to 4 times the mean value).

The average price for a staff member is almost 60 thousand guilders ($30,000) per year. On average nursing personnel and paramedical personnel are paid somewhat less than the other personnel, since the management is included in the latter group.

The average annual variable cost of hospitals in the sample is about 80 million ($40 million) guilders. Variation is substantial: the cost of the largest hospital is over 50 times the cost of the smallest hospital.

4.5 Empirical Results

4.5.1 *Introduction*

Sixteen empirical models are analyzed. These alternative specifications incorporate different approaches to production, behavioral assumptions and estimation techniques. This section presents the most robust and striking results of these models. The full set of results can be found in Blank, Eggink and Merkies (1998b).

4.5.2 *Economic Behavior and Substitution Possibilities*

The empirical analyses suggest that hospitals behave as cost minimizers: they take the level of production as given and minimize cost. This conclusion is robust over both the final production (or treated cases) and the intermediary production (procedures and patient days) approaches (see Blank, Eggink and Merkies 1998b). Although the hospitals do react to service delivery prices, their reaction does not comply with the hypothesis of revenue maximization. For a revenue maximizing hospital the theory prescribes that an increase in the service delivery price of a certain product is an incentive to produce more of that service. In that case, the service delivery prices (or the budget parameters) could be instruments for the government to influence hospital policy, for instance, to minimize waiting lists. However, the empirical findings indicate the reverse. In practice, when the revenue of a certain patient group increases, hospitals treat fewer patients from that group, since with the higher prices is easier to meet the budget. The remaining resources can perhaps be used to treat more patients of another type. The same holds for the procedures and patient days.

Hospitals have a preference for nursing and other personnel, i.e. they hire more nursing and other personnel than required from a financial (cost minimizing) standpoint. The preference for nursing personnel ensues from the desire to offer more care to patients

(more 'hands per bed'). The preference for other personnel can probably be attributed to the desire of the management to make the staff as large as possible in order to gain prestige and power.

The findings indicate that there are ample possibilities for substitution between the variable resources. In particular, nursing personnel and material supplies show large substitution possibilities, as do paramedical and nursing personnel, and other personnel and material supplies. Therefore, when the wages of personnel are increased a shift to other resources such as material supplies may occur. This may not be consistent with the policy objective to attract more personnel by improving the terms of employment.

Since the cost minimization model was found to be the most appropriate specification (see Blank, Eggink and Merkies 1998a), the remaining results are based upon this model.

4.5.3 *Technical Change*

In the period of study (1985-1995) new technology gave rise to new treatments and more complex procedures. Moreover, the intensity of a patient day increased due to a decrease in the length of stay. This gave rise to a cost increase of about 5% per year in producing the intermediary products (procedures and patient days). However, no autonomous cost increase in the treatment of patients was observed. The cost increase in the production of medical procedures was offset by an adjustment in the treatment by physicians: a decrease in the number of procedures per patient and a decrease in the length of stay. Therefore, even though the procedures and patient days became more expensive, this was not the case for the treatment of patients.

4.5.4 *Capital and Medical Research*

The analyses include variable resources, i.e. personnel and material supplies, whereas capital is seen as a fixed resource. The regression analyses indicate that there is a small capital surplus, especially in the larger hospitals and the teaching hospitals.[9]

Medical research occupies a strikingly large place in a hospital. A sizeable proportion of the costs are related to this activity. Even though the costs cannot be allocated exactly over the various groups of patients (or the procedures and patient days) and medical research, an inspection of the marginal costs indicates the importance of medical research. Especially in large hospitals, the amount of medical research is large; however,

the marginal costs are lower there. This indicates the presence of scale effects in the area of medical research (see Section 4.5.5). This also indicates that a rise in the number of treated cases (or procedures) does not necessarily lead to a proportionate rise in costs. This fact will be explored further in the policy simulations in Section 4.5.8.

4.5.5 *Economies of Scale*

The results of both the regression and the DEA analyses indicate that the scale effects are small. Some economies of scale are found in the small hospitals, but they are not very large. The increase in scale in the recent decades did not have a major effect on costs. The diseconomies of scale in the intermediary production of procedures and patient days, are offset to some extent by the behavior of the physicians. They decide what procedures are performed for each patient, and seem to opt for fewer procedures, or less expensive ones, to cope with the inefficiencies of the hospital.

4.5.6 *The Behavior of Physicians*

By including both final and intermediary production in the study, and comparing the results, the behavior of physicians is studied implicitly (see arrow B in Figure 4.2).[10] Efficiency is found to be somewhat smaller in the intermediary production models than in the final production models (1 to 3%). This implies that physicians offset the inefficiencies in the production of procedures and patient days to a small extent, by choosing less expensive treatments for patients. Since the cost minimization model is preferred in both the final production approach and in the intermediary production approach, physicians do not frustrate the economic aims of a hospital's management. In fact they even seem to compensate for managerial inefficiencies.

4.5.7 *Economic Efficiencies*

A cost efficiency score is calculated for each of the hospitals using each of the models. The mean efficiency varies from 80% to 85%. Theoretically, this means that the cost could be decreased by 15% to 20% if all hospitals were to be fully efficient. However, this does not imply that large efficiency improvements (or cost decreases) are attainable. In fact, the differences in efficiency can only be partly explained by characteristics of the hospitals, and the characteristics can only be partly affected by the management of the hospitals or even by the government.[11]

Table 4.3 summarizes the effects of the various characteristics included in the analyses (see Table A4.3 for definitions of the variables used). The table distinguishes characteristics with a strong or a weak relationship to efficiency. Since four different (cost-minimization) models are analyzed based on the two approaches to production and the two empirical methods, different effects are found. Where all models yield the same result for a certain variable, this is noted as strong proof. Where the results are mostly the same, this is noted as weak proof. Moreover, some variables may show no relationship with efficiency at all, and the models may even yield contradictory results.[12]

Table 4.3 indicates that the cost efficiency is ***strongly related*** to the experience of personnel, cooperation with other health care institutions, the length of the working week of paramedical personnel, the amount of absence due to illness and the presence of an outpost clinic.

The experience of personnel is an has a large impact on the efficiency of hospitals. More highly trained and experienced personnel will increase the cost of personnel, thereby decreasing efficiency. Highly trained personnel may yield more service deliveries than other personnel; however, these extra service deliveries do not offset the extra costs. This emphasizes the importance of correctly measuring quantities of resource usage.

Table 4.3

Summary of the relationship between the characteristics and efficiency[a]

strong proof		*weak proof*		*no proof/ contradiction*	
experience of personnel	-	type of management[b]	-	consultants	c
cooperation	+	competition	-	religious affiliation	•
working week paramedical personnel	-	functional-budgeting	-	after merger	•
absence due to illness	-	before merger	-	occupancy of operating theaters	c
outpost clinic	-	working week of nursing personnel	-	occupancy of delivery rooms	c
		working week of other personnel	+	occupancy of X-ray rooms	c
		occupancy rate of beds	+	physicians on payroll	c
		contracting out	-	physicians in training	c
		contracting in	-	financial reserves	•

a *+: positive relationship; -: negative relationship; •: no relationship; c: contradictory results.*
b *Foundation vs. other types.*

Source: Social and Cultural Planning Office

If personnel was measured in terms of fte's rather than in terms of volumes, this effect would have pointed in the opposite direction. Of course, the quality of services may improve by hiring highly trained and experienced personnel. Even though the experience of personnel has a large impact on efficiency, a hospital cannot easily change its workforce. In the Netherlands personnel cannot readily be dismissed, and personnel is scarce. In addition, dismissing highly trained personnel is not beneficial to society.

Cooperation with other institutions, both in the domestic area and in the field of medicine, yields cost savings. Such cooperation becomes more present throughout the years.

The length of the working week for paramedical personnel, inversely related to the fraction of part-time workers, decreases efficiency. This is the result of the requirement that social security contributions be paid only on salary exceeding some threshold. Moreover, hiring part-timers will hardly cause the overhead to increase since each worker, part-time or full-time, will work full shifts.[13] The negative effect of the absence due to illness is obvious, since replacements have to be paid. The presence of an outpost clinic adversely affects the efficiency of the hospital. This is a consequence of the extra administrative, security and maintenance personnel, and of the extra cost of transportation.

Some other characteristics show only a *weak relationship* with efficiency. This is the case for the variables: type of management (foundation vs. other), rate of competition, functional budgeting, before merger, working week of nursing and other personnel, occupancy rate of beds and the amount of work contracted out or in. A striking result in this row of variables is that in contrast to market situations, competition here is to the detriment of efficiency. This is because, in the Netherlands, the prices in hospitals are fixed, so there is no price competition. Instead, in order to attract patients hospitals compete on the basis of quality.

No effects could be found for religious affiliation, past mergers and the financial reserves of the hospital.[14] Contradictory results were found for the number of consultants, the occupancy rates of operating theaters, delivery rooms, and X-ray rooms, the fraction of physicians on the payroll (instead of self-employed) and the fraction of physicians in training. The fact that no robust conclusions can be drawn for 6 of the 25 variables used here illustrates the importance of using and comparing the results of different models.

4.5.8 *Policy Simulations*

This section presents a number of useful policy simulations to quantify the effect on efficiency of some possible scenarios. Three groups of simulations are presented: the effects of demographic changes, changes in scale and changes in background characteristics. All simulations were performed on the basis of the situation in 1995.

Demographic Developments

The Dutch population is growing larger and is ageing (see Statistics Netherlands 1996). This will affect the demand for hospital care, particularly since older people have a higher demand for care than younger people. This simulation forecasts the effect of the demographic developments up to 2005. Several situations are studied:

1. *Demographics*

   In the first simulation, the population grows and the composition according to age groups changes accordingly. The per capita demand in each age group is assumed to remain constant.
2. *Demographics and trend in demand*

   In the second simulation, in addition to the demographic developments of the first simulation, the observed trend in per capita demand continues. In particular, the shift from clinical to outpatient treatment is taken into account.
3. *Demographics and change in composition of patients*

   The third simulation includes the demographic changes, as in simulation 1, and the fact that types of illness may differ between younger and older people. Thus, the simulation incorporates the relationship between ageing of the patients and the composition of the patient groups according to medical specialty.
4. *Demographics and medical research*

   In the fourth variant, medical research is assumed to grow as rapidly as the number of patients. Similar to the first simulation the demand per age group is assumed to remain constant.
5. *Demographics and trend in demand and medical research*

   The final variant focuses on the growing extent of research in hospitals, relative to the number of patients. Therefore, alongside a growing and ageing population (second simulation), the observed trend in the volume of medical research is extrapolated into the future.

Appendix A4.3 contains the growth rates used as the inputs of the simulations. The simulations are based on the outcomes of the regression analysis where the treated cases (or patients) are seen as the service delivery. Figure 4.3 depicts the percentage change in cost between the simulated situation and the actual situation in 1995. In the simulation, even though capital is assumed to grow in accordance with the number of patients, the

extra capital costs are not included in the results since a variable cost model is used. However, the use of extra variable resources corresponding to the increase in capital is included. The technological change in the cost function (0.2% per year) is also assumed to continue in the future.

Figure 4.3

Cost Effects of Demographic Changes

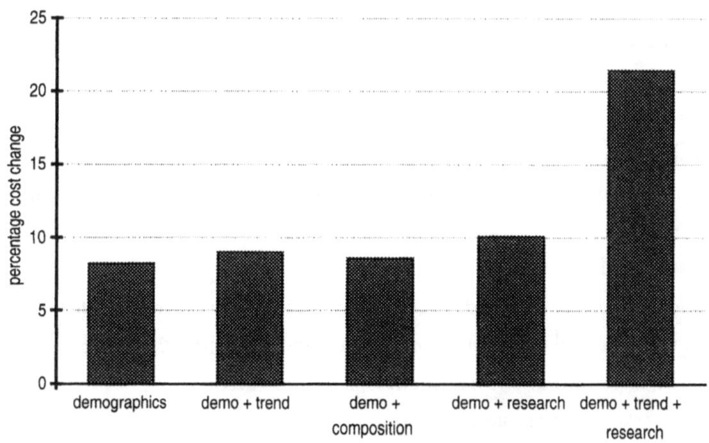

Figure 4.3 indicates that the growth in service deliveries of 8.5% between 1995 and 2005 yields almost the same growth in costs (8.2%), even given the diseconomies of scale in the smaller hospitals. This is a consequence of the fact that the extent of medical research is kept constant. Increasing this variable in proportion to the growth of treated cases yields an additional 2% cost increase (10.1%). Including either the rising demand per capita (trend in demand), or the shifts between the types of patients, would incur very little further increase in costs (0.8 and 0.4% respectively). However, when the trends in demand and in medical research (per patient) are extrapolated, an increase of more than 20% is found. This indicates that extrapolation of the increasing trend in medical research would cause costs to grow exorbitantly.

These simulations show that, if the present trends continue, demographic changes may place a large burden on hospital care. This holds for medical research in particular. If the cost of capital is included in the simulations, the effects become even larger.

Changes in the Scale of Service Deliveries

The government regulates the capacity of the hospitals by approving the number of beds and the number of physicians per hospital. As mentioned in Section 4.2.1 many closures and mergers took place, in the period from 1985 to 1995, thereby increasing the scale of service deliveries in the remaining hospitals. The analyses indicated that there are diseconomies of scale in the large hospitals, and that the increase in scale went too far from the standpoint of efficiency.

The following simulations assess the effect of the ongoing increase in scale. The simulations are performed by altering the number of hospitals and the number of patients per hospital. Two simulations are performed:

1. *Continuing increase in scale*
   In the first simulation the smallest hospitals (fewer than 8,000 discharges) are clsed down and their capacity and service deliveries are divided equally over the remaining hospitals. This means 18% of the hospitals are closed down.
2. *Reversing increase in scale*
   In the second simulation part of the capacity and service deliveries of the largest 25% of hospitals is divided equally over the smallest 50% of hospitals. This is done in such a way that the maximum number of patients would be the maximum of the 75% smallest hospitals. Thus the size of the hospitals becomes more equal.

Again the regression version of the final production approach is used. The results are shown in Figure 4.4.

Figure 4.4

Cost Effects of Changes in the Scale of Output

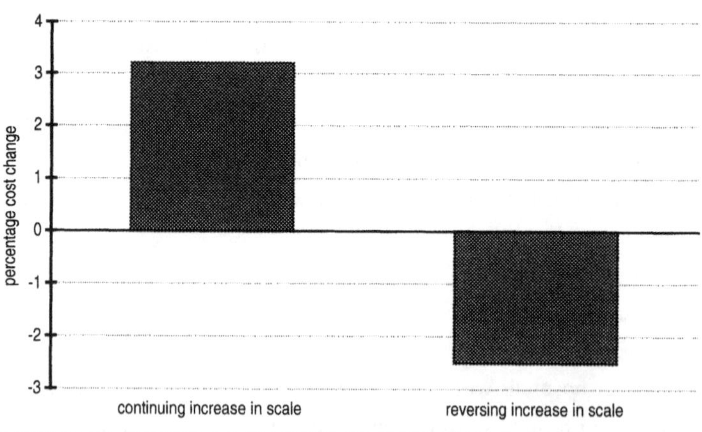

Figure 4.4 indicates that, even though the negative scale effects are found to be small, continuing the process of mergers and closures may increase variable costs substantially. Closure of the smallest 18% of hospitals and redistributing their capacity and service deliveries may lead to a cost increase of 3.2%. Alternatively, reversing the scale increase by dividing service deliveries more equally over the hospitals may decrease costs. This takes advantage of the economies of scale in the smaller hospitals. The above simulations do not take into account the fact that competition also changes when hospitals are closed or the scale is changed. In Section 4.5.7, a decrease in competition was shown to increase efficiency. This effect will offset the costs of the diseconomies of scale to some extent. Thus, this simulation suggests that growth in service deliveries, for instance, corresponding to the demographic changes, should be implemented in the smaller hospitals. More mergers and closures should not be propagated.

Effect of Background Characteristics

The effect of policy measures concerning the variables that influence efficiency are quantified. A simulation is performed for four background characteristics with a strong relationship with efficiency. These were the following characteristics: cooperation, working week of paramedical personnel, absence due to illness and outpost clinics. In addition to the four characteristics mentioned also the experience of personnel was found to have a strong relationship with efficiency. However, it is only useful to study the effect of characteristics that can be influenced by the management of hospitals or by the government. As mentioned in Section 4.5.7, this characteristic is hardly an option for policy making, and the corresponding simulations are omitted.

The object is to identify the potential savings due to a change in these variables. This is accomplished by means of simulations in which favorable but realistic values for the characteristics are assumed. The other variables are fixed at their actual values. The following simulations are performed.

1. *Cooperation*

   In the first simulation it is assumed that all hospitals cooperated with another health care institution.

2. *Working week of paramedical personnel*

   In the second simulation, the mean length of the working week of paramedical personnel is taken to be 66% of the full-time working week (26.4 hours a week), which is the minimum in the dataset.

3. *Absence due to illness*

   In the third simulation the effect of absence due to illness is simulated by assuming that the absence rate does not exceed the average rate in the dataset. This means that nothing changes for hospitals with a low absence rate, whereas the mean rate of absence is assumed for hospitals that currently have a relatively high absence rate.

4. *Outpost clinics*

   Finally, in the fourth simulation it is assumed that none of the hospitals have an outpost clinic.

The effects of these measures are calculated for both the regression method and the DEA method in order to estimate the possible effects. The final production model is employed here. The decreases of costs corresponding to the four policy measures are given in percentages in Figure 4.5.

Figure 4.5

COST EFFECTS OF CHANGES IN BACKGROUND CHARACTERISTICS

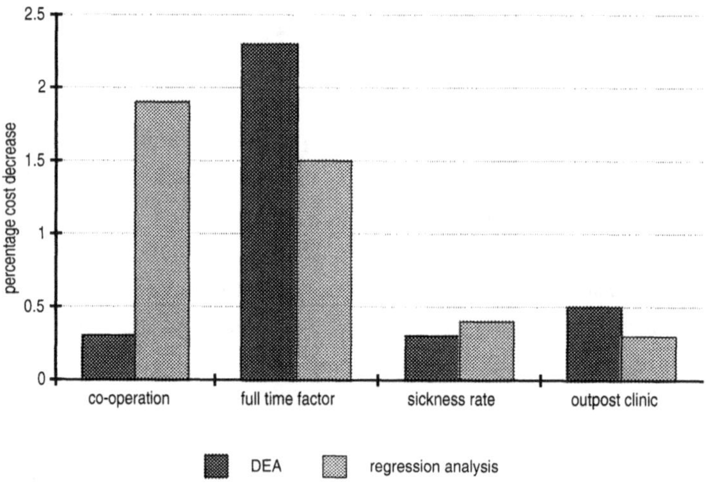

As mentioned in Section 4.5.6 cooperation with other institutions increases efficiency. The estimates in Figure 4.5 indicate that cooperation may lead to average savings of 0.3% and 1.9% of variable costs in the two methods respectively. Employing paramedical personnel for an average of 26.4 hours per week (a mean working week of 66%) may yield another 1.5% or 2.3% reduction in costs for the sector as a whole. Adopting an effective sickness reduction policy, ensuring that absence due to illness does not exceed the average level, may reduce the costs by about 0.3%. Finally, disposing of the outpost clinics may increase efficiency with 0.3%-0.5%. In practice, not all these reductions can be attained. Moreover, the results depend on the method chosen. Roughly speaking, implementing policy measures that affect the characteristics in the figure, may result in a reduction of variable costs by 3.4 to 4.1%.

4.6 SUMMARY AND CONCLUSIONS

This chapter describes a study of the production structure and the efficiency of hospitals in the Netherlands. Two approaches to production are adopted here, based on treated cases and on procedures and patients days respectively, and the results of the approaches are compared. Furthermore, economic behavior was investigated empirically and the robustness of the results is studied by employing both a regression method and DEA.
From previous research it was concluded that hospitals behave as cost minimizers rather than as revenue maximizers (see Blank, Eggink and Merkies 1998b). Based on this behavioral assumption it is shown in this chapter that the possibilities to improve efficiency are modest. There are small diseconomies of scale, which indicates that extension of production should take place in smaller hospitals rather than in larger ones. Technical changes have hardly any effect on costs. By using day-care instead of clinical care, and by decreasing the length of stay, patients are treated more efficiently. However, the procedures become more complex and patient days become more intensive. These two developments offset one another. It must be noted that both developments will have increased the quality of care.

Hospitals have some instruments to improve efficiency. In particular, cooperation with other health care institutions, hiring fewer full-time nurses and paramedical personnel, lowering the rate of absence due to illness and disposing of outpost clinics may improve efficiency. Collectively, such measures could at best lead to a savings in variable costs of about 4%. Furthermore, there are indications that competition decreases efficiency. Allowing new private suppliers to enter the hospital market, currently a subject of political debate, will thus be accompanied by an increase in the costs of hospital production.

The figure of 4% improvement of efficiency should not be used as a target for policy makers. Budget cuts should not be imposed on each hospital in equal measure (the so-called cheese-slicer method). This would mean that efficient hospitals would also have to decrease their costs, which can only be achieved by decreasing the volume or the quality of service delivery. It is more important to provide hospitals with guidance on ways to improve efficiency. Finally, it must be noted that the extent of medical research grows rapidly and the corresponding costs are substantial. Concentration of medical research in larger hospitals may limit this cost increase.

There are some more general conclusions. First, it is shown that employing different definitions of service delivery yields more insight into the production structure of the

objects of research, in this case Dutch hospitals. The conversion of intermediary products into final products may also be very informative for policy makers. Moreover, the definition of other variables in the analyses, such as resources or prices, should be considered carefully. The results of the efficiency analyses may be very sensitive to the definition of the variables in the model. Chapter 6 illustrates the same point.

Second, various empirical models are analyzed, differing not only in measures of service deliveries, but also in assumptions on economic behavior and empirical methods. The robustness of the results can be ascertained by comparing the results of those models. The fact that some explanations of efficiency are contradictory in the various models, stresses the importance of considering alternative empirical specifications. Without such a test, the efficiency results could be misleading and suggest incorrect policy measures.

A final conclusion is that knowing the mean level of efficiency in a sector is not a sufficient basis for policy making. As has been mentioned above, the level of inefficiency should not be used as a target, since imposing a budget cut on each unit in a sector equally also forces the efficient units to cut their costs. This will probably affect their service deliveries adversely. Moreover, a part of the inefficiency is caused by aspects that cannot be controlled by policy makers or managers. An important goal of research should be to learn more about the causes of inefficiency and prescriptions for improving efficiency.

APPENDIX

A4.1 INFORMATION ON THE EMPIRICAL METHODS

Regression Analysis

A flexible functional form is assumed in the regression models, the so-called translog function (Lau 1986: 1516-1566). This implies that no assumptions are made on features such as scale effects. See for Blank, Eggink and Merkies (1998b) for explicit formulations of the models. Each model contains a "main" equation, such as the cost or revenue function and a set of corresponding input or output share equations. The input share equations are derived from Shephard's Lemma; the output share equations are derived from Roy's identity (see for instance Shephard 1970 and 1974 and Chapter 2). In addition to the usual variables on resources, service deliveries and prices, the main equations also contain several characteristics of the hospitals (see Table 4.2). These characteristics directly measure economic efficiency. The models are estimated as a multivariate regression system. The joint density of the system of equations is assumed to be a normal distribution and the systems are estimated by Zellner's Seemingly Unrelated Regression (Zellner 1962) using TSP-software.

Given a certain form of economic behavior, the hospital may have other objectives as well. For instance, the hospital management may affect the quality of the care delivered by the hospital (see e.g. Dor, Duffy and Wong 1997, or Gertler and Waldman 1992 and Blank and Eggink 1996 for the case of nursing homes). Such considerations may also be incorporated into the models. In this setting, several methods can be used to incorporate such deviating types of economic behavior. The best-known methods are the theory on resource preference and shadow prices, and that of quality maximizing behavior. See Chapter 2 for a discussion on these methods. In this chapter the assumption of quality maximization is used. In a cost function setting this implies that the hospital deviates from the optimal resource combination for reasons of quality. In practice, this was included in the analyses by correcting the service delivery level for differences in quality. In this study, this method is performed on the model found to be the most appropriate for the behavior of Dutch hospitals.

Data Envelopment Analysis

In the DEA method, the four assumptions on economic behavior and economic constraints can be imposed. This is discussed extensively in Färe, Grosskopf and Lovell (1994). Formal descriptions of the four models are given there. See Blank, Eggink and Merkies (1998a) for empirical application of these models. The DEA models were calculated using linear programming software (SIMPLEX).

In the DEA setting the efficiency can be decomposed into scale efficiency, allocative and technical efficiency (see Chapter 2). It is not possible to directly account for the differences in the characteristics that affect efficiency. If the sample can be divided into subgroups, the DEA method can be applied to the groups separately. For continuous variables, such as the rate of absence due to illness, this is not possible (see Burgess and Wilson 1998). The DEA-efficiency scores are analyzed in a second step. Using regression analysis the relationship between the background characteristics of the hospitals and the efficiency scores are obtained. Specifically, since the efficiency scores are censored from above at 1, the Tobit method was used (see for instance Maddala 1990: 151). The dependent variable in the Tobit analysis is defined as (1-score)/score. Burgess and Wilson (1998) use an adapted a version of DEA in which this censoring problem does not occur. Their method yields similar results for the estimated relationships with the background variables. In the DEA setting no quality adjustment methods are available.

A4.2 Construction of Variables

Final Production

The day care patients form a separate service delivery indicator. The discharges of inpatients in the 30 medical specialties distinguished in the surveys, are aggregated into 5 categories. The aggregation is done on the basis of two characteristics:

1. the cost of a discharge per specialty as calculated by the Institute for Health Care Management and used in the reimbursement scheme;
2. the number of patient days per patient per specialty (average length of stay).

In order to use the various groups in statistical analyses, the number of discharges per group must be sufficiently large. This results in the groups of medical specialties given in Table A4.1. For the sake of completeness the first group (day-care patients) is also included in the table.

Table A4.1
Groups of Medical Specialties

	medical specialties
1	day-care patients
2	cardiology, allergology, plastic surgery, gynecology, ENT specialties, ophthalmology, oral/dental surgery, outpatient deliveries
3	pediatrics, urology, radio therapy, neurology, psychiatry
4	surgery, orthopedics neurosurgery, cardio pulmonic surgical specialties
5	internal medicine
6	lung specialties, rheumatology, gastroenterology, clinical geriatrics, dermatology and other specialties

Intermediary Production

Five types of patient days are distinguished: day-care days, patient days on a psychiatric ward, patient days in intensive care, patient days on the children's ward and other patient days.

The medical procedures included in the analyses comprise almost alle procedures performed in the hospital. Eight groups of medical procedures are constructed, based on the type of procedures. Severity of cases plays only a secondary role. The categories are listed in Table A4.2.

Table A4.2
Groups of Medical Procedures

	medical procedures
1	X-ray procedures
2	lab procedures (including the pathological lab procedures, nuclear therapy and function examinations)
3	reactivation sessions (including physiotherapy and speech therapy)
4	inpatient surgery
5	outpatient surgery
6	outpatient deliveries
7	hemodialysis and heart catheterizations
8	special surgery (open-heart surgery, kidney transplants and neurological surgery)

Construction of Medical Research Variable

The volume of medical research is approximated by the time spent by physicians not related to patient care. In order to calculate this, the number of physicians (measured in terms of full-time equivalents) is related to the number of discharges for the various groups of medical specialties. The number of special surgical procedures, also used in the technology index, is included in this relationship as well since these procedures require extra time and attention on the part of physicians. By using a so-called *stochastic frontier analysis*, the maximum number of patients per physician is determined (see e.g. Aigner, Lovell and Schmidt 1977 or Chapter 2). When physicians in a certain hospital treat fewer patients than this maximum number, it is assumed that they devote the remaining time to research and educational purposes.

Construction of Case-mix Index

The case-mix index measures the degree of specialization in the hospital. The index is based on the information theory (see e.g. Butler 1995). It measures the concentration of physicians of certain medical specialties in certain hospitals. The idea is that complexity is associated with concentration. A hospital with many physicians for a certain specialty have a larger number of sub-specialties than a hospital with a small number of physicians. These hospitals are thus more able to treat patients with a higher severity than other hospitals, and the mean severity of cases is higher in those hospitals. The construction of the case-mix index follows Butler (1995: 76-82).

Construction of Capital Index

The quantity of capital is divided into two categories: buildings and medical inventory. An indirect measure of capital had to be constructed for the two categories. Separate regression models are constructed that relate the initial capital expenses per capital category to the age of the capital and indicators of the volume and quality of capital. The age of the capital is calculated as the average of the cumulated depreciations divided by the depreciations in the present year. The number of beds, the number of operating theaters, the number of delivery rooms and the number of X-ray rooms are used as the indicators for the volume of both buildings and medical inventories.

The quality of capital in buildings is measured by the availability of a first aid department, a CT-scanner and a radio therapy department. The quality of capital in medical inventories is measured by the availability of a first aid department and a CT-scanner. Several other models are analyzed but do not yield better results. The quantity of capital is then calculated by "forecasting", ignoring the age and the quality indicators in the regression analyses. The resulting measures of capital are homogenous for the different hospitals. In the analyses the capital in buildings and medical inventories are aggregated into a single volume of capital by weighting them with their mean cost share (0.8 and 0.2 respectively).

Construction of Resource Prices

The unit value is calculated for the three personnel categories by dividing the cost of the category by the corresponding number of full-time equivalents. Since there is no natural unit of measurement for material supplies, its unit value is not derived by dividing the cost by the corresponding price. Instead a price index of material supplies is constructed directly. The cost of material supplies is divided into the categories: food, energy (heating, etc.), furniture and maintenance, medical care, transport and communication, and development and recreation. A price index for each of these categories is published by Statistics Netherlands. A different price index can be constructed for each hospital by taking the share of each category in the total cost of material supplies as a weight.

For each of the personnel resources and material supplies, a price variable is constructed that varied across region and time period only. This is done by regressing the unit values on the regions and time periods. A variable was also included that indicates whether there is a high or a low population density in the neighborhood of the hospital. The forecasts of these regressions are used as resource prices in the empirical analyses. Thus, the resource prices are the same for observations in the same region, the same time period and the same population density.

Overview of Background Characteristics

Table A4.3

Overview of background characteristics

	definition	*mean*	*standard deviation*
Service delivery characteristics			
experience of nursing personnel	deviation price of nursing personnel from regional price nursing personnel	100	5
experience of paramedical personnel	deviation price of paramedical personnel from regional price paramedical personnel	100	9
experience of other personnel	deviation price of other personnel from regional price other personnel	101	13
consultants	fraction of consultant physicians in the hospital (in %)	6	9
Institutional features			
type of management	foundation vs. other (in %)	76	-
religious affiliation	religious vs. not religious (in %)	54	-
competition	Herfindahlindex for the number of hospitals in the region	20	14
type	general vs. teaching hospital (in %)	6	-
functional budget funding (FB)	reimbursement by FB system or other (in %)	64	-
cooperation	cooperation with other hospital, nursing home etc.(in %)	17	-
before merger	hospital merges in the following year (in %)	4	-
after merger	hospital has merged in previous year (in %)	2	-
Process characteristics			
working week of nursing personnel	number of nurses divided by the number of fulltime equivalents	80	8
working week of paramedical	personnel number of paramedical personnel divided by the number of fulltime equivalents	80	7
working week of other personnel	number of other personnel divided by the number of fulltime equivalents	71	9
occupancy of beds	average number of inpatient days per bed	235	30
occupancy of operating theaters	average number of surgeries per operating theater	1,303	463
occupancy of delivery rooms	average number of deliveries per delivery room	283	99
occupancy of X-ray rooms	average number of X-rays per X-ray room	5,930	2,070
absence through illness	fraction of personnel cost paid during absence through illness (in % of costs)	5	2
contracting out	fraction of cost for outsourcing services (in % of costs)	7	4
contracting in	fraction of cost received for services for other parties (in % of costs)	2	3
physicians on payroll	fraction of physicians paid by the hospital (in %)	25	19
physicians in training	fraction of physicians in training (in %)	18	12
outpost clinic	the presence of an outpost clinic (in %)	32	-
financial reserves	percentage of financial reserves in variable costs (in % of costs)	1	6

Source: Social and Cultural Planning Office

A4.3 Growth Rates of Demographic Changes

Table A4.4

Growth Rates for Simulation of Demographic Changes, 1995-2005

		percentage changes between 1995 and 2005
demographics		8.5
trend in discharges		-5
trend in day-care days		225
trend in research		61.5
group 1	(day-care patients)	0
group 2	(cardiology, allergology, plastic surgery, gynecology, ENT, ophthalmology, oral/dental surgery, outpat. deliv.)	-3.7
group 3	(pediatrics, urology, radio therapy, neurology, psychiatry)	7.4
group 4	(surgery, orthopedics, neurosurgery, cardio pulmonic surgical specialties)	-0.8
group 5	(internal medicine)	-2.3
group 6	(lung specialties, rheumatology, gastroenterology, clinical geriatrics, dermatology and other specialties)	10.9
technological change		3.9

Source: Social and Cultural Planning Office, Statistics Netherlands (1996)

Notes

[1] The authors thank Hal Fried for his useful comments on an earlier draft of this chapter.

[2] Even though the study is written in Dutch, an extensive English summary has been provided which describes the Dutch hospital sector, the main features of the analyses and the results.

[3] For instance, in the rehabilitation clinics, the length of stay is much longer than in the other types of hospital. In 1995, the mean length of stay in special hospitals was 30 days, whereas in general and teaching hospital the mean length of stay was only 10 days.

[4] This so-called Functional Budget system (FB system) was introduced at all general hospitals in 1988. Until 1987 a similar system was used for about 30% of the budget. The remaining 70% was reimbursed on the basis of fixed budgets set in 1985. The teaching hospitals are reimbursed fully on the basis of a fixed budget.

[5] This indicator is used in many empirical studies of hospitals. See e.g. Jensen and Morrisey (1986), Gaynor and Anderson (1993), Byrnes and Valdmanis (1996), Ferrier and Valdmanis (1996) and Parkin and Hollingsworth (1997).

[6] This intermediary production approach can be found in the literature as well. See e.g. Banker, Conrad and Strauss (1986), Cowing and Holtman (1983), Grosskopf and Valdmanis (1987, 1993), Register and Bruning (1987), Morey, Fine and Loree (1990), Vitaliano en Toren (1996), Webster, Kennedy and Johnson (1998) and Guisset, Sicotte and D'hoore (1998). Mixtures of the final and intermediary production approach are also common in the literature. Examples are Granneman, Brown and Pauly (1986), Zuckerman, Hadley and Iezzoni (1994), Valdmanis (1992).

[7] In Blank, Eggink and Merkies (1998a) the appropriate model is chosen on indicators derived from the regression analyses. The choice criteria are the statistical properties (explained variance (R^2) and significance of parameters), the plausibility of the results (scale elasticities, technical change and economic efficiency), and theoretical properties (monotonicity and convexity or concavity of the estimated functions). No significance levels or explained variance are available in the DEA setting. In that case the efficiency levels must be the criteria of choice. The model should yield high efficiency scores, since the hospitals try to reach the corresponding frontier. In this article the final production approach is used. Blank, Eggink and Merkies (1998b) apply these methods for the intermediary production approach as well. Here the results from that study are used.

[8] The costs of physicians (wages) are not included in the cost or price variables either.

[9] This finding is based on checking the so-called envelope condition (see Koutsoyiannis 1987).

[10] Guisset, Sicotte and D'hoore (1998) use a similar approach in a DEA setting. However, they model the behavior of physicians more explicitly, since they do not study the production of treated cases given the input of labor and material supplies as is done here. Instead, they use the intermediary products as resources for the production of treated cases. The results are similar to the findings of the comparison between the final and intermediary production models in this study.

[11] In the regression analyses the relationship between the efficiency scores and the background characteristics is estimated directly. In the DEA analyses only the scores result directly and the relationship is assessed in a second step regression analysis.

[12] The following criteria for the parameters are used here

Strong proof:	the same sign in all models and significant at least once.
Weak proof:	the same sign in most models, significant at least once, and no significant inverse sign.
No effect:	nowhere significant.
Contradiction:	significant positive and significant negative effects.

[13] This is in contrast to for example administrative personnel, who may work for only a few hours each day.

[14] The insignificant effect of the financial reserves may be caused by measurement problems. Further study is needed on this issue.

References

Aigner, D.J., C.A.K. Lovell, and P. Schmidt. 1977. Formulation and Estimation of Stochastic Frontier Production Function Models. *Journal of Econometrics* 6 (1): 21-3.

Ball, R., E. Roberts, and R.Thomas. 1999. Lessons Learned for DEA Practice from Health Care Applications in the UK. In: Westerman, G. (ed.), *Data Envelopment Analysis in the Service Sector*. Wiesbaden: Gabler.

Banker, R.D., R.F. Conrad, and R.P. Strauss. 1986. A Comparative Application of Data Envelopment Analysis and Translog Methods: an Illustrative Study of Hospital Production. *Management Science* 32 (1): 30-44.

Blank, J.L.T., and E. Eggink. 1996. *A Quality-Adjusted Cost Function in a Regulated Industry: The Case of Dutch Nursing Homes (paper presented at the Productivity workshop 1996 Athens, Atlanta)*. Rijswijk: Sociaal en Cultureel Planbureau.

Blank, J.L.T., E. Eggink, and A.H.Q.M. Merkies. 1998a. *The Choice of an Appropriate Economic Model for the Structure of Production: the Case of Dutch General Hospitals*. Den Haag: Sociaal en Cultureel Planbureau (Werkdocument 49).

Blank, J.L.T., E. Eggink, and A.H.Q.M. Merkies. 1998b. *Between Bed and Budget (Tussen bed en budget: een empirisch onderzoek naar de doelmatigheid van algemene en academische ziekenhuizen in Nederland)*. Den Haag: Sociaal en Cultureel Planbureau.

Breyer, F., and P. Zweifel. 1992. *Gesundheitsökonomie*. Berlin: Springer-Verlag.

Burgess J.F., and P.W. Wilson. 1998. Variation in Inefficiency among US Hospitals. INFOR 36 (3): 84-102.

Butler, J.R.G. 1995. *Hospital Cost Analysis*. Dordrecht: Kluwer Academic Publishers.

Byrnes, P., and V. Valdmanis. 1996. Analyzing Technical and Allocative Efficiency of Hospitals. In Charnes A., W. Cooper, A.Y. Lewin and L.M. Seiford (eds.), *Data Envelopment Analysis: Theory, Methodology and Applications*. Norwell, MA: Kluwer Academic Publishers, 129-144.

Cowing, T.G., and A.G. Holtmann. 1983. Multiproduct Short-run Hospital Cost Functions: Empirical Evidence and Policy Implications from Cross-section Data. *Southern Economic Journal* 49 (3): 637-653.

Dor, A., S. Duffy, and H. Wong. 1997. Expense Preference Behavior and Contract-Management: Evidence from U.S. Hospitals. *Southern Economic Journal* 64 (2): 542-554.

Färe, R., and D. Primont. 1995. *Multi-Output Production and Duality: Theory and Applications*. Dordrecht: Kluwer Academic Publishers.

Färe, R., S. Grosskopf, and C.A.K. Lovell. 1994. *Production Frontiers*. Cambridge: Cambridge University Press.

Feldstein, M.S. 1968. *Economic Analysis for Health Service Efficiency*. Chicago: Markham Publishing Company (Econometric Studies of the British National Health Service).

Ferrier, G.D., and V. Valdmanis. 1996. Rural Hospital Performance and Its Correlates. *The Journal of Productivity Analysis* 7: 63-80.

Ganley, J.A., and J.S. Cubbin. 1992. *Public Sector Efficiency Measurement; Applications of Data Envelopment Analysis.* Amsterdam: North-Holland.

Gaynor, Martin, and Gerard F. Anderson. 1993. *Uncertain Demand, The Structure of Hospital Costs, and the Cost of Empty Hospital Beds.* Cambridge: National Bureau of Economic Research (Johns Hopkins Department of Economics Working Paper).

Gertler, Paul J., and Donald M. Waldman. 1992. Quality-Adjusted Cost Functions and Policy Evaluation in the Nursing Home Industry. *The Journal of Political Economy* 100 (6) (1232-1256) (1992).

Granneman, W., R.S. Brown, and M.V. Pauly. 1986. Estimating Hospital Costs: A Multiple-Ouput Analysis. *Journal of Health Economics* 5 (3): 107-127.

Grosskopf S., and V. Valdmanis. 1987. Measuring Hospital Performance: A Non-Parametric Approach. *Journal of Health Economics* 6: 89-197.

Grosskopf S., and V. Valdmanis. 1993. Evaluating Hospital Performance With Case-Mix-Adjusted Outputs. *Medical Care* 31 (6): 525-533.

Guisset, A.-L., C. Sicotte, and W. D'hoore. 1998. Efficience des hôspitaux québecois: test d'un modèle emboîté. *Presented at the INFORMS-meeting*, Montreal, April 1998 .

Harris, J.E. 1977. The Internal Organization of Hospitals: Some Economic Implications. *The Bell Journal of Economics*: 467-482.

Institute for Health Care Management. 1986. *De intramurale gezondheidszorg in cijfers per 1 januari, 1986.* Utrecht: Nationaal Ziekenhuisinstituut.

Institute for Health Care Management. 1996. *De intramurale gezondheidszorg in cijfers per 1 januari, 1996.* Utrecht: Nationaal Ziekenhuisinstituut.

Jensen, G.A., and M.A. Morrisey. 1986. The Role of Physicians in Hospital Production. *Review of Economics and Statistics* 68 (3): 432-442.

Koutsoyiannis, A. 1987. *Modern Microeconomics.* London: MacMillan Press, 1987.

Lau, L.J. 1986. Functional Forms in Econometric Model Building. *Handbook of Econometrics* III: 1515-1566.

Maddala, G.S. 1990. *Limited-Dependent and Qualitative Variables in Econometrics.* Cambridge: Cambridge University Press (Econometric Society Monographs 3).

Morey, R.C, D.J. Fine, and S.W. Loree. 1990. Comparing Allocative Efficiencies of Hospitals. Omega, *International Journal of Management Science* 18 (1): 71-83.

Parkin, D., and B. Hollingsworth. 1997. Measuring Production Efficiency of Acute Hospitals in Scotland, 1991-94: Validty Issues in Data Envelopment Analysis. *Applied economics* 29: 1425-1433.

Register, C.A., and E.R. Bruning. 1987. Profit Incentives and Technical Efficiency in the Production of Hospital Care. *Southern Economic Journal*: 899-914.

Shephard, R.W. 1970. *Theory of Cost and Production Functions*. Princeton, NJ: Princeton University Press.

Shephard, R.W. 1974. *Indirect Production Functions*. Meisenheim am Glan: Verlag Anton Hain (Mathematical Systems in Economics, no 10).

Statistics Netherlands. 1996. *Bevolkingsprognose 1996*. Voorburg/Heerlen: CBS.

Statistics Netherlands. 1999. *Statistisch Jaarboek 1999*. Voorburg/Heerlen: CBS.

Valdmanis, V. 1992. Sensitivity Analysis for DEA Models. An Empirical Example Using Public vs. NFP Hospitals. *Journal of Public Economics* 48: 185-205.

Vitaliano, D.F., and M. Toren. 1996. Hospital Cost and Efficiency in a Regime of Stringent Regulation. *Eastern Economic Journal* 22 (2): 161-175.

Webster, R., S. Kennedy, and L. Johnson. 1998. *Measuring the Efficiency and Productivity of Australian Private Hospitals*. (Methodology Advisory Committee Paper).

Zellner, A. 1962. An Efficient Method of Estimating Seemingly Unrelated Regression and Tests of Aggregation Bias. *Journal of the American Statistical Association* 57: 500-509.

Zuckerman, Stephen, Jack Hadley, and Lisa Iezzoni. 1994. Measuring Hospital Efficiency with Frontier Cost Functions. *Journal of Health Economics* 13 (3): 255-280.

5 Remedying Excess Capacities in French Surgery Units by Industry Reallocations: The Scope for Short and Long Term Improvements in Plant Capacity Utilization

Benoît Dervaux, Kristiaan Kerstens and Hervé Leleu[1]

5.1 Introduction

Several studies have documented the existence of important overcapacities in the French hospital sector, and especially in specialized care. About 34,000 beds would be redundant in terms of short stay capacity. This amounts to about 15% of total capacity (Guyomar 1995). Despite this surplus in supply, large regional inequalities persist with regard to both the health status of the population and the availability of specialized equipment (Coca 1995).

A plausible underlying mechanism leading to the documented overcapacities is the nature of competition in the health industry. The strategy of hospitals is to enlarge their capacities and their range of services, thereby in general privileging specialized care. This quality competition leads to overcapacities at the industry level (Wooley and Frech 1989).

The recent interest in evaluating efficiency has undoubtedly improved our understanding of plausible mechanisms inhibiting the functioning of public sector institutions. But, most frontier or "best practice" studies have limited themselves to documenting the various inefficiencies in an accurate way. One disadvantage documented in the literature is that production units often object to certain policy decisions made on the basis of these performance results (Epstein and Henderson 1989). Almost inevitably, efficient units remain unaffected, though their performance is only relatively good, while inefficient units are "punished" by, for instance, cutting budgets or being forced to adjust their inputs and outputs to certain observed best practice targets.

This study, by contrast, investigates structural remedies at the industry level. In particular, the study not only documents inefficiencies at the firm level, in our case hospitals. It also looks at remedies by considering reallocations of outputs and inputs among the firms within the same sector. Furthermore, it studies the ideal industry configuration by reporting on the optimal combination of firms meeting actual sector demands and compatible with currently available inputs.

Most performance studies focus on the firm level and use meanwhile well-known, standard frontier methodologies. The evaluation of industry efficiency is less well developed. This contribution extends the seminal work of Färe, Grosskopf, and Li (1992) who develop long-run industry production frontiers. In these long-run industry models, all inputs and outputs can be reallocated with the goal of achieving an optimal industry structure. These models complement the short-run industry production function earlier developed in Johansen (1972). Recognizing that capital is a fixed factor and that most investments take the form of capital goods of a certain vintage, the latter author includes capacity constraints in the industry production function. This implicitly assumes that ex ante, i.e., before the investment decision is taken, the technological choices facing the firm allow for large substitution possibilities between inputs. But once investments have been made (ex post), none or only few substitution possibilities remain in the short run. This is called the putty-clay tradition. The models of Färe, Grosskopf, and Li (1992), by contrast, ignore any capacity limitations. These models can be helpful in reorganizing the health care industry in general. Furthermore, they tend to find solutions that imply savings for the public budget.

These industry production frontiers are extended in several ways. Basically, we modify the models to handle multiple outputs and we eliminate recourse to ad-hoc specifications of production capacity. An ad-hoc specification is, e.g., found in Førsund, Hjalmarsson, and Summa (1996) who apply the short-run Johansen (1972) model. Instead, we adopt the methodology proposed in Färe, Grosskopf, and Kokkelenberg (1989) that provides an operational definition for the plant capacity definition first introduced by Johansen (1968).[2] We are aware of only a handful of applications of these models in general (e.g. Li 1992), including one in the health care industry in particular (Li and Ng 1995).

As stressed by Førsund and Hjalmarsson (1987: 141), the economic relevance of these sector models is that they allow for both a positive and normative interpretation. In their words, these models can either be positively used to "simulate industry behaviour under decentralised decision making" or normatively applied to show "how to organise the industry in the most efficient way".

When considering reallocations among hospitals, we obviously have a normative social planning perspective in mind. In particular, we presuppose that in the short run current market demand can be planned and reallocated among hospitals such that existing capacities are optimally utilized. Industry performance increases mainly by an elimination of technical inefficiencies and by an improvement of capacity utilization rates. In a long run perspective, existing capacities no longer constrain the planning process.

Indeed, according to several specialists the rationalization of supply would allow to: limit the phenomenon of supply-induced demand, exploit more fully returns to scale and economies of scope, economize by grouping reserve capacities, and preserve the quality of care. With respect to the latter, one should notice that guaranteeing a minimal critical mass of interventions allows to fully exploiting learning effects. This has been documented in the medical literature by, for instance, Luft, Bunker, and Enthoven (1979) and Luft, Hunt, and Maerki (1987), among others. In this sense, concentration of activities beyond a minimum level is a necessary condition to assure good quality of care. For instance, a recent French policy measure in this spirit is to close down all maternity hospitals with fewer than 300 births a year.

Since a critical number of interventions is a condition to attain minimal quality levels, we ideally would like to guarantee a minimum level of quality of care in the whole industry. However, no information on quality of care is available. Therefore, the explicit objective of our models is to maximize the savings in all types of surgery rooms. Negatively stated, preventing the use of underutilized surgery rooms cannot decrease the quality of surgical care. Furthermore, as it is widely believed that medical doctors (MDs) are a critical resource, we verify whether or not our primary goal is compatible with the current industry allocation of variable inputs.

This study complements preliminary results with these models reported in Dervaux, Kerstens, and Leleu (1999) in two respects. First, we include personnel information. Earlier, we assumed that the personnel constraint was not binding at the industry optimum. Now we can simply check whether this is the case or not. This is important since a numerus clausus in France, effective since 1978, restrains the supply of MDs in general terms in an effort to limit the supply of health care, and consequently supply-induced demand. Neglecting eventual different needs between disciplines, this quantity constraint recently seems to create shortages for certain specialists. It is a widespread opinion that this shortage explains the inability to fully exploit existing capacities (see, e.g., de Pouvourville 1996). Anticipating our empirical results, we show that the current allocation of MDs is sufficient to face demand.

Second, outputs are slightly aggregated in an effort to counterbalance the reduction in the number of organizations evaluated. The latter is a consequence of a shift in the level of analysis from surgery units to hospitals. We carefully control whether our empirical results remain qualitatively in line with this earlier research. Lacking other studies using similar models, this yields at least one point of comparison to assess the plausibility of our outcomes.

This paper unfolds as follows. The next section provides some background information, including some statistics, on the French health care sector. Section 5.3 starts with a discussion of the firm and industry efficiency models using Data Envelopment Analysis (DEA) techniques. Section 5.4 contains a detailed description of the samples utilized and develops the technology specification for the hospitals. All empirical results are presented in Section 5.5. A final section concludes, formulates critical notes, and suggests some extensions.

5.2 HEALTH CARE IN FRANCE: BACKGROUND AND STATISTICS

In 1997, health care expenditures represented 9.6% of French GDP, reaching a level of $2,250 in per capita terms. France has among the highest share of health expenditures in GNP compared to OECD countries, only preceded by the United States, Germany and Switzerland. About 47% of the total health budget is devoted to hospital care. In this field, the public sector delivers 70.1% of the industry capacity, conceived in terms of the number of beds. Among private hospitals, a clear distinction should be made between hospitals participating to the public health service (denoted PSPH), counting for 9.1% of the industry capacity, and not-for-profit or for-profit hospitals, respectively 4.3% and 16.3% of the industry capacity.

Despite differences in ownership status, PSPH hospitals are very similar to public hospitals, because they face the same constraints (e.g., obligation to admit any patient at any time) and share the same financing rules. From 1983 to 1996, public and PSPH hospitals have operated under a global budget to cover their operating costs, including MDs wages. Since 1996, a prospective payment system based on Diagnostic Related Groups (DRGs) has been introduced. Private hospitals receive a per diem payment to cover nursing care, medication and catering. They also receive a lump sum for each surgery or delivery so as to finance their capital investments. In contrast to public hospitals, private hospitals are organized as doctors' workshops, since most MDs are independent workers that are paid on a fee for service basis.

This coexistence of two financing schemes leads to huge differences in economic incentives between public and private providers. This undoubtedly adds to the perceived influence of ownership status. Private hospitals are induced to invest in activities where productivity gains are possible though a strategy of specialization and standardization (like surgery and obstetrics). Table 5.1 highlights this phenomenon by presenting the repartition of beds by activity and according to ownership status.

Our study concentrates on the Nord/Pas-de-Calais region. Narrowing the geographical perspective does not change the main features described so far. One exception is that the market share of the private sector in this region is slightly below the national level (29.8% versus 35.4%).

Table 5.1

Repartition of Beds by Activity and Ownership Status (percentages)

	public sector	PSPH	private sector not-for-profit	private sector for-profit
medicine	78.9	7.9	2.1	11.0
surgery	46.2	7.4	3.6	42.8
obstetrics	59.4	4.9	2.7	33.0
rehabilitation care	42.4	21.6	14.9	21.0
long term care	91.8	3.8	3.4	0.9
psychiatry	68.3	15.2	2.6	13.9
total	64.7	10.6	5.1	19.7

5.3 Firm and Industry Efficiency Models

This section first depicts the intuition behind the firm and industry models used in this contribution. Then, these basic models are extended to adopt them to the planning context of the French health care system. Basically, this section provides a non-technical summary of the underlying notions. Appendix A5.1 furnishes a rather detailed numerical example illustrating the basic concepts behind the different models. Details on the required computations are outlined in Dervaux, Kerstens, and Leleu (1999) and can be found in a technical appendix available from the authors on request.

5.3.1 Methodological Intuition

A unified methodological framework for measuring firm and industry efficiency been introduced in Sengupta (1991) and refined by Färe, Grosskopf, and Li (1992). These authors bring together two strands of literature. First, they build upon the by now popular literature focusing on assessing firm performance. This literature is associated with the seminal contribution of Farrell (1957) and is reviewed by Lovell (1993). Second, a somewhat smaller series of articles and books concentrate on modeling industry production functions and sector efficiency. The main contributions here are probably the Aigner and Chu (1968) and Seitz (1971) articles and the Johansen (1972) and Førsund and Hjalmarsson (1987) books.

These technologies of firms and industries are constructed from data on organizations or decision making units (DMUs) by imposing appropriate assumptions. Reasoning in a multi-input multi-output framework, a combination of different types of inputs is used to produce a combination of different types of outputs. A production possibility or transformation set represents the set of all feasible input-output combinations (see Section 2.2 of this volume). For our application, it is essential to distinguish between fixed and variable inputs.

At the firm level, one of the basic technologies is a constant returns to scale model. This technology, that actually imposes some additional conditions, has been presented in Appendix A2.1 and is also repeated in an Appendix available upon request (or see Färe, Grosskopf, and Lovell (1994)). The main motivation for this model is that it is linked to the long run competitive ideal of a zero-profit equilibrium. Industry specifications of technologies are derived from the above firm-based technologies by considering all inputs and outputs available in the sector as simultaneously reallocatable. This contrasts with firm efficiency models whereby none of the observed inputs and outputs of the individual firms can be reallocated among other firms.

Frontier methods require not only a specification of technology, but also some concept of distance to the frontier. The output measure of efficiency is defined as a factor indicating the largest feasible extension of the current outputs, while maintaining constant input levels. It is positive and bounded below by unity. The unit value is obtained when the observation is producing efficiently. A value greater than unity indicates the possible percentage increase in outputs given current input levels. It also has a straightforward revenue interpretation.

This output oriented efficiency measure is mainly introduced to explain the notions of plant capacity and plant capacity utilization below. In the empirical section, we employ both output- and input-oriented efficiency measures (see Section 2.2 of this volume). The first to determine current hospital plant capacity utilization and the technology at full capacity, the latter to operationalize our main objective of minimizing the use of surgical rooms.

We now turn to the notion of plant capacity at both the firm and the industry level. Following the definition of Johansen (1968), plant capacity is the maximal amount that can be produced per unit of time with existing plant and equipment without restrictions on the availability of variable production factors.[3] Plant capacity utilization (PCU) is then measured as a ratio of current, but technically efficient, outputs to the maximal outputs according to the above definition. To avoid confusing technical efficiency and incomplete plant capacity utilization, Färe, Grosskopf, and Kokkelenberg (1989) have devised a procedure using efficiency measures to separate both notions in a frontier context. An observation is first evaluated in terms of technical efficiency to a standard technology. Then, the same observation is compared to a plant capacity technology, i.e., the same technology but without restrictions on the availability of variable inputs. Details on the required computations are outlined in Dervaux, Kerstens, and Leleu (1999).

Having defined all the elements for our analysis, our modeling strategy can be summarized as follows. First, starting from observed inputs and outputs at the firm level we determine the plant capacity for each firm using an output oriented efficiency measure. These computations are instrumental to determine the maximum production capacities at the firm level, and, implicitly, at the industry level. Then, switching to an industry perspective, we look for a reallocation of the total amount of fixed inputs and outputs in the industry relative to a full plant capacity technology in two ways. First, we minimize proportionally the use of fixed inputs in a short run perspective, that is, given existing plant capacities. Second, we pursue the same objective function but disregarding existing plant capacities. This amounts to allowing a rescaling of existing plant capacities. All these steps in our procedure are summarized in Table 5.2. Observe that variable inputs never constrain the solution of the optimization problems, but are part of the decision variables. We check afterwards whether there are sufficient variable inputs to satisfy the demands resulting from the optimal solutions.

The Appendix A5.1 develops two numerical examples to illuminate the basic notions underlying our analysis.

Table 5.2

Summary of Models

	orientation	*outputs*	*fixed inputs*	*variable inputs*
firm PCU	output	firm allocation	firm allocation	free
industry: current PCU	input (fixed)	industry allocation	industry allocation	free
industry: rescaling of PCU	input (fixed)	industry allocation	industry allocation	free

5.3.2 *Model Refinements: Adaptation to French Planning Context*

To account for the planning context of French health care, our model diverges from the above described classical ones in three ways.[4] First, we include an emergency constraint when determining plant capacities at the firm level. The emergency nature of certain surgical interventions tends to decrease technical efficiency as well as capacity utilization, since these cases are very demanding in terms of organizational flexibility (Friedman and Pauly 1981). Of course, a minimum number of hospitals (most often public ones) should remain open at night to assure proper treatment of the uncertain demand related to emergency cases. In our model, we impose that the reference hospital at the frontier treats at least as many emergency patients as the evaluated hospital.

Second, we take into account the organizational structure of supply and demand in the Nord/Pas-de-Calais region. The northern hospital regulation agency (ARH) distinguishes between four areas ("bassins de vie") in this region: Littoral, Lille, Hainaut and Artois. Each of these areas defines a geographical entity with homogenous socio-cultural characteristics and a great similarity in the demand for health care. The long-term goal of the ARH is to organize health care supply in such a way that population needs in each area can be covered. Therefore, in the industry models we impose that local demand is met in each of the four areas of the Nord/Pas-de-Calais region.

Third, having mentioned the problem of regional disparities in health care access in the introduction, an open question is whether there are also phenomena of suppressed demand in the four sub-regions of Nord/Pas-de-Calais. As a matter of fact, the Nord/Pas-de-Calais region is known to be under-equipped compared to national standards (see Coca 1995). This is important given the policy objective of ARH to refine the structure of health care supply so as to cater for geographically heterogeneous population needs. Therefore, we also estimate a theoretical demand for surgical interventions using national data. Details of the procedure are outlined in Section 5.5.2.

5.4 Surgery in French Hospitals

5.4.1 *Sample Description*

We use two data sets for this study. Our principal data source is a census of all surgery units in France operating in 1992 conducted by the French social security. A total of 1,605 hospitals have been surveyed. Our study concentrates on a sub-sample of 92 hospitals situated in the northern part of France (Nord/Pas-de-Calais region). This census gathered information at four distinct tiers: the hospital level; all surgical wards; all surgery units within each hospital; and a sample of surgical interventions drawn for each surgery unit. This data set still offers the most comprehensive picture of surgical activity and equipment available today in France. Unfortunately, data on personnel (MDs, nurses, nursing aids,...) is not available. To overcome this drawback, we extract information on staff numbers from an administrative survey of all private and public hospitals annually organized by the Ministry of Health since 1994. In this second data source, personnel numbers are, however, only available at the hospital level, not at any of the tiers underneath.

Clearly, there is a discrepancy of two years between both data sources. This forces us to assume that no drastic personnel changes have occurred between 1992 and 1994. Furthermore, due to the need to match two data sets at different levels of aggregation, we have been obliged to work at the level of the hospital. Our earlier study focused instead at the surgery unit. Since fewer units of analysis remain, we are also forced to aggregate somewhat our definition of outputs. This is the only way to escape the curse of dimensionality inherent in non-parametric frontier analysis. Unfortunately, no standard test procedures are available to guide us in this modeling effort (though see Banker (1996) for some recent proposals).

The two data sets have been matched by looking at the number of surgical beds available in each hospital in 1992 and 1994. The concatenation of both data sources has been impossible for four small hospitals, which reduces the sample size from 92 to 88 hospitals. Furthermore, surgical activities are not equivalently defined in the two data sets. To maintain consistency, we have reallocated staff from obstetrical to surgical wards according some specific rules. Details on the matching procedure are given in Appendix A5.2.

A final challenge is that personnel need not exclusively be dedicated to surgical interventions as such. MDs, but especially nurses and nursing aids may have other activities in, for instance, surgical wards. In the second data source, there is no information on the precise repartition of time between surgical interventions and surgical wards. We have to find

ways to account for the time devoted in surgical wards so as to avoid specification bias. First, we include the total number of inpatient days in surgical wards as an additional output. Second, we also include the number of beds in surgical wards as an additional input. Thus, our specification of technology no longer narrowly focuses on surgical interventions, but is enlarged to include related activities of personnel in surgical wards.

5.4.2 *Model Specification*

Turning to a description of surgical technology, we sequentially devote our attention to its outputs, and then its fixed and variable inputs. Starting with the outputs, we distinguish surgical acts by their nature and complexity. Regarding their nature, acts are grouped according to a successive combination of two criteria: first, classical surgery versus exploratory acts, and subsequently bone surgery versus visceral surgery. Regarding their complexity, acts are classified according to a French index based on surgical standards (denoted ICR). This index summarizes the complexity of each surgical act and is constructed from judgements of medical experts regarding the necessary resources under ideal circumstances. We partition surgical activity into heavy, medium and light surgical acts.[5] Theoretically, the combination of the classifications according to nature and complexity yields 12 (4x3) categories. However, in practice certain categories are empty and so we end up with six types of surgical activities. In addition to these outputs, we have isolated endoscopic activity, since this is the most frequently observed surgical activity. In fact, this sole act accounts, on average, for about 17% of total surgical interventions. Finally, for reasons explained in the previous subsection, we also included total inpatient days in surgical wards. This leads to a total of eight output categories. Note that regrettably we have no information on the quality of surgical interventions or on the quality of care in surgical wards.

There are three kinds of inputs in our model: the number of surgery rooms per kind; the number of beds; and personnel per category. Surgery rooms are considered as fixed inputs, since they imply huge investments and most, if not all, of these costs are sunk. Beds and personnel, by contrast, are variable input categories, since they can be allocated in different ways at little cost. Before discussing each of these inputs in turn, we first contrast our view of beds as variable inputs to the tradition to regard them as a fixed input when analyzing hospital efficiency. The goal of our analysis is first and foremost to evaluate surgical activity within hospitals. Therefore, surgery rooms, due to their sunk costs, are the fixed factors on which our analysis sheds light: should one maintain a surgical activity in this hospital or not, and if so at what level? Beds in surgical wards

depend only partially upon the activity of the surgery unit, since they can be easily recycled for other purposes. Therefore, they are variable dimensions for our purpose.

Surgery rooms differ mainly in their fixed equipment and, consequently, in the types of acts that can be performed while guaranteeing minimal quality standards. The survey distinguishes between seven types of rooms: from nursing rooms to highly specialized rooms dedicated to cardiology and neurology (precise definitions are found in Table A5.2). We aggregate type D and E rooms, since only one hospital in the region has a type E room. Thus, our analysis ends up with six room types.

Regarding the variable inputs, we include the number of beds in surgery wards. Furthermore, full time equivalent personnel information is classified according to three types: MDs, nurses, and nursing aids. Descriptive statistics on outputs, surgery rooms, beds and personnel categories are found in Table 5.3.

Clearly, most interventions (about 60%) are classified under the medium visceral and light surgery categories. Few hospitals (only 28%) engage in heavy exploratory acts. Turning to the inputs, 67% of fixed inputs fall under the B and C categories, representing the standard surgery rooms. Rooms of type G are sparsely available in the region. Note also that the distributions of some of the inputs and outputs have a wide range and are highly skewed, exemplifying the size differences between the various hospitals.

5.5 FIRM AND INDUSTRY EFFICIENCY: EMPIRICAL RESULTS

5.5.1 *Industry Level Results*

First, the current situation is characterized in terms of plant capacity utilization. For each firm (hospital), we compute the plant capacity utilization model.[6]

Table 5.3 also summarizes the results for this PCU firm model. It turns out that, at full capacity, hospitals could raise their outputs by 15% on average. Moreover, 59 of the 88 hospitals (67 %) operate at full capacity. We also compare initial and optimal numbers of rooms as well as the initial and optimal outputs. In line with previous results (see Dervaux, Kerstens, and Leleu 1999), inefficiencies are mainly found in surgery rooms with below average equipment (type A rooms) and with rather standard surgery acts (light surgery and medium visceral surgery). When operating at full capacity, the surgery industry could raise its activities by at least 10%.

Table 5.3

DESCRIPTIVE STATISTICS ON OUTPUTS AND INPUTS AND PCU RESULTS

	mean	st. dev	max.	% zeros	total	PCU	%[a]
outputs							
heavy surgery	220	640	5,930	10	19,400	22,400	15
heavy exploratory acts	90	490	4,510	72	8,010	8,830	10
medium bone surgery	660	820	6,400	0	57,700	65,600	14
medium visceral surgery	1,780	1,840	15,500	0	156,000	195,000	25
light exploratory acts	420	980	8,720	9	36,600	41,800	14
light surgery	1,540	1,610	12,400	0	135,000	172,000	27
endoscopy	970	2,180	15,300	16	85,400	97,300	14
inpatient days	25,500	28,200	252,000	0	2,240,000	2,550,000	12
fixed inputs							
room B	1.7	4.6	41	43	145	145	0
room C	3.5	4.3	38	10	306	306	0
room D, E	0.8	1.6	10	59	71	70	1
room A	0.6	0.8	4	57	49	34	29
room F	0.9	2.4	20	63	79	74	6
room G	0.3	0.8	5	83	23	22.7	1
total rooms	7.6	12.0	113	0	673	653	3
variable inputs							
medical doctors	10.3	14.9	138	0	905	949	-5
nurses	44.4	89.9	827	0	3,910	3,920	0
nursing aids	52.3	99.9	919	0	4,610	4,530	2
beds	94.1	128	1,120	0	8,280	8,670	-5

a Excess or shortage capacities as a percentage of initial endowments.

Turning now to the industry analysis, Table 5.4 provides results for the input efficiency measure computed on the fixed inputs solely for both the short-term Johansen model and the long-term model. Compared with the current situation, it reveals the excess capacities in the fixed inputs. In both models, the optimal number of rooms is estimated by a maximal contraction of the input vector, eventually diminished by the slack variables. The main conclusions are as follows. First, at the industry level, rooms can be reduced by about 15% in the short-run and even by 29% in the long-term. Like in the PCU firm model, the largest overcapacities are situated in surgery rooms with below average equipment (rooms of type A and F).

Table 5.4

Available and Optimal Outputs and Inputs in the Industry Models

		short-run model		long-run model	
	total	*number*	*%* [a]	*number*	*%* [a]
outputs					
heavy surgery	19,400	20,400	5	19,400	0
heavy exploratory acts	8,010	8,600	8	9,200	15
medium bone surgery	57,700	59,700	3	58,500	1
medium visceral surgery	156,000	175,000	12	198,000	26
light exploratory acts	36,600	38,800	6	53,000	44
light surgery	135,000	152,000	13	191,000	41
endoscopy	85,400	87,200	2	100,000	18
inpatient days	2,240,000	2,240,000	0	2,240,000	0
fixed inputs					
room B	145	126	13	107	26
room C	306	266	13	225	26
room D, E	71	62	13	52	26
room A	49	30	39	28	44
room F	79	68	14	52	34
room G	23	19	16	17	26
total rooms	673	571	15	480	29
variable inputs					
medical doctors	905	833	8	777	14
nurses	3,910	3,500	10	3,020	23
nursing aids	4,610	4,000	13	3,040	34
beds	8,280	7,590	8	7,230	13

a *Excess or shortage capacities as a percentage of initial endowments.*

These estimated overcapacities for the Johansen model are exactly the same as in Dervaux, Kerstens, and Leleu (1999), but they are about 18% less in the long-term model. The different level of aggregation in both studies probably explains this difference. In the first article, these models were solved at the surgery unit level, while the present work has been obliged to focus on the hospital level. While outputs have been aggregated, this does not seem to have counterbalanced the above shift in perspective.

Results for the optimal variable inputs are also presented in Tables 5.3 and 5.4. For all models, we calculate optimal quantities of personnel and beds for hospitals at full capacity utilization. Starting with the PCU firm model in Table 5.3, we notice that the initial allocation of nurses and nursing aids is just sufficient to meet the optimal number of variable inputs necessary at full capacity. By contrast, the actual numbers of MDs and beds in the sector are insufficient to respond to demand at full capacity. Turning to the short- and long-term models, we see that, by contrast, the actual allocation for the industry is sufficient (Table 5.4). Allowing for reallocation, the demand could be largely met with current variable input endowments. At the very minimum in the short-run model, there is even an 8% excess of MDs and beds. Though its is equally clear that the latter two variable input dimensions appear to be, relatively speaking, the most constraining in all three models.

5.5.2 *Results for the Four Sub-regions*

Table 5.5 reports the structure of results for short- and long-run models for the four areas, i.e., the optimal values for the activity vector. Accounting for the organizational structure of supply and demand in the Nord/Pas-de-Calais region, local demand is met in each area in the optimization problems. In the short-run model, 72% of hospitals operate at full capacity instead of the 67% currently. 16% produce below full capacity and 12% have a zero optimal activity. The latter implies stopping activities in the surgery units of these hospitals. Closing down hospitals is not at stake. Results are rather homogeneous for these four areas.

In the long-run model, scaling surgery units within hospitals up or down is allowed. The latter obviously encourages reallocations between hospitals. Now only 23% (= 17% + 6%) of hospitals, or rather scaled up or down versions of current units, are needed to produce the industry output. The latter results show the normative aspect of the long-term model. In each area, more than two-thirds of hospitals transfer their surgical activity to these scaled units. A detailed look at the empirical results reveals that indeed the surgery units with a mix similar to the industry mix are scaled up, while more specialized units are either eliminated or used to complete the industry needs.

A problem is that surgical activities in some hospitals are scaled up rather drastically. For instance, one unit in the sample is scaled up by a factor smaller than 18. This scaling of a current full capacity technology raises a methodological issue concerning the feasibility and usefulness of the resulting optimal activity vector. First, taken literally as scaling up the surgical activities of an existing unit, it raises questions about the tech-

nological and medical feasibility. Second, a more plausible interpretation is that it indicates the optimal duplication of this existing plant. In the latter case, about 18 other units (not in table) should be structured and organized similar to this optimal unit. Since all cannot possibly be attached to the same hospital, their exact location remains an open issue. By contrast, the short-run plant capacity model has the advantage of imposing less assumptions, since only a scaling down or up of activities within a unit with existing capacity is feasible.

Table 5.5

Structure of Reallocations in the Four Areas

		short-run model			long-run model		
	number of hospitals	% full capacity	% below full capacity	% zero capacity	% of scaled hospitals		% zero capacity
					up	down	
Littoral	18	67	28	5	22	0	78
Lille	26	70	15	15	15	4	81
Artois	22	82	14	4	18	9	73
Hainaut	22	68	9	23	14	9	77
total	88	72	16	12	17	6	77

For these areas, we actually consider three models: in addition to the classical Johansen and long-term models, we also develop a long-term model with an estimated instead of observed demand. To estimate population needs in surgery, we employ the observed national rates of surgical interventions in each of our 7 output categories per age and per sex. These rates are used to compute a theoretical demand in each area according to its own population structure in terms of age and sex. The final output, inpatient days, is corrected by regressing this variable on a constant and the total number of surgical interventions at the national level. Results for these computations are reported in Table A5.3. Summarizing, the deviations between estimated and current outputs reveal that the Lille sub-region has a surgical activity largely exceeding its needs. By contrast, Artois and Hainaut seem to be deprived of interventions in many of the output categories.

Table 5.6 presents the optimal number of rooms in Littoral, Lille, Artois and Hainaut. In the short-run model, Lille and Artois, with the largest current endowments, show the least overcapacities in rooms (respectively 12% and 11%). The reverse results are, however, obtained in the long-run model (respectively 31% and 33%). The two others regions have about the same amount of excess rooms under both models (between 21% and

24%). The structure of results changes drastically when we consider estimated instead of observed demand. Overcapacities in Lille reach 42%, since the observed demand is larger than the estimated one. Results are less pronounced for Littoral, though there are still overcapacities. For Hainaut and Artois, results change both quantitatively and qualitatively. While they experienced overcapacities with observed demand, the new results clearly indicate that they are just about sufficiently equipped or even lack a few rooms.

Table 5.6

Available and Optimal Number of Rooms for the Four Areas

	initial situation	short-run model		long-run model observed demand		long-run model estimated demand	
		number	%[a]	number	%[a]	number	%[a]
Littoral	124	97	21	95	23	107	14
Lille	284	250	12	197	31	164	42
Artois	149	133	11	100	33	153	-3
Hainaut	116	91	22	88	24	118	-2
total	673	571	15	480	29	543	19

a Excess or shortage capacities as a percentage of available rooms.

Table 5.7 presents optimal variable inputs for the estimated demand model. MDs seem to be the critical resource to respond to estimated demand. While abundantly available in Lille, they are a scarce resource for Artois and Hainaut.

Table 5.7

Initial and Optimal Number of Variable Inputs: Estimated Demand Model

	medical doctors			nurses			nursing aids			beds		
	obs.	opt.	%[a]	obs.	opt.	%[a]	obs.	opt.	%[a]	obs.	opt.	%[a]
Littoral	183	170	7	686	584	15	945	650	31	1,580	1,310	17
Lille	412	295	28	1,810	916	49	1,170	921	48	3,100	2,350	24
Artois	169	232	-37	803	584	27	1,020	1,080	-6	2,080	2,310	-11
Hainaut	141	170	-21	611	526	14	866	675	22	1,520	1,340	12
total	905	868	4	3,910	2,610	33	4,610	3,330	28	8,280	7,310	12

a Excess or shortage of staff and beds as a percentage of initial endowments.

These areas need an additional 37% respectively 21% of their current allocation of MDs to respond to population needs. At the regional level, MDs are just sufficient to cover surgical needs. Furthermore, only Artois has an insufficient number of beds in the long run. Again, Lille shows structural excess capacities for all types of personnel and beds.

5.6 Conclusion

For policy purposes, the following tentative conclusions can be formulated. Our empirical results highlight existing overcapacities in French surgery. The use of our industry reallocation models at full plant capacity reveals that about 15% of surgery rooms can be closed down while satisfying the regional demand for health care in the short run. In the long run, overcapacities reach even up to 29%. The number of MDs is sufficient at the regional level to meet current and even estimated health care demand, at least after reducing surgical capacities in specific hospitals. A comparison with our previous work shows that these results appear quite robust with regard to both the definition of outputs and inputs, and the level of analysis.

As to geographical distribution, overcapacities exist in all areas within the Nord/Pas-de-Calais region and mainly affect rooms with below average equipment. They appear slightly more important in Littoral and Hainaut. This picture changes sharply when considering the needs of the population instead of observed demand. In this case, most of the overcapacities are concentrated in Lille and, to a lesser extent, in Littoral. Other areas seem, on balance, to be justly equipped. To cope with these equity concerns, it is necessary to rather drastically reallocate medical labor between areas.

The data coming from two data sets, we have taken great care to match information on personnel with the main census. But, measurement errors on variable inputs cannot be excluded. Nevertheless, these limitations must not be overemphasized, since the variable inputs do not constrain the computations of plant capacities. As a matter of fact, personnel is always a decision variable at the firm level. It is only at the industry level that personnel becomes critical in verifying whether the optimal amounts needed can be met with the actual allocation of resources. Optimal versus actual endowments are checked for both the region and the sub-regions. This comparison shows that concerns about a scarcity of MDs seem premature.

From a methodological point of view, the short-run plant capacity model has the advantage of imposing the least hypotheses. The usefulness of the long-term model requires further investigation, since it assumes that a scaling of a current full capacity

technology is feasible and useful. As indicated in the text, there is a question of interpretation of the optimal activity bundle.

It is, however, good to point out two caveats. First, reallocation of inputs and outputs may imply certain adjustment costs. For instance, if one closes down a surgery unit, then this implies writing off large amounts of sunk costs. Equally so, personnel that may be reallocated to other units may face both monetary costs and social stress. Adjustment costs are, however, not explicitly modeled at this stage. A counter-argument is that at least the short-run industry models need not imply irreversible decisions. E.g., if the model results dictate that one should no longer use a specific surgical room, this decision can be undone at a later stage.

Second, concentration of activities, an almost unavoidable result in these models, could be interpreted to imply a shift from social to private costs. That is, a reduction of the public sector health budgets is counterbalanced by increased costs for private users. For instance, a user pays more transportation costs, as average distances increase due to concentration, to undergo a surgical intervention, as will his family members paying him a visit. But, such eventual shift in costs should be weighted against the private and social benefits in terms of improved quality of care. Balancing these gains and losses ideally requires a full-fledged cost benefit analysis, which is for example done by Merkies in Chapter 8. However, this is beyond the scope of our contribution. Furthermore, we believe these costs are no major drawback, since surgery in most cases does not correspond to a recurrent need.

In conclusion, we believe these empirical results underscore the huge policy potential of these industry reallocation models. We hope to apply similar policy-oriented models shortly to the Picardie region. Let us finish by pointing out some further refinements. Clearly, in reality some demands should be met at the local level, while others could better be treated at the regional or even national level. This requires the introduction of nested or hierarchical levels. This is not impossible in a frontier context, as shown in Athanassopoulos (1998) and illustrated in Dervaux, Kerstens, and Leleu (1999). Finally, the link between our long-term plant capacity model and the more traditional Färe, Grosskopf, and Li (1992) models remains to be explored.

APPENDIX

A5.1 NUMERICAL EXAMPLES

This subsection develops two numerical examples to illuminate the basic notions underlying our analysis. We start by explaining the plant capacity concept: first at the firm level and then at the industry level. Then we develop the industry frontiers with and without capacity constraints.

The plant capacity concept is explained in Figure A5.1 using a simple numerical example (see also Table A5.1).

Figure A5.1
PLANT CAPACITY AT THE FIRM AND INDUSTRY LEVEL

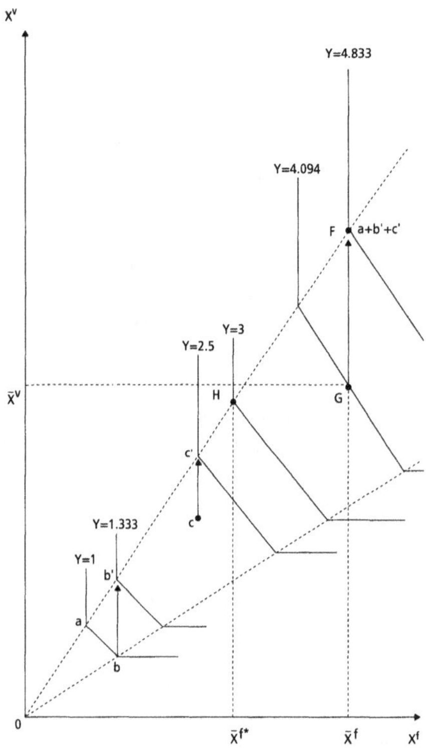

The sector includes three firms, represented by points a, b and c, producing a same output from different endowments in variable and fixed factors. The x-axis represents fixed inputs, the y-axis variable inputs. The technology exhibits constant returns to scale. This shows up by the rays from the origin to the efficient points on the frontier. This ray implies that the efficient point can be completely scaled up or down without restrictions. Thus any proportional increase in all outputs can be obtained by simply increasing all inputs by that same proportion. Firms a and b are efficient, whereas firm c uses too much of each input. It can be calculated that its output efficiency score is 0.48.

Table A5.1
NUMERICAL EXAMPLES

	output	*fixed input*	*variable input*	*te*	*% pcu*
numerical example 1					
firm a	1	3	3.5	1	100
firm b	1	4	2	1	75
firm c	1	7.5	5.5	0.48	84
total industry	3	14.5	11	0.73	85

	variable input/ output	*fixed input 1*	*fixed input 2*	*te*	*% pcu*
numerical example 2					
firm a	1	1	6	1	100
firm b	1	5	1	1	100
firm c	1	3	2	1	100
firm d	1	3	5	0.73	100
total industry	4	12	14	0.84	100

The technology used by firm a can be seen to be the least intensive in terms of the fixed factor compared to the processes employed by firms b and c. When the firms are unconstrained in terms of their variable input, it is firm a and all of its extensions along the ray oa that guarantee a maximal output. Thus, it is somewhere along this ray oa that all firms reach their full plant capacity. Firm a then operates at full capacity by definition. Firm b, by contrast, could produce a greater quantity of output given its current level of fixed input. Moving point b to the ray oa (i.e., vertically onto point b') yields an output equal to 1.33 (larger than the current unit output). Having determined the maximal output at full plant capacity, one can compute the plant capacity utilization rate of firm b as the ratio of its current output divided by this output at full plant capacity. Firm b thus currently produces at 75% of its plant capacity (=1/1.33). Computations for firm c proceed along the same lines, but with the complication that we first have to remove the technical inefficiency. First, we project this observation onto the frontier into the output direction (not visible on Figure A5.1). When being efficient, point c could produce 2.09 output units. Second, moving this frontier projection vertically towards c' yields an optimal capacity output of 2.5 units. Thus, the plant capacity utilization rate of firm c is about 83.8% (=2.09/2.5). Notice that capacity utilization rates are indeed computed at the frontier, to avoid any confusion with the technical efficiency notion.

To determine the capacity for the whole industry, we first observe that the industry allocation simply equals the sum of all inputs and outputs of the firms in the industry (point G). Industry capacity output now corresponds to moving this allocation onto the ray oa (i.e., point F). This full capacity output equals 4.83 (=1+1.33+2.5). This number corresponds exactly to the sum of the individual firm's capacity output levels. Notice that, geometrically speaking, point F is simply the sum of the line pieces oa, ob' and oc'. Furthermore, line piece GF equals the sum of the line pieces bb' and cc'. The latter simply correspond to the additional variable inputs necessary, both at the individual and industry level, to be able to produce at full plant capacity. This summing property illustrates that the Johansen measure of plant capacity is consistent for sector aggregation. Current industry output equals 3 (= 1+1+1). If one eliminated the inefficiency of firm c, the industry could at best attain an output level of 4.09 (= 1+1+2.09). Thus, the capacity utilization rate at the industry level is 84.7% (= 4.09/4.83).

To illuminate the difference between firm models and the short-run industry frontier, we carefully indicate their production possibilities using the same numerical example. This reveals what one can expect when one

either cannot or can reallocate inputs and outputs among firms in an industry. Since our empirical study considers demand as exogenous, we seek to find the minimal quantity of fixed inputs compatible with the actual industry output and using the full capacity production plans a, b' and c'. In Figure A5.1, the solution of this problem is located at the intersection (point H) between ray oa and the isoquant corresponding to the current industry output.

Notice that, as in the PCU firm model, variable inputs do not constrain the optimal solution, since they are decision variables. However, we may face scarce resources at the industry level. In fact, two situations can occur. On the one hand, this solution point may be compatible with the current total endowment of variable inputs in the industry. Hence, current fixed inputs and optimal variable inputs can produce the optimal industry output. On the other hand, optimal variable inputs may exceed the available industry allocation. In this case, if one cannot procure the necessary additional variable inputs, then it is impossible to produce the optimal industry output from the full capacity production plans.

Turning to the distinction between short- and long-term industry models, we need to introduce a second numerical example where four firms use two fixed inputs as well as one variable input to produce a single, identical output (see again Table A5.1). Production possibilities at full capacity are presented in Figure A5.2.

Figure A5.2
SHORT- AND LONG-RUN INDUSTRY MODELS

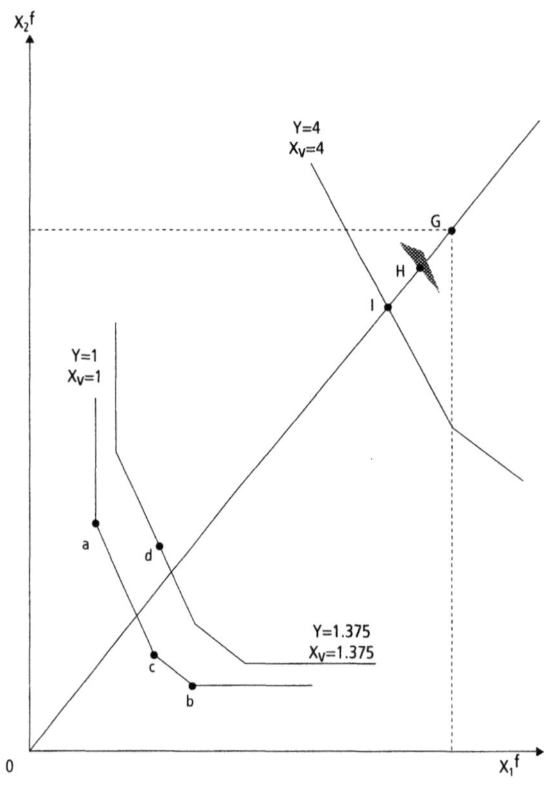

Twelve units of fixed input 1 and 14 units of input 2 produce 4 units of output. The current industry allocation for both fixed inputs is again indicated by point G. In the long run, efficient production with industry output 4 is feasible. A projection of point G onto the industry isoquant yields intersection point I, with an efficiency score of 0.84. The short-run Johansen production model has a feasible set represented by the grayed region. This set is determined by all combinations of the four individual firms yielding an industry output of 4, while respecting the capacity constraint. Minimizing the use of fixed inputs, the projection of point G onto this feasible set leads to point H with efficiency score 0.91. Clearly, for the same output level, the short-run model requires more of both fixed inputs, resulting in less optimal savings at the industry level. Intuitively stated, the mechanism behind this result is that the long-run model can scale both up and down initial capacities in its search for an industry optimum. The short-run model, by contrast, can only maintain existing full capacities or scale them down.

Readers who are interested in details on the required models and computations are referred to a technical appendix that is available from the authors upon simple request.

A5.2 DEFINITIONS AND DESCRIPTIVE STATISTICS

The two data sets have been matched by looking at the number of surgical beds available in each hospital in 1992 and 1994. Most regional hospitals show no change in capacity during this time period. On average, the number of surgical beds in the Nord/Pas-de-Calais region decreased only by 3% between both years. The concatenation of both data sources has been impossible for four small hospitals, because they closed down surgical units during this two years period. But, these four hospitals represented altogether less than 1.5% of the industry capacity in the region. This reduces the sample size from 92 to 88 hospitals. Furthermore, surgical activities are not equivalently defined in the two data sets. In the census, part of the obstetrical activity (like caesarian sections) is considered as surgical, while the administrative survey uses a more traditional etiological distinction between obstetrics and surgery. To maintain consistency, we reallocate staff from obstetrical to surgical wards in proportion to the percentage of obstetrical beds considered as surgical in the census.

Table A5.2

DEFINITION OF SURGERY ROOMS

1. type b rooms: simple equipment for basic surgical acts
2. type c rooms: standard equipment suitable for the most common surgical acts
3. type d rooms: aseptic rooms with high air quality control
4. type e rooms: specialized rooms devoted to cardiology and neurology
5. type a rooms: nursing rooms
6. type f rooms: rooms for endoscopy
7. type g rooms: radiological intervention rooms

Table A.5.3

Observed and Estimated Demand for Surgical Activities for the Four Areas

	Littoral			Lille			Artois			Hainaut			total region		
	obs.	est.	%	obs.	est.	%	obs.	est.	%	obs.	est.	%	obs.	est.	%
heavy surgery[a]	3.9	2.0	49	6.2	11.1	-79	6.0	3.7	37	4.0	2.5	37	20.1	19.4	4
heavy exploratory[a]	0.9	0.6	34	1.4	5.9	-316	1.4	1.2	12	0.9	0.3	73	4.6	8.0	-73
medium bone surgery[a]	11.9	13.6	-15	19.2	20.9	-9	17.4	12.7	27	11.7	10.5	10	60.2	57.7	4
medium visceral surgery[a]	32.9	31.3	5	53.0	57.7	-9	47.8	39.1	18	32.3	28.3	12	165.9	156.4	6
light exploratory[a]	8.7	5.9	32	14.0	20.6	-48	12.7	5.1	60	8.6	5.0	42	43.9	36.6	17
light surgery[a]	27.5	28.0	-2	44.7	50.9	-14	38.8	28.8	26	26.2	27.3	-4	137.2	135.1	2
endoscopy[a]	22.1	16.8	24	35.1	38.9	-11	33.3	24.0	28	22.4	5.7	75	112.9	85.4	24
inpatient days[b]	440	421	4	828	676	18	583	612	-5	433	415	4	2,284	2,111	8

a *Demand expressed per 1,000 acts of surgery.*

b *Demand expressed per 1,000 inpatient days.*

Notes

[1] Keywords: firm efficiency; plant capacity; short and long run industry efficiency; reallocation; hospitals. We thank H. Allemand (CNAMTS) for authorizing the use of the "GAIN Chirurgie" survey as well as B. Colladon and B. Vincke (ENSM) for their medical expertise at an early stage of this study. Our conclusions do not reflect official positions of CNAMTS nor ENSM. This research benefitted from financial support from the Ministry of Work and Social Affairs (MIRE convention N° 31/95). The comments of J. Blank are gratefully acknowledged.

[2] See Färe, Grosskopf, and Valdmanis (1989) for an application in the health care sector.

[3] This technical or engineering notion of capacity utilization can be contrasted to other cost-based notions of capacity utilization (see Johansen 1968 or Morrison 1985).

[4] The reader can consult Dervaux, Kerstens, and Leleu (1999) for technical details on the first two issues.

[5] We retain two thresholds: one at 25 and another at 100 ICR points.

[6] We follow the standard two-stage procedures of Ali and Seiford (1993). First, the plant capacity is determined by optimizing a radial output efficiency measure. Second, the sum of residual slacks is maximized, taking account of the optimal adjustment of outputs from the first stage.

References

Aigner, D., and S. Chu. 1968. On Estimating the Industry Production Function. *American Economic Review* 58 (4): 826-839.

Ali, A., and L. Seiford. 1993. The Mathematical Programming Approach to Efficiency Analysis. In: Fried H., C.A.K. Lovell, and S. Schmidt (eds.), *The Measurement of Productive Efficiency: Techniques and Applications*. Oxford: Oxford University Press, 120-159.

Athanassopoulos, A. 1998. Decision Support for Target-Based Resource Allocation of Public Services in Multiunit and Multilevel Systems. *Management Science* 44 (2): 173-187.

Banker, R. 1996. Hypothesis Test Using Data Envelopment Analysis. *Journal of Productivity Analysis* 7 (2-3): 139-159.

Coca, E. 1995. *Les inégalités entre hôpitaux. Obstacle à l'efficacité et à l'équité de la maîtrise des dépenses hospitalières*. Paris: Berger-Levrault.

de Pouvourville, G. 1996. L'héritage. In: Contandriopoulos, A-P., and Y. Souteyrand (eds.), *L'hôpital stratège: Dynamiques locales et offre de soins*. Paris: John Libbey, 271-283.

Dervaux, B., K. Kerstens, and H. Leleu. 1999. Rationalisation de l'offre de soins en chirurgie: Réduction des surcapacités et réallocation sectorielle. *Revue Economique* 50 (3): 645-655.

Epstein, M., and J. Henderson. 1989. Data Envelopment Analysis for Managerial Control and Diagnosis. *Decision Sciences* 20 (1): 90-119.

Färe, R., S. Grosskopf, and E. Kokkelenberg. 1989. Measuring Plant Capacity, Utilization and Technical Change: A Nonparametric Approach. *International Economic Review* 30 (3): 655-666.

Färe, R., S. Grosskopf, and S-K. Li. 1992. Linear Programming Models for Firm and Industry Performance. *Scandinavian Journal of Economics* 94 (4): 599-608.

Färe, R., S. Grosskopf, and C.A.K. Lovell. 1994. *Production Frontiers*. Cambridge: Cambridge University Press.

Färe, R., S. Grosskopf, and V. Valdmanis. 1989. Capacity, Competition and Efficiency in Hospitals: A Nonparametric Approach. *Journal of Productivity Analysis* 1 (2): 123-138.

Farrell, M. 1957. The Measurement of Productive Efficiency. *Journal of the Royal Statistical Society Series A: General* 120 (3): 253-281.

Førsund, F., and L. Hjalmarsson. 1987. *Analyses of Industrial Structure: A Putty-Clay Approach*. Stockholm: Almquist & Wiksell.

Førsund, F., L. Hjalmarsson, and T. Summa. 1996. The Interplay between Micro-Frontier and Sectoral Short-Run Production Functions. *Scandinavian Journal of Economics* 98 (3): 365-386.

Friedman, B., and M. Pauly. 1981. Cost Functions for a Service Firm with Variable Quality and Stochastic Demand: The Case of Hospitals. *Review of Economics and Statistics* 63 (4): 620-624.

Guyomar, C. 1995. Comment mesurer les capacités excédentaires dans les hôpitaux? *Solidarité Santé* 4: 9-16.

Johansen, L. 1968. Production Functions and the Concept of Capacity, Namur, Recherches Récentes sur la Fonction de Production (Collection «Economie Mathématique et Econometrie», n° 2) reprinted in Førsund, F.R. (ed.), *Collected Works of Leif Johansen*, Volume 1, Amsterdam, North Holland, 359-382].

Johansen, L. 1972. *Production Functions: An Integration of Micro and Macro, Short Run and Long Run Aspects*. Amsterdam: North Holland Publishing Company.

Li, S-K. 1992. Three Essays on Efficiency Measurement. Carbondale (IL): Southern Illinois University (PhD dissertation).

Li, S-K., and Y.C. Ng. 1995. Measuring the Productive Efficiency of a Group of Firms. *International Advances in Economic Research* 1 (4): 377-390.

Lovell, C.A.K. 1993. Production Frontiers and Productive Efficiency. In: Fried, H., C.A.K. Lovell, and S. Schmidt (eds.), *The Measurement of Productive Efficiency: Techniques and Applications*. Oxford: Oxford University Press, 3-67.

Luft, H.S., J.P. Bunker, and A.C. Enthoven. 1979. Should Operations Be Regionalized? The Empirical Relation between Volume and Mortality. *New England Journal of Medicine* 301 (25): 1364-1369.

Luft, H.S., S.S. Hunt, and S.C. Maerki. 1987. The Volume-Outcome Relationship: Practice-Makes-Perfect or Selective-Referral Patterns? *Health Services Research* 22 (2): 157-182.

Morrison, C. 1985. Primal and Dual Capacity Utilization: An Application to Productivity Measurement in the U.S. Automobile Industry. *Journal of Business and Economic Statistics* 3 (4): 312-324.

Seitz, W. 1971. Productive Efficiency in the Steam-Electric Generating Industry. *Journal of Political Economy*, 79 (4): 878-886.

Sengupta, J.K. 1991. Farrell Efficiency: Some Generalizations and Econometric Implications. In: Kaul, T., and J. Sengupta (eds.), *Economic Models, Estimation, and Socioeconomic Systems: Essays in Honor of Karl A. Fox*. Amsterdam: Elsevier, 125-140.

Wooley J.M., and H.E. Frech. 1989. How Hospitals Compete: A Review of the Literature. *Journal of Law and Public Policy* 2 (1): 57-79.

6 Efficiency of Dutch Nursing Homes: The Sensitivity of DEA-scores to Different Sets of Resource Prices

Evelien Eggink and Jos L.T. Blank[1]

6.1 Introduction

The Dutch nursing home industry is currently the subject of a great deal of public interest because expenditures on nursing home care are large, have increased significantly over the past 20 years, and are expected to continue increasing in the future because the population is growing older. Real expenditure on nursing home care in the Netherlands increased from 2.8 billion guilders in 1974 (0.7% of GNP) to 5.9 billion guilders in 1997 (0.9% of GNP).[2] Most important, the homes are reimbursed almost entirely from government sources.

Nursing home efficiency is important to both policy makers and management. Policy makers can use this information to introduce more incentives in the reimbursement scheme or change regulations in order to cut budgets without affecting the level and quality of services. Management of nursing homes can benefit from efficiency analyses by changing the production process in order to save costs and assure continuity.

The primary objective of this chapter is to examine the efficiency of the Dutch nursing home industry, using the method of Data Envelopment Analysis (DEA). Efficiency scores are calculated for each nursing home. Mean efficiency scores and variation in efficiency scores are presented. Further, differences in efficiency scores are analyzed by viewing the relationship with background characteristics of individual nursing homes. Such relationships are of interest to managers and policy makers since they yield insight into the causes of inefficiency. Stability of such relationships is required in order to take effective policy measures. These relationships are estimated in a second step using regression analysis.

In studies of productivity and cost-efficiency, resource prices play an important role. For instance, when calculating the cost-efficiency of firms, resource prices are used to compare the mix of the resources employed by the different firms. Different resource prices lead to a different optimal mix of resources. The way resource prices are measured may therefore play a crucial role in identifying inefficiencies. Most of the empirical applications of the DEA method do not address the choice of resource price measures specifically (see for instance, Nyman and Bricker 1989, Fizel and Nunnikhoven 1993 and Kooreman 1994 for DEA-studies of nursing homes). In empirical research, price measures are often chosen on practical grounds, such as the availability of data, whereas conceptual matters may be important as well. *Therefore, the second objective of this chapter is to evaluate the impact of using different resource price measures on cost-efficiency, calculated by using DEA.*

In many industries, firms do not all face the same prices for their resources. It is clear that resource prices may vary over time. In addition, apart from intertemporal variation in resource prices there may also be inter-regional variation in resource price levels due to differences in market conditions, such as scarcity at the factor markets. There may even exist inter-firm price differences due to differences in ownership or scale of production. These three points of view give rise to three different measures of resource prices. The first resource price measure only reflects the intertemporal variation and is referred to as national prices, the second also includes the inter-regional variation (regional prices). The third measure includes the inter-firm resource price differences as well (firm-specific prices).

The efficiency scores resulting from the three different price measures are compared and analyzed in detail. Three main issues will be addressed in this paper. First, the mean efficiency may be affected by the price measures. Second, the ranking of the firms according to their efficiency scores may change when using another price measure. Formal tests will be performed to assess the significance of the differences between the sets of scores. A similar approach can be found in Chapter 11. Hughes and Yaisawarng also apply these methods in analyzing the effects of three different measures of capital. The final question addressed here is the stability of the relationship between the efficiency scores and characteristics of the firms. Therefore we study the impact of different resource price measures on these relationships.

This chapter is organized as follows. In Section 6.2 we will describe the Dutch nursing home industry. Some quantitative information on the nursing homes will be presented. In Section 6.3, the DEA-model used in the calculations is presented and the different definitions of the resource prices are described. Section 6.4 describes the empirical setting of this chapter, i.e. the data used in the calculations. Then, in Section 6.5, the

empirical results are given. Section 6.6 summarizes the results and presents some conclusions and policy recommendations.

The nursing home industry is a rather popular service among efficiency researchers. The literature on this subject is extensive. The results presented in this chapter are derived from the study *Zuinig op zorg* (*Careful with Care*, Blank and Eggink 1996). The DEA study of Kooreman (1994) also concerned Dutch nursing homes. Other DEA studies on nursing home efficiency were performed by Nyman and Bricker (1989), Fizel and Nunninkhoven (1993), Chattopadhyay and Heffley (1994), Chattopadhyay and Ray (1996). Parametric analyses were conducted by Ullman and Holtman (1985), Bekele and Holtman (1987), Tuckman and Chang (1988), McKay (1988 and 1989), Dor (1989), Gertler (1992), Gertler and Waldman (1992), Hofler and Rungeling (1994) and Vitaliano and Toren (1994). All of them use cost functions to analyze efficiency. Most of the literature on this subject refers to U.S. nursing homes.

Amongst researchers there is consensus that nursing homes are producing care. The number of patient days seems, therefore, to be the most appropriate measure of service delivery. All researchers recognize the impact of the severity of illnesses and the quality of services on efficiency, so much attention is paid to the measurement of illness severity and quality and to the trade-off between efficiency and quality. Differences among researchers relate mostly to the aim of their studies. Some of them focus on scale effects, others on financing and ownership.

6.2 The Dutch Nursing Home Industry

In general, Dutch nursing homes provide care to post-operative, or chronically ill residents and residents requiring home-type care with basic nursing, therapy and social activities. The nursing home industry is mainly a service to the elderly. In 1997 about 93% of the patients were older than 65 years; about 61% were even older than 80 years (Institute for Health Care Management 1998).

New patients are assigned to nursing homes by independent assessment boards, operating in 27 health regions. This means that nursing homes have no say about the access of patients, and their selection possibilities are very limited.

Nursing homes are almost fully reimbursed by the central government on a prospective basis. Patients contribute only a small part of the costs. Budgets are based on local agreements on the numbers of in-patient days, day care patient days, and external activities,

the so-called extramural care. They are not based on the quality of services or the disability (care need) of patients. Although any surplus does remain available to the nursing home, it is not allowed to make any profits. Over the long term, revenues and costs have to stay in equilibrium. Since wages are regulated as well, the possibilities of the management to increase their own salaries are very limited. Hence, in this case, not-for-profit can not be a matter of 'for-profit in disguise'. Capacities in terms of beds and day care places are determined by the Council of Hospital Services according to rules formulated by the central government. Thus, the managements of Dutch nursing homes do not control capital usage.

In spite of these regulations, nursing homes are still in some competition. Although the assessment boards play an important role in assigning patients to nursing homes, patients or their families can always refuse such a referral or request a transfer to another home. Nursing homes with low occupancy rates will be subjected to budgets cutbacks and finally closed down.

Table 6.1 summarizes the most important features of the Dutch nursing home industry in the years 1985 and 1997.

Table 6.1

SOME CHARACTERISTICS OF THE DUTCH NURSING HOME INDUSTRY, 1985 AND 1997

	1985	1997
number of nursing homes	326	331
in-patient days (x million)	17.6	20.2
day care patient days (x million)	0.6	1.1
variable cost (in NLG billion)[a]	3.5	6.1
personnel (in ftes x thousand)	52.8	69.2

a 1 NLG ≈ USD 0.5 (*exchange rate 1998*), *prices of 1995.*

Source: Institute for Health Care Management (1988, 1998)

The number of nursing homes in the Netherlands exceeds 300. This number is rather stable in time. Nursing homes 'produced' almost 18 million in-patient days and 600 thousand day care patient days in 1985. Because of the aging population the government had to increase nursing homes' capacities. The number of in-patient days grew to

over 20 million and the number of day care patient days almost doubled (1.1 million in 1997). Nevertheless, the 'production' of day care patients days is relatively unimportant. The cost of the hospitals increased from 3.5 billion in 1985 to 6.1 billion guilders in 1997 (+75%). Only 20% of the cost increase was due to increasing prices (not in table). In 1985 more than 52 thousand full time equivalents were employed in the nursing home industry. This number grew to 69 thousand in 1997 (+31%).

6.3 Data Envelopment Analysis and Measurement of Resource Prices

6.3.1 *Data Envelopment Analysis*

This section deals with the DEA model adopted in this chapter. An extensive body of literature on DEA is available, consisting of descriptions, extensions and applications of the DEA method (see Chapter 2; Seiford 1996; Coelli, Rao and Battese 1998). This section is therefore limited to a short discussion of the choice of the DEA specification.

Efficient nursing homes can be detected by the method of linear programming. These efficient nursing homes form an envelope (frontier or best practice) for all nursing homes in the dataset. The distance from each individual home to the frontier is a measure of efficiency. In Chapter 2 of this volume Lovell discusses this method in more detail. He also stresses the choice of the specification of the model.

Several choices have to be made when developing a DEA model. First, an orientation has to be chosen. Since this chapter focuses on cost-efficiency, the resource-oriented DEA is used (see for instance Chapter 2; Lovell 1993; Ali and Seiford 1993): to what extent can resources be diminished without affecting the level of production in order to attain minimum costs? In the empirical setting of the Dutch nursing home sector this is an appropriate orientation, since in the Dutch nursing home sector services are exogenous (see Section 6.2). Furthermore, assumptions have to be made about the returns to scale. Here, variable returns to scale are adopted. This assumption implies that the input-service ratio can be decreased by changing the level of production.

Cost efficiency can be decomposed into allocative, pure technical and scale efficiency. Allocative efficiency refers to the mix of various resources. By changing the resource mix, it may be possible to diminish cost without affecting services. Pure technical efficiency indicates to what extent all resources can be diminished proportionally keeping the level of service delivery fixed. Scale efficiency measures the impact of a change in the level of production on the required resources.

6.3.2 Measurement of Resource Prices

As noted in the introduction, in the empirical analyses in this chapter, three measures of resource prices are introduced:

1. *National resource prices.* These resource prices are the same for all firms observed in the same year, and show only intertemporal variation. This resource price can be calculated as the aggregated national cost of a resource divided by the corresponding aggregated national quantity.
2. *Regional resource prices.* These resource prices are the same for firms in the same region and the same year. This resource price can be calculated using the aggregated regional cost and aggregated regional quantity.
3. *Firm-specific prices.* These resource prices differ between all firms, calculated as average cost per unit of resources used by a firm, the so called unit values.

The *national resource price* is the average resource price for all firms in the sector, describing the situation countrywide. The only variation in the national prices is the intertemporal variation. In the Netherlands, resource price differences between firms are not likely to be large: wages are generally determined by collective labor agreements and, since the geographical distances in the country are small, factor mobility may be relatively high. The national resource prices may therefore be an adequate representation of the actual resource prices faced by the firms. Using these resource prices in DEA calculations implies that any deviation from minimum cost is seen as inefficiency, since prices are the same for all firms in a given year.

However, some *inter-regional price differences* may be present. If firms are free to choose the region they are operating in, they can choose the region with the lowest resource prices. Then, any increase in cost resulting from a different set of resource prices should in fact be interpreted as inefficiency. In many situations, especially in the public sector, firms cannot choose to operate in a region with favorable resource prices. In that case, the resource prices are exogenously given and the inefficiency resulting from the resource price differences between the regions should be filtered out. This can be done by using the regional resource prices rather than the national resource prices which are the same for all firms in a given year.

Moreover, there may be other characteristics of firms that affect the prices paid for the resources. For example, a public firm may pay a higher wage rate than a competitive firm. The scale of the firm may also affect the resource prices. There may be pecuniary benefits for large firms, especially for resources such as material supplies. In these cases

firm specific resource prices may be more appropriate. The calculation of the three price measures is described more formally in the appendix.

When heterogeneity within a resource is present, the quantity, such as the number of fte's for labor, may not be an accurate measure. Instead, the volume of a resource is to be preferred. The volume is a measure that corrects quantity for heterogeneity. The volume of a resource is calculated by dividing cost by the corresponding price. So the volume measured also depends on the measurement of prices. These volumes are used as measures of resources in the analyses.

6.4 DATA

The empirical analyses will be performed using a panel data set of Dutch nursing homes covering the years 1984, 1987-1993. Even though it leaves a gap in the time periods, the data for 1984 are included in the analyses in order to add more data to the analyses, and to ensure that there is enough intertemporal price variation. The data were collected by the Institute for Health Care Management and are derived from several surveys, such as surveys on financial, patient and personnel variables. These files contain information on all nursing homes, i.e. over 300 observations each year. Various consistency checks were run on the data to ensure that changes in average values and distribution of values across time were not extreme. After eliminating observations containing incomplete or inconsistent values, observations for 1,939 nursing homes over eight years of study remain (260-310 homes per year).

From the extensive data set we chose four quantitative *services*: the number of in-patient days in two categories (patients with physical disabilities or with psycho-geriatric disability) and the number of day care patient days in the same two categories. *Resources* include (skilled) nursing personnel, other personnel (management, maintenance, secretarial etc.) and material supplies. As we are only concerned with the variable cost, capital and nurse trainees are included in the fixed resources.[3] Capital is measured by an index consisting of the number of beds, the number of day care places, the number of single rooms, double rooms, rooms with three, four or five beds, and rooms with more than five beds.

As mentioned in Section 6.1, a number of *background variables* of the nursing homes are used to explain the differences in cost-efficiency among homes. A distinction is made between three groups of characteristics related to production, institutional environment and the production process.

The first group of characteristics are concerned with aspects of *production* that are not included in the DEA- analysis itself. These include the disability index, measuring the mean care need of nursing home patients in the region[4], the fraction of male patients and various quality measures. The disability index measures the extent to which patients can perform domestic chores and personal care. The fraction of male patients is also included to reflect differences in need for care. These characteristics were carefully chosen from a larger set of characteristics relating to the need for care, like age, the origin (home for the elderly, independent home, hospital) and the fraction of deceased patients. In addition, several quality measures are included in the analysis. These include the quality of the housing, measured by the capital cost per bed, and the presence of a pharmacy and nurse trainees in the nursing home. In addition the extramural activities a nursing home produces are included in te characteristics, measured by the subsidy amount the nursing home receives for those activities.[5]

The *institutional characteristics* are characteristics of the nursing homes that cannot be controlled by the management, but may have an impact on the efficiency. Since the size of the homes is decided by the government, the number of beds in the home is included. A measure of market competition in the region is also included, by means of a competition ratio.[6] Further the management type, religious affiliation and the type of home (for physically disabled, psycho-geriatric patients or both) are included.

The third group, the *process characteristics*, can be controlled by the management. Important characteristics included in this group are the occupation rate, the ward size, the amount of work contracted out or contracted in, and the absence through illness rate of the personnel.

Table 6.2 contains the descriptives of the variable costs, the fixed resources, the services, and the resource prices. The variable resources are described by their costs and prices; the fixed resources and the services are represented by their volumes. An overview of all background variables and their statistical descriptives is presented in the Appendix (Table A6.1).

Table 6.2 indicates that the average annual variable cost of the nursing homes in the sample equals 11.5 million guilders, or about 5.75 million US dollars. The variation is rather large (a standard deviation of 5.1 million guilders), due to large differences in size among nursing homes.

Table 6.2

DESCRIPTIVE STATISTICS

	mean	*standard deviation*	*minimum*	*maximum*
resources				
variable cost (x NLG million)[a]	11.5	5.1	1.2	35.4
cost of nurses (x NLG million)[a]	4.7	2.1	4.7	16.0
cost of other personnel (x NLG million)[a]	4.1	1.9	2.2	13.8
cost of material supplies (x NLG million)[a]	2.7	1.2	2.5	9.4
capital[b]	100	57	10	344
nurse trainees	31	18	0	122
services				
patient days (x 1,000)	64.0	26.7	8.9	354.0
in-patient days, physical disability (x 1,000)	31.0	26.0	0	133.0
in-patient days, psycho-geriatric disability (x 1,000)	30.4	22.6	0	101.0
day care patient days, physical disability (x 1,000)	1.3	1.8	0	10.0
day care patient days, psycho-geriatric disability (x 1,000)	1.3	1.6	0	8.0
resource prices				
national prices				
nurses (x NLG 1,000)	51.1	4.6	42.6	58.4
other personnel (x NLG 1,000)	55.4	4.7	48.4	62.8
material supplies[c]	1.00	0.05	0.94	1.16
regional prices				
nurses (x NLG 1,000)	51.1	4.7	41.2	61.1
other personnel (x NLG 1,000)	55.5	5.5	44.7	66.6
material supplies[c]	1.00	0.05	0.94	1.11
firm-specific prices				
nurses (x NLG 1,000)	51.3	6.4	19.8	98.5
other personnel (x NLG 1,000)	56.5	11.1	24.1	139.8
material supplies[c]	1.00	0.05	0.91	1.16

a *1 NLG ≈ USD 0.5 (prices of 1998).*

b *The volume of capital used is a constructed variable, measuring both quantitative and qualitative aspects of capital. Its value cannot be directly interpreted.*

c *The price of material supplies is an index number. Its mean value is set at 1 and the numbers in the table only have a relative interpretation.*

Source: Social and Cultural Planning Office

The cost of nurses per home equals on average 4.7 million guilders, while the number of other personnel is a little lower (4.1 million guilders). Again, the variation is large since the size of the homes differ considerably. The cost of material supplies also shows a large variation amongst homes. The minimum and maximum values indicate that the largest home uses almost four times as much material supplies than the smallest home. The amount of capital used in the homes in the sample also shows a large variation. When the mean value of this constructed variable is set at 100, a minimum value of 10 and a maximum value of over 300 arise. These differences are partly due to wide qualitative differences in the housing provided by the homes (not in Table 6.2). Some homes have five and six-bed rooms, while others only have rooms for one or two patients. Finally, the number of nurse trainees varies from 0 to 122. On average, a nursing home employs about 30 trainees.

On average the homes in the sample 'produce' 64,000 patient days, with a minimum of almost 9,000 for the smallest home in the sample, and a maximum of 354,000 patient days for the largest one. The number of in-patient days is much larger than the number of day care patient days; the latter only forms 4% of the total number of patient days. About a quarter of the homes in the sample do not have a day-care center, and have no day-care patients, mostly in the earlier years in the sample (not in Table 6.2).

The national prices show that the price of skilled nurses equals about 51,000 guilders (or USD 25,500) per full-time equivalent. Since managers and paramedical staff are included in the other personnel, the national price for that category of personnel is somewhat higher, at around 55,400 guilders per full time equivalent. There is considerable variation in national prices for personnel as a consequence of some general wage rises in collective labor agreements in the period 1984-1993. The price of supplies is a price index, measuring the price deviation from the mean price in 1990. Therefore, its level has no meaning. The variation over time appears not to be very large.

The mean values for the regional prices are practically the same as for the national prices, though, as expected, the variation is a little higher. This is especially the case for the price of other personnel.

Finally, the mean values of the firm-specific prices for both nurses and for other personnel are higher than the national and regional prices.[7] Also, the variation is considerably higher; the maximum values for the two labor resources equal twice the more aggregated maximum values.

6.5 EMPIRICAL RESULTS

6.5.1 *Efficiency Scores*

For each of the three price measures, the cost-efficiency is calculated for each of the nursing homes in the sample. The calculations use the DEA-model discussed in Section 6.3 and the data described in Section 6.4. Table 6.3 summarizes the resulting efficiency scores.

Table 6.3

MEAN DEA COST-EFFICIENCY SCORES PER YEAR, 1984-1993

	national prices	*regional prices*	*firm-specific prices*
1984	0.82	0.83	0.80
1987	0.81	0.82	0.78
1988	0.79	0.80	0.76
1989	0.78	0.78	0.74
1990	0.76	0.76	0.72
1991	0.75	0.76	0.72
1992	0.75	0.76	0.71
1993	0.74	0.75	0.71
Mean	0.78	0.78	0.74

Source: Social and Cultural Planning Office

Table 6.3 shows that the mean cost-efficiency using national prices equals 0.78. This indicates that theoretically on average the cost of the homes could be decreased by 22% without decreasing the provision of services. Moreover, the cost-efficiency decreases over time, from 0.82 in 1984 to 0.74 in 1993. This is not very surprising since the disability index (measuring the care need) has increased in that period. In the analysis the measure of services is not corrected for this change; therefore the increased care need is included in the inefficiency here.

The empirical results in Table 6.3 show that the mean efficiency is affected by the definition of resource prices chosen. The mean efficiency for the whole sample using the firm-specific prices is smaller than the mean efficiency using the national prices, which in turn is a tiny bit smaller than the efficiency using the regional prices (not visible in table). The differences are 4% and 0.8%, respectively. In addition, the efficiency scores

may differ substantially for individual homes. Using the regional or firm-specific prices instead of the national prices also yields decreasing cost-efficiency over time.

As mentioned before in Section 6.3, the cost efficiency can be decomposed in allocative, pure technical and scale efficiency. Allocative efficiency refers to the mix of various resources. Pure technical efficiency relates to a proportional decrease in all resources. Scale efficiency refers to the necessary extent of proportional increase in resources to accomplish a proportional increase in services. Table 6.4 shows the mean values of the cost efficiency and its components for each of the three price measures.

Table 6.4

DECOMPOSITION OF COST EFFICIENCY (MEAN SCORES)

	national prices	*regional prices*	*firm-specific prices*
allocative efficiency	0.93	0.93	0.90
pure technical efficiency	0.89	0.89	0.87
scale efficiency	0.94	0.94	0.94
cost efficiency	0.78	0.78	0.74

Source: Social and Cultural Planning Office

Table 6.4 shows that for all three price measures the main source of inefficiency is pure technical (11%-13%). The same service delivery can thus be achieved by a joint reduction of all resources by 11% to 13%. By changing the mix of resources a reduction of cost of 7 to 10% is possible (allocative efficiency). Inefficiencies related to the scale of services amount to 6%. In general (small) homes can benefit from scale effects by increasing their services (not in Table 6.4). As a result, cost efficiency varies between 74% and 78%.

National and regional prices based mean efficiencies correspond very well. As can be expected the firm specific prices model leads to higher allocative inefficiencies. It follows that in a world with more price differentiation the possibility of an incorrect allocation of resources increases. Technical efficiency in the firm specific price model is also lower.

Formal tests are performed to test whether the means of the cost efficiency scores differ significantly and whether the scores of the three sets differ pairwise.[8] According to the means tests, the means of the efficiency scores calculated with national prices and with regional prices do not differ significantly. The mean score calculated with the firm-specific prices differs significantly from the means for the two other sets of scores.

According to the pairwise comparison test, the three sets of scores do differ significantly. Thus, even though the means of the scores may not differ significantly, the sets of scores do differ for each of the pairs.

6.5.2 *Ranking of Homes by Their Efficiency Scores*

In the previous section it is shown that the sets of scores corresponding to the three price measures differ significantly. Now the question arises what happens to the ranking of the homes according to their respective efficiency scores when using another price measure. For each of the three models, the homes are ranked by the scores in ascending order. Then the three rankings are compared. In Figures 6.1 to 6.3 the differences between the rankings of the homes are illustrated for each pair of models.

Figure 6.1

DIFFERENCE IN THE RANKING OF THE HOMES USING YEAR VALUES VERSUS REGIONAL VALUES

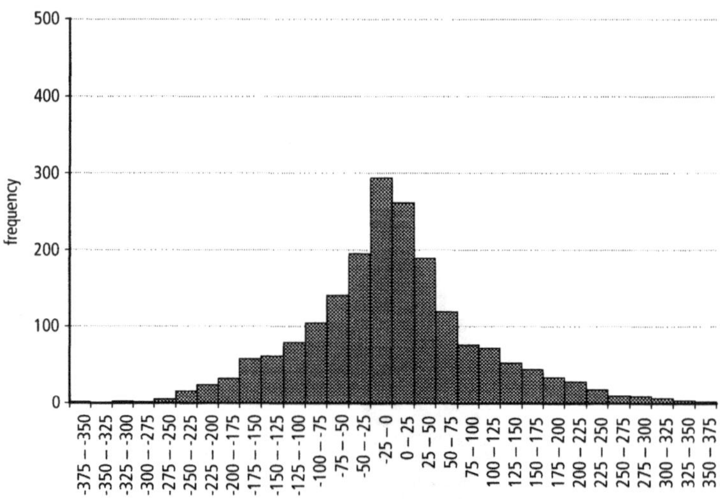

Figure 6.2

DIFFERENCE IN THE RANKING OF THE HOMES USING YEAR VALUES VERSUS FIRM-SPECIFIC VALUES

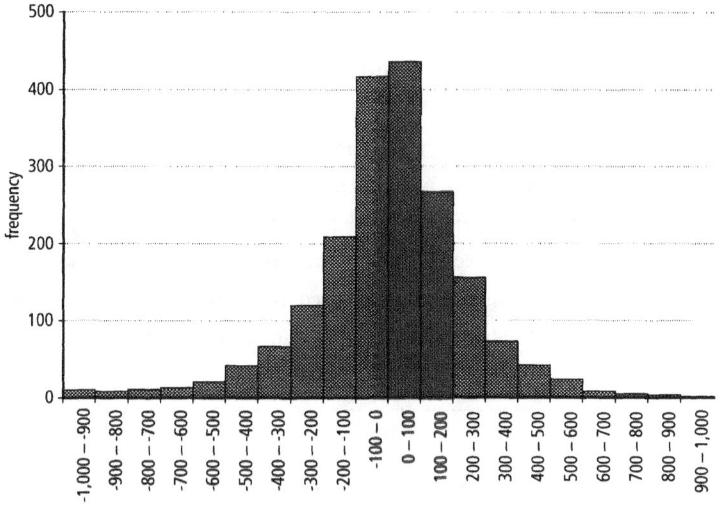

Figure 6.3

DIFFERENCE IN THE RANKING OF THE HOMES USING REGIONAL VALUES VERSUS FIRM-SPECIFIC VALUES

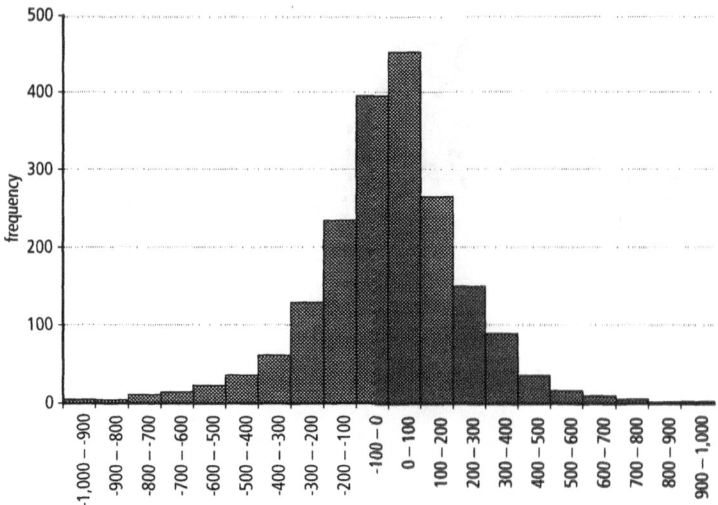

Figure 6.1 shows that, even though many homes may be ranked almost the same in the separate models, for some homes the ranking changes dramatically when another price measure is chosen in the DEA calculations. For the efficiency calculated with national and regional prices the rankings of about 60% of the homes does not change by more than 100 (up or down), while less than 10% changes by more than 250 (see Figure 6.1). However, the differences between the rankings calculated with the firm-specific prices and the national and regional prices respectively are much larger (see Figures 6.2 and 6.3). Even though the ranking of about a third of the homes does not change by more than 100, about a third of the rankings changes by more than 250. The difference may be as high as 1,700 for a few homes (not in figures). This shows that using DEA to detect the cost-efficient homes in terms of ranking is very sensitive to the definition of the prices. This statement is confirmed by a formal test that shows that the rankings of the efficiency scores differ significantly among the different sets of efficiency scores.[9] See Appendix A6.3 for the test results.

However, research is not often concerned with finding the efficiency ranking of homes, but rather with explaining the differences in cost-efficiency. Therefore the next step in this chapter is to seek explanations for the differences in efficiency, not only among the homes but also between the sets of efficiency scores for the three sets of prices.

6.5.3 *Explanations for Differences in Efficiency*

The relationship between the efficiency scores and characteristics of the homes is of interest to managers and policy makers. They yield insight into the background and causes of inefficiency. Using this information effective policy measures can be taken. However, this requires that the relationship between efficiency and characteristics is robust. In the setting of this chapter this implicates that the relationship does not change when another price measure is used in the calculation of efficiency.

In order to investigate the stability of such relationships empirically, they are studied in a second step analysis, using regression analysis.[10] In Section 6.4 various characteristics are mentioned, grouped into the three types of characteristics described earlier: production, institutional and process characteristics. These are all included in the regression analysis.

Table 6.5 summarizes the results of the regression analyses for the scores resulting from the three different models. For each set of scores, corresponding to a price measure, the sign of the relationship between the characteristics and the efficiency score is given. In

addition, "++" and "--" indicate that the effect is significantly positive or negative. The explanatory power is also presented. This describes the fraction of the variance in the efficiency scores that can be explained by differences in the variable at hand.[11] A high explanatory power indicates that the variable is an important determinant of efficiency. Finally, the explained variance of the models as a whole is given.

Table 6.5 shows that the estimates of the three models are much alike. First we will discuss Table 6.5 in general terms. The small differences among the models will be discussed further on. The characteristics included in the analyses can explain a reasonable portion of the total variance in the efficiency scores: about 30%.

In the *characteristics of production* the disability index, the fraction of male patients and extramural activities play an important role. As can be expected, the higher the disability of patients, the lower the efficiency. It seems that men in nursing homes need relatively less care than women. Extramural activities, care provided outside the nursing home, are not included in the measure of services. Since they do lead to extra cost they are measured as inefficiency here. The other variables mentioned show in almost all cases a not significant effect on efficiency.

The type of home is the most important *institutional characteristic*. Homes providing care solely to physically disabled patients have a higher efficiency and those providing care exclusively to psycho-geriatric patients a lower efficiency than so called combined homes. As type of patients is already incorporated in the analyses, one should be careful in interpreting the result. In the case of homes for physically disabled, some cost reductions may occur because of specialization in one type of patients. One can also conclude that on average the care need of physically disabled patients in these specialized homes is lower than in combined homes. Further, it seems that homes for psycho-geriatric are specialized in providing care to severe psycho-geriatric patients. Less important in terms of explanatory power, but nevertheless in most cases significant is the effect of competition. The competition ratio measures the extent of competition amongst nursing homes in a region. It shows that increasing competition deteriorates efficiency. This is a rather common result in existing literature on health economics (see Gertler 1992, Gertler and Waldman 1992). In general, health institutions do not compete by service prices but by quality of services. The extra cost for quality of service is here interpreted as inefficiency.

Finally, the occupation rate seems to be the most important characteristic in the *process characteristics*. The higher the occupation rate, the more efficient the home. Large wards increase efficiency, and contracting in and out, and absence through illness do not show significant effects on cost efficiency.

Table 6.5

Results of the Regression Analyses of Background Characteristics[a]

	national prices		regional prices		firm-specific prices	
	coefficient	expl. power	coefficient	expl. power	coefficient	expl. power
characteristics of production						
disability index	-	0	- -	2	- -	2
fraction male patients	++	4	++	2	++	2
extramural activities	- -	6	- -	5	- -	2
fraction nurse trainees	+	0	+	0	-	0
capital per bed	+	0	+	0	- -	2
pharmacy	+	0	-	0	-	0
institutional characteristics						
number of beds	-	0	+	0	+	0
competition ratio	- -	1	- -	1	-	0
not religiously affiliated	+	0	+	1	+	1
protestant	+	0	+	1	-	0
management type[b]	++	1	++	1	-	0
home for physically disabled	++	43	++	43	++	44
home for psycho-geriatric disabled	- -	30	- -	29	- -	38
process characteristics						
occupation rate	++	13	++	14	++	7
cost of contracting out	-	1	-	1	-	0
revenue from contracting in	-	0	-	0	-	0
ward size	++	1	+	1	+	0
absence through illness[c]	+	0	-	0	-	0
explained variance (in %)	31		31		30	

a - : *negative non-significant relationship with efficiency score*
- - : *negative significant relationship with efficiency score*
+ : *positive non-significant relationship with efficiency score*
++: *positive significant relationship with efficiency score*

b *Foundation vs. other types.*

c *The absence through illness is not known for 1984. Therefore the coefficient for this variable is calculated separately for 1987-1993 only. The other estimates are only marginally affected.*

Source: Social and Cultural Planning Office

Table 6.5 also shows that many variables have a similar effect on the three different sets of efficiency scores. The relationship between most of the characteristics and the efficiency score has the same sign and significance in the three models. Also, the explanatory powers are very similar. Only the differences in the effects of capital per bed and management type are striking. They show contradictory results in terms of sign and significance. The different effect of capital per bed is puzzling, but the different effects of management type can be easily explained. Foundations pay fewer pension premiums for their employees than other homes. This is reflected in higher efficiency. As premiums are a part of wage cost they are incorporated in the firm specific prices. As they are corrected in the national and regional resource prices these differences are not included in the efficiency score anymore.

Summarizing, in all three models the main causes for differences in efficiency between the homes are the fraction of male patients, the extramural activities, the type of home (for psycho-geriatrically disabled, physically disabled or both types of patients) and the occupation rate. Except for the occupation rate these variables can not be used as instruments for management of nursing homes to improve efficiency. The estimated effects of the background features are rather insensitive to the choice of measurement of resource prices.

6.6 Conclusions

In this chapter the cost-efficiency of nursing homes in the Netherlands was analyzed. The method of Data Envelopment Analysis was used to assess cost efficiency scores. A dataset consisting of 1,939 Dutch nursing homes in the period 1984-1993 (excluding 1985 and 1986) was used in the analysis. Special attention was paid to the impact of the measurement of resource prices on DEA cost-efficiency.

The yearly mean cost inefficiency varies between 17% to 29%, depending on the year of observation and the way resource prices are measured. The measured cost efficiency has decreased in the period 1984-1993, also due to an increasing disability of patients. Averaged over the whole sample cost inefficiency varies between 22% and 26%, depending on the measurement of resource prices. About 6% inefficiency is related to the size of production, 7-10% to a suboptimal mix of resources and 11-13% to technical issues. The scale efficiency shows that by enlarging the scale of production of small homes some savings can be achieved. The suboptimal mix of resources may not be a matter of efficiency only. It was shown in Blank and Eggink (1996) that there is a systematic preference for nurses which can be interpreted as quality of service. This line of reasoning is

not uncommon. Gertler and Waldman (1992) also use in their research on homes in New York State a method to correct for quality aspects. McKay (1988) completely leaves the nurses out of her analyses of Texas nursing homes, since in her view every extra nurse implies more quality and can therefore not contribute to inefficiency.

A second step analysis shows that only one third of the variation in inefficiency can be explained by background characteristics. Only a very small part of the inefficiency can be reduced to managerial or process variables. The main conclusion is that only small efficiency gains can be achieved. The largest part of inefficiency is still unexplained and most of the explanatory variables are out of control of the government or management. More information on process and managerial features may contribute to a better insight in the backgrounds of inefficiency. The Dutch Department of Health has only recently acknowledged the importance of this information and has assigned some consultancy offices to improve the provision of relevant data and conduct some benchmarks.

In this chapter, three definitions of resource prices are introduced: national, regional and firm-specific prices. The first resource price measure only reflects the intertemporal variation (national prices), the second measure also includes the inter-regional variation (regional prices). The third measure includes the firm related resource price differences as well (firm-specific prices). From the comparison of the results of the three models three conclusions follow. The first conclusion is that the mean efficiency is lower when the price measure shows a large variation, like the firm-specific prices. Second, the ranking of the firms according to their efficiency scores changes when using another price definition. Statements like home A is more efficient than home B can therefore not be justified. The final conclusion is that the relationship between cost-efficiency and characteristics of firms is hardly affected by using different definitions of prices. So, the conclusion is that DEA should not be used to find an absolute measure of cost inefficiency, or to rank the firms according to their efficiency scores, since this depends on the price definition chosen. The choice of price definition should therefore be considered carefully and should be based on prior knowledge or beliefs about the price variations among the firms. However, the explanation of cost-efficiency seems to be only marginally affected by using another price definition. Thus, for the purpose of finding the explanation for inefficient behavior, this method yields robust results. This implies that the DEA method can be used as a basis for policy measures, independent of the price measure chosen in the DEA cost efficiency calculations.

APPENDIX

A6.1 CALCULATION OF THE PRICE MEASURES

This appendix gives a short description of the calculation of the three different measures of resource prices used in this chapter.

Firm-specific Resource Prices

As usual, firm-specific prices are calculated by costs divided by resource quantities for each individual home. Then the resource prices of nursing personnel and other personnel are based on the so-called unit values: costs divided by the number of personnel. The firm-specific resource price of supplies is based on price indices for food, medical supplies, heating, electricity and so forth, published monthly by Statistics Netherlands. The price indices of these components of materials are then aggregated to form a price of material supplies by weighing with the corresponding cost shares. Since the cost shares may differ between nursing homes, reflecting differences in the composition or the quality of the material supplies, the resultant price is different for each home.

Regional Resource Prices

The regional prices of a resource are the same for homes in the same region and year of observation. In order to calculate regional prices, the aggregated regional costs are divided by the aggregated regional quantities of the resource for each region and year. For both nursing and other personnel the regional prices above can be straightforwardly derived as the aggregated costs divided by the aggregated number of full-time equivalents. The regional price for material supplies is calculated in a different way, since no natural quantity measure is at hand. The price indices used in the calculation of the unit values again form the basis. Here, these price indices are weighed by the mean cost shares of the different components in the region, instead of the cost shares of the individual home. The resultant price differs only between regions and over time periods.

National Resource Prices

The national prices of a resource are the same for homes in the same year. The prices are calculated in a similar way to the regional prices. For personnel aggregated national costs are divided by aggregated national full-time equivalents for all cases in the same time period. For material supplies, the weights used in the construction of the price index are calculated as the mean cost share for all homes in the time period. This results in the national prices for nursing personnel, other personnel and material supplies.

A6.2 Descriptives of the Background Characteristics

Table A6.1

Descriptive Statistics Background Characteristics, 1984, 1987-1995

	mean	*standard deviation*	*minimum*	*maximum*
characteristics of production				
disability index[a]	8.13	0.50	6.40	9.40
fraction male patients	0.26	0.08	0	1
extramural activities	24.60	36.80	0	106.60
fraction nurse trainees	0.24	0.10	0	0.54
capital per bed	1.06	0.25	0.34	1.64
pharmacy[b]	0.20	0.40	0	1
institutional characteristics				
number of beds	171	70	25	430
competition ratio	0.08	0.05	0.04	0.50
management type[b]	0.85	0.35	0	1
protestant[b]	0.24	0.43	0	1
not religiously affilliated[b]	0.45	0.50	0	1
home for physically disabled[b]	0.27	0.44	0	1
home for psycho-geriatrically disabled[b]	0.25	0.43	0	1
process characteristics				
occupation rate	0.98	0.04	0.57	1.24
cost of contracting out (in % of variable cost)	0.06	0.04	0	0.38
revenue from contracting in (in % of variable cost)	0.01	0.02	0	0.22
ward size	30	7	8	86
absence through illness (in % of cost of personnel)[c]	6	3	1	21

a The disability index is only available at regional level. The standard deviation will therefore be relatively small.
b This variable is a dichotomous variable.
c This variable is not known for the year 1984.

Source: Social and Cultural Planning Office

A6.3 Mean Value, Correlation and Rank Correlation Tests

Table A6.2 contains the test values for the t-test on the mean scores of cost efficiency, and the Pearson correlation test on the sets of the scores.

Table A6.2
Test Results for the T-tests of the Mean Value and the Pearson Correlation-tests

	national prices	*regional prices*
t-test of mean value		
regional prices	-2.27	-
firm-specific prices	-9.59	-11.88
Pearson correlation-tests		
regional prices	-4.00	-
firm-specific prices	-0.04	-9.65

Source: Social and Cultural Planning Office

The critical value for the t-tests (3,876 degrees of freedom = 2*number of observations-2) equals 2.57 at the 1% level. Thus, Table A6.2 indicates that the means of the efficiency scores calculated with national prices and with regional prices do not differ significantly. The mean score calculated with the firm-specific prices differs significantly from the means for the two other sets of scores. The critical value for the Pearson correlation test (1,937 degrees of freedom = number of observations-2) also equals 2.57 at the 1% level. Thus, even though the means of the scores may not differ significantly, here the hypothesis that the sets of scores are the same is rejected for each of the pairs of prices.

Table A6.3 shows the test values of the Spearman rank correlation test.

Table A6.3
Test Results for the Spearman Rank Correlationtests

	national prices	*regional prices*
regional prices	-5.41	-
firm-specific prices	-13.30	-12.55

Source: Social and Cultural Planning Office

The critical value for the Spearman rank correlation test (1,937 degrees of freedom = number of observations-2) equals 2.57 at the 1% level. Thus, indeed the hypothesis that the rankings of the efficiency scores are the same is rejected for each of the pairs of prices.

A6.4 Estimation Results of the Tobit Analyses

The dependent variable in the Tobit analysis is defined as (*1-efficiency score)/efficiency score*. Therefore a variable with a positive coefficient in the Tobit estimation has to be interpreted as having a negative relationship with the efficiency score.

The estimates of the Tobit analyses for the cost efficiency scores using the three different prices are given in Table A6.4.

Table A6.4

Results of the Tobit Analyses, 1984, 1987-1995

	national prices		*regional prices*		*firm-specific prices*	
	coefficient	*t-value*	*coefficient*	*t-value*	*coefficient*	*t-value*
constant	1.23	10.11	1.05	9.15	0.91	5.88
chracteristics of production						
disability index	0.01	1.14	0.03	3.06	0.03	2.94
fraction male patients	-0.24	-4.67	-0.17	-3.55	-0.24	-3.75
extramural activities	0.00	5.63	0.00	5.02	0.00	3.34
fraction nurse trainees	0.00	-0.01	-0.05	-1.23	0.00	0.09
capital per bed	-0.02	-1.26	-0.01	-0.46	0.07	3.01
pharmacy	0.00	-0.17	0.00	0.09	0.01	0.56
institutional characteristics						
number of beds	0.00	0.37	0.00	-0.21	0.00	-0.66
concentration ratio	-0.16	-2.49	-0.16	-2.79	-0.09	-1.08
not religiously affiliated	-0.01	-0.84	-0.01	-1.59	-0.02	-1.50
protestant	-0.01	-0.61	-0.02	-1.55	0.01	0.79
management type	-0.02	-2.22	-0.02	-2.27	0.01	0.79
home for physically disabled	-0.15	-13.55	-0.14	-13.39	-0.18	-13.32
home for psycho-geriatric disabled	0.12	11.67	0.12	11.52	0.17	12.80
process characteristics						
occupation rate	-0.87	-9.14	-0.85	-9.47	-0.79	-6.53
cost of contracting out	0.20	1.90	0.18	1.78	0.06	0.45
revenue from contracting in	0.13	0.55	0.23	1.03	0.35	1.13
ward size	0.00	-2.35	0.00	-1.84	0.00	-1.66
absence through illness[a]	-0.03	-0.15	0.01	0.03	0.32	1.22

a *The absence through illness is not known for 1984. Therefore the coefficient for this variable is calculated separately for 1987-1993 only. The other estimates are only marginally affected.*

Source: Social and Cultural Planning Office

Notes

[1] A paper with more technical details has been presented at the Informs Meeting in Montreal in April 1998 and can be obtained on request from the authors Evelien Eggink or Jos Blank (ee@scp.nl or jo@scp.nl) or from the SCP-website http://www.scp.nl/users/production/articles/deaprice/deaprice.htm.

[2] 1 guilder is about USD 0.5.

[3] The number of nurse trainees is assumed to be a fixed resource. Since the government policy is to phase out the in-service training of nurses, the numbers of nurse trainees have decreased considerably over the years. The proportion of nurse trainees in total staff decreased from 20% in 1984 to 13% in 1993.

[4] For reasons of privacy, the disability index was not available at the level of individual nursing homes. Therefore the disability index used in the analysis is calculated for each of the 27 health regions, representing the mean care need of the patients in the region. Thus, the disability index is the same for all nursing homes in the same health region. The index is obtained from the SIG, an organization gathering information on health care institutions in the Netherlands.

[5] Subsidies for extramural activities have been distributed to Dutch nursing homes since 1990. Since there was no information about the money amount of the subsidy for each nursing home, the total national subsidy was assumed to be divided over the nursing homes proportionally to the variable cost.

[6] The competition ratio is the reciprocal of the Herfindahl index (Kwoka 1985). The Herfindahl index is a measure of concentration of firm activities. For interpretation convenience we have used the competition ratio. The Herfindahl index has been based on the number of in-patient days.

[7] This is a consequence of the fact that the regional and national prices are calculated as aggregated cost divided by aggregated full-time equivalents. Thus the regional and national price is in fact the corresponding mean unit value, weighted with the number of full-time equivalents in the home. The mean unit value weights small homes the same as large homes. Thus, the fact that the mean unit value exceeds the mean regional and national price indicates that small homes pay higher prices than large homes.

[8] A t-test is performed to test the hypothesis that the means of the efficiency scores differ significantly (Mood, Graybill and Boes 1974: 435). Moreover, it is tested whether the sets of scores as a whole differ significantly. Pearson correlation coefficients are therefore calculated for each pair of efficiency scores (see e.g. Hanke and Reitsch 1991). Table A6.2 gives the test-values for the t-tests and the Pearson correlation test.

[9] A formal test, based on the Spearman rank correlation is used to determine if the differences in the rankings of the cost-efficiency are significant (Hanke and Reitsch 1991). The results are given in Table A6.3.

[10] In particular, since the efficiency scores are censored from above at 1, the Tobit method is used (see for instance Maddala 1987: 151). The estimates of the coefficients are given in the appendix in Table A6.4.

[11] The explanatory power of a characteristic is calculated as follows. The variance of the forecast using the characteristic at hand only (the characteristic multiplied by the corresponding coefficient) is divided by the overall variance of the forecast. The resulting explanatory powers are then corrected by the total explained variance in order to find the fractions explained by each variable. In fact, the effect of the correlation between the characteristics is divided among the characteristics.

References

Ali, A.I., and L.M. Seiford. 1993. The Mathematical Programming Approach to Efficiency Analysis. In: Fried, H.O., C.A.K. Lovell, and S.S. Schmidt (eds.), *In: The Measurement of Productivity Efficiency: Techniques and Applications*. Oxford: Oxford University Press, 120-159.

Bekele, G., and A.G. Holtmann. 1987. A Cost Function for Nursing Homes: Toward a System of Diagnostic Reimbursement Groupings. *Eastern Economic Journal* 13 (2):115-122.

Blank, J.L.T., and E. Eggink. 1996. *Zuinig op zorg: een empirisch onderzoek naar de productiestructuur van verpleeghuizen in Nederland (Careful with Care)*. Rijswijk Sociaal en Cultureel Planbureau (Cahier 131).

Chattopadhyay, S., and D. Heffley. 1994. Are For-Profit Nursing Homes More Efficient? Data Envelopment Analysis with a Case-Mix Constraint. *Eastern Economic Journal* 20 (2):171-186.

Chattopadhyay, S., and S.C. Ray. 1996. Technical, Scale, and Size Efficiency in Nursing Home Care: A Nonparametric Analysis of Connecticut Homes. *Health Economics* 5 (4):363-373.

Coelli, T., D.S.P. Rao, and G.F. Battese. 1998. *An Introduction to Efficiency and Productivity Analysis*. Boston: Kluwer Academic Publishers.

Dor, Avi. 1989. The Costs of Medicare Patients in Nursing Homes in the United States: A Multiple Output Analysis. *Journal of Health Economics* 8: 253-270.

Fizel, J.L., and T.S. Nunnikhoven. 1993. The Efficiency of Nursing Home Chains. *Applied Economics* 15: 49-55.

Gertler, P.J. 1992. Medicaid and the Cost of Improving Access to Nursing Home Care. *The Review of Economics and Statistics* 74 (2): 338-345.

Gertler, P.J., and D.M. Waldman. 1992. Quality-Adjusted Cost Functions and Policy Evaluation in the Nursing Home Industry. *The Journal of Political Economy* 100 (6): 1232-1256.

Hanke, J.E., and A.G. Reitsch. 1991. *Understanding Business Statistics*. Homewood, Ill.: Irwin Inc.

Hofler, R.A., and B. Rungeling. 1994. U.S. Nursing Homes: Are They Cost Efficient? *Economic Letters* 44 (3): 301-305.

Institute for Health Care Management. 1988. *De intramurale gezondheidszorg in cijfers per 1 januari, 1988*. Utrecht: Nationaal Ziekenhuisinstituut.

Institute for Health Care Management. 1998. *De intramurale gezondheidszorg in cijfers per 1 januari, 1998*. Utrecht: Nationaal Ziekenhuisinstituut.

Kooreman, P. 1994. Nursing Home Care in the Netherlands. *Journal of Health Economics* 13 (3):301-316.

Kwoka, J.E. 1985. The Herfindahl Index in Theory and Practice. *The Antitrust Bulletin* 30 (4): 915-947.

Lovell, C.A.K. 1993. Production Frontiers and Productive Efficiency. In: Fried, H.O., C.A.K. Lovell, and S.S. Schmidt (eds.), *The Measurement of Productive Efficiency*. New York: Oxford University Press, 3-67.

Maddala, G.S. 1987. *Limited-Dependent and Qualitative Variables in Econometrics*. Cambridge: Cambridge University Press.

McKay, N.L. 1988. An Econometric Analysis of Costs and Scale Economies in the Nursing Home Industry. *Journal of Human Resources* 23 (1):57-75.

Mood, A.M., F.A. Graybill, and D.C. Boes. 1974. *Introduction to the Theory of Statistics*. Tokyo: McGraw-Hill.

Nyman J.A., and D.L. Bricker. 1989. Profit Incentives and Technical Efficiency in the Production of Nursing Home Care. *Review of Economics and Statistics* 71 (4):586-594.

Seiford, L.M. 1996. Data Envelopment Analysis: The Evolution of the State of the Art (1987-1995). *The Journal of Productivity Analysis* 7 (2):99-137.

Tuckman, H.P., and C.F. Chang. 1988. Cost Convergende between For-profit and Not-for-profit Nursing Homes. Does Competition Matter? *Quarterly Review of Economics and Business* 28 (4): 50-65.

Ullmann, S.G., and A.G. Holtmann. 1985. Economies of Scope, Ownership, and Nursing Home Costs. *Quarterly Review of Economics and Statistics* 25 (2): 83-94.

Vitaliano, D.F., and M. Toren. 1994. Cost and Efficiency in Nursing Homes: A Stochastic Frontier Approach. *Journal of Health Economics* 13 (3): 281-300.

Part Three: Educational Services

7 Input Regulations and Allocative Efficiency in U.S. Public Schools

Shawna Grosskopf, Kathy J. Hayes, Lori L. Taylor and William L. Weber

7.1 Introduction

Public primary and secondary education in the United States is under considerable pressure to improve performance. There is near universal consensus that public schools must do a better job of educating America's children. This general opinion is also supported by results of recent research into school performance. Although the U.S. spends more on education than ever before, achievement is falling. Eric Hanushek's 1986 survey of the literature on educational production functions found no evidence of a systematic correlation between educational expenditures and student achievement gains.

There are a number of possible explanations for public school inefficiency. Regulations imposed by state or federal authorities might lead to inefficient decisions by local public school officials. For example, in Texas the state legislature mandates maximum class sizes for some grade levels and sets the salary structure for classroom teachers, despite research that has shown that class sizes and teacher salaries have little correlation with student achievement (Hanushek 1986; Bridge, Judd and Moock 1979). Alternatively, school district administrators may simply not know how to behave efficiently. (After all, economists have been able to provide little guidance about the characteristics of efficient schools.) Finally, the current institutional structure of the public school system may provide incentives for administrators to opt for the status quo. The market for primary and secondary educational services in the United States is at best an imperfectly competitive market.

Although public sector primary and secondary education in the United States is highly decentralized (there are almost 18,000 U.S. school districts), public schools generally cannot choose their students, and their ability to choose the mix and level of resources

to use may also be limited as mentioned above. We are interested in determining whether giving schools more control over their input choices, even within their given budgets, could yield improved performance. This is one of the rationales for proposals to introduce site-based management, for example.

In this chapter, we use recently developed models of performance to simulate the potential gains in efficiency from reducing input regulations, as would occur under site-based management, for example. This technique was first used in Grosskopf, Hayes, Taylor and Weber (1999a) to simulate input deregulation in 244 urban Texas school districts operating in 1989. Here, we expand the sample to 302 school districts and update to 1997. Following Grosskopf, Hayes, Taylor and Weber (1999a), we first determine the performance relative to the best-practice frontier for Texas school districts given the observed allocations of labor and non labor inputs, i.e., those consistent with the status quo. We simulate the introduction of choice by allowing schools to choose their allocation of labor and non labor inputs with no restriction other than a budget constraint (input prices are also given as they would be in a competitive market situation). The difference between the status quo and deregulated best-practice frontiers represents the potential gains from reallocation of school district inputs. Put differently, this difference captures allocative inefficiency. If input choices are distorted under the current regulatory regime, allowing input choice should result in improved performance from the same total resources currently used in education. The techniques we use are exceptionally useful for analyses of the public sector in general, providing a natural performance measure for public sector service provision.

The chapter is organized as follows. In Section 7.2 we provide a brief description of some of the institutional characteristics of U.S. and Texas public schools. Section 7.3 reviews related research. Section 7.4 includes a description of our data. Section 7.5 presents the outcomes of the analysis. Section 7.6 concludes this chapter.

7.2 PRIMARY AND SECONDARY EDUCATION IN THE U.S.

As mentioned in the introduction, most primary and secondary education (grades kindergarten through 12) is delivered from the public sector: nine out of ten students attend public schools in the United States. There is, however, considerable variation in how schools are financed and administered since education is the responsibility of the states rather than the central government. In practice, provision of primary and secondary education occurs largely at the local level – only Hawaii has explicitly centralized education to the state level.

As with administration, financing varies by state, but generally speaking, in most states there is a mix of financing from both the state (often through foundation grants to the school districts) and from the local level. Again, generally speaking, states rely on sales and income taxes whereas school districts or the local governments often rely on property taxes for revenues.

In terms of regulations that affect input choices of schools, again, these are generally determined at the state level, and therefore vary by state. Our application is to Texas public schools. According to the Texas Education code (subchapter B Section 16.054 of the Texas Education Code), each school district must maintain an average ratio of at least one teacher for each 20 students, and may not enroll more than 22 students per class in grades K-4. School districts that receive guidance program funds must hire a certified counselor for each elementary school and at least one counselor for every 500 elementary school students. In addition, all teachers must be assigned to the state-developed career ladder, and are subject to the state set minimum salaries, based on years of experience. These are some of the constraints on input choice faced by Texas school districts.

7.3 Models of School Production and Productivity

Over the years, economists have used a variety of techniques to evaluate school performance. Most researchers have focused on estimating single-output, average production functions for schooling. Although a few recent studies have examined monetary returns to schooling (Betts 1995 and Card and Krueger 1992a, 1992b), the most common measures of educational outputs have been test scores (for examples, see Berger and Toma 1994, Eberts and Stone 1987, Wahlberg and Fowler 1987 and the literature surveyed in Hanushek 1986). Generally, researchers assume that schools produce these educational outputs using inputs related to school personnel, per-pupil expenditures, and family background. The production functions yield estimates of the marginal products of the inputs, and allow researchers to infer which inputs would have the greatest marginal impact on achievement. Most researchers using this approach have found that inputs within school district control (such as expenditures or class sizes) have little or no marginal impact on test scores (Hanushek 1986). Card and Krueger (1992a, 1992b) find evidence that school inputs have a positive effect on the monetary returns to schooling, but their analysis is based on state-level data about school characteristics and may be subject to aggregation bias (see Hanushek, Rivkin and Taylor 1996). Using less aggregate data, Betts (1995) finds no evidence of marginal effects.

Recently, some researchers have modified production function analysis to incorporate allocative and technical inefficiencies, and multiple measures of educational output. Most of the researchers using this generalized approach have relied on nonstochastic techniques like data envelopment analysis (e.g., Bessent and Bessent 1980; Bessent, Bessent, Kennington and Reagan 1982, Bessent, Bessent, Elam and Long 1984; Färe, Grosskopf, and Weber 1989, and Grosskopf, Hayes, Taylor, and Weber (1999a). However, a few researchers, such as Cooper and Cohn (1997), and Deller, and Rudnicki (1993), have used stochastic techniques.[1] Ray and Mukherjee (1998), Ruggiero (1996), McCarty, and Yaisawarng (1993), and Ray (1991) combine DEA and regression analysis in partially stochastic two-step procedures that incorporate multiple outputs. Grosskopf, Hayes, Taylor and Weber (1997) use another modified production function to examine the consequences of budgetary reforms when school districts are inefficient.[2] Like the production-function analyses, these studies generally find evidence of substantial school inefficiency.

Analyses of educational cost functions yield similar results. Barrow (1991) estimated a cost function frontier for schools in England and found that actual costs were 4% to 16% above the minimum estimated cost for the schools in his sample. Callan and Santerre (1990) found evidence that school districts in Connecticut produce primary and secondary education using inefficiently large quantities of capital and transportation services. Jimenez (1986) concluded that schools in Bolivia and Paraguay used excessive amounts of capital and that many of the schools in Bolivia exhibited diseconomies of scale. Eberts and Stone (1986) found that rent extraction in the form of higher teacher salaries adds between 7% and 15% to education costs in unionized school districts in the United States.

An alternative way of modeling educational technology is to allow for endogenous output and fixed budgets – in contrast to cost functions which take output as exogenous and do not include a budget constraint.[3] Glass, McKillop and O'Rourke (1998) also used this model to look at productivity change in U.K. universities. This approach was also employed by Grosskopf and Moutray (forthcoming) to assess the site-based management reforms introduced (and later abandoned) in the Chicago public school system. As mentioned above, Grosskopf, Hayes, Taylor and Weber (1999a) also employ a similar model to simulate the potential impact of deregulation in Texas school districts for 1989. We employ that approach here to an expanded and updated sample of Texas school districts.

We use the status quo and deregulated (simulated post reform) models discussed above.[4] As in Grosskopf, Hayes, Taylor and Weber (1999a), we estimate our models using DEA. For general discussion of DEA we refer to Section 2.4. The technical details of our approach are discussed in Appendix A7.1.

7.4 DATA AND SPECIFICATION

We use DEA to estimate the performance of a sample of 302 Texas school districts operating in 1997 for our two models: the status quo and with simulated deregulation of input allocations.[5] Our proxies for outputs are based on a procedure used by Hanushek and Taylor (1990). Instead of using raw test score data from a single year as a measure of output, they argue that test scores should be adjusted for home produced inputs and previous attainment, providing a measure of the value added by the school to educational achievement. Our data set includes math test scores for students in grades six and 10.[6] We estimate the value added by the school district using a regression approach. We regress a set of variables which include previous scores from the standardized state test in 1995 as well as proxies for home-produced inputs on the standardized math scores for 1997 – see the appendix for details. The proxies we use for home produced inputs include ethnicity, the fraction of the cohort that is not receiving free or reduced-price lunches, as well as changes in these variables between 1995 and 1997. Of course, we would prefer to have information on the home environments of the students (parent schooling, etc.), but this information was not available.

We use the residuals from our regression to construct our output measures. Estimating school outputs as equation residuals generates output measures that represent deviations from the state average. It follows that school districts which add less value than the state average have negative output measures. We transform the value-added residuals into tractable output measures by adding the mean of the posttest scores to the corresponding value-added residuals and scaling the resulting variable by the relevant enrollments to get a total output measure.

As in Grosskopf, Hayes, Taylor and Weber (1999a), the variable inputs which we consider to be within school district control are limited to instructional and administrative personnel. We define the quantity of instructional inputs per pupil as the weighted average of the number of teachers and teacher aides per pupil. The quantity of administrative inputs per pupil is the weighted average of the number of administrators and support personnel per pupil. In both cases, we derive weights from the average wages paid for the personnel categories in each metropolitan area. To generate measures of total instructional (INST) and administrative (NINST) inputs, we multiply these per-pupil measures of variable input by the sum of the enrollments in grades six and 10 ($ENROLL_i = ENROLL_{i6} + ENROLL_{i10}$).

There are several important school inputs that are beyond school district control, at least in the near term. As in Grosskopf, Hayes, Taylor and Weber (1999a), we focus on

Table 7.1

DESCRIPTIVES

	N	Mean	St. Deviation	Min.	Max.
Variable Inputs					
NINST	302	13.0	21.6	0.73	198
INST	302	83.9	132	8.36	1,250
Variable Input Price					
NINST PAY (USD)	302	53,500	2,650	48,100	57,800
INST PAY(USD)	302	32,400	1,090	30,300	34,500
Fixed Inputs					
$STUINPUT_{10}$	302	44,800	70,600	3,740	627,000
$STUINPUT_{6}$	302	52,800	88,700	4,410	912,000
CAPINPUT	302	1,290	2,690	77.6	33,300
Outputs (Value added test scores by grade)					
XAG_{10}	302	44,600	72,500	3,880	657,000
XAG_{6}	302	52,400	90,300	4,270	933,000
Costs and Enrollment					
C (x USD 1,000)	302	3,440	5,700	315	54,900
$ENROLL_{6}$	302	668	1,150	54	12,000
$ENROLL_{10}$	302	595	965	52	8,770

two: non labor school inputs and family inputs. Unfortunately, we have no direct measures for either of these inputs. Because the quantity of non labor inputs should be highly correlated with expenditures on library books, furniture and equipment, physical plant maintenance and general maintenance and operations, we use a principal components index of per-pupil expenditures in these four categories, multiplied by $ENROLL_i$, as our proxy for the quantity of non labor inputs (CAPINPUT). We use the predicted values from our output regression equation multiplied by the corresponding grade-level enrollments ($ENROLL_{ig}$) to measure the contribution of home production at each grade level ($STUINPUT_{ig}$), yielding an index that depends on past achievement test scores, the ethnic and socioeconomic composition of the school district and the change in cohort characteristics. There are two measures of fixed student inputs corresponding to the primary and secondary grade levels for each school district. Table 7.1 presents descrip-

tive statistics for the data used in this analysis. For a more detailed discussion of the input and output measures, see Grosskopf, Hayes, Taylor, and Weber (1999a and 1999b).

7.5 Estimation and Results

Next we calculate the performance of each school district in our sample relative to our two best-practice frontiers: the status quo and the deregulated model. For the status quo model, we specify given inputs as the observed levels of the two types of personnel as well as our three fixed inputs. Outputs are our measures of value-added for grades six and 10. For our simulated reform model we allow each school district to hypothetically choose the levels of the two types of personnel, subject to the budget constraint. This budget constraint is the total personnel expenditure in the school district. Input prices are assumed fixed based on the average salary levels in the metropolitan area. For both models, a school district is judged efficient (i.e., their students are reaching best-practice achievement levels given their resources) if the value of the efficiency measure is one. Inefficient school districts will have measures greater than one. These school districts are not reaching best-practice achievement levels.

Table 7.2

Summary of Results

	Mean	St. Deviation	Min.	Max.
All School Districts (n = 302)				
Status Quo	1.06	0.04	1.00	1.17
Post Reform	1.07	0.04	1.00	1.18
Gain	1.01	0.01	1.00	1.13
GAIN = 1 (n = 62)				
Status Quo	1.04	0.05	1.00	1.16
Post Reform	1.04	0.05	1.00	1.16
Gain	1.00	0.00	1.00	1.00
GAIN > 1 (n = 240)				
Status Quo	1.06	0.04	1.00	1.14
Post Reform	1.07	0.04	1.00	1.18
Gain	1.01	0.02	1.00	1.13

We report summary statistics in Table 7.2. For the sample as a whole, we find that on average, output could be increased under the status quo by about 6%. After our simulated reform the average potential increase in outputs grows to 7%. Thus the average potential gain from allowing school districts to choose variable inputs subject to budget constraints rather than taking their initial levels as fixed is about 0.7%. That is, on average, school districts could increase value added by almost 6% if they used their initial input bundle efficiently and by an additional 0.7% if they could reallocate inputs efficiently. These results are similar to those we estimated based on the 244 urban school districts from 1989.[7]

We note that not all of the school districts in the sample would benefit from our simulated reform: 62 of the 302 school districts would realize no additional improvement in performance through reallocation, i.e., GAIN=1 for these school districts. For the 240 school districts which could gain from reallocation, we find that on average they could realize an increase of more than 7% in value added if they had allocated their inputs efficiently under reform. This is largely due to technical inefficiency, however. On average these school districts could increase achievement by about 6% by eliminating technical inefficiency. Thus the 'gain' due solely to improving allocative efficiency through our simulated deregulation appears to be rather small for these school districts, i.e., only about 1% on average. To put these numbers in perspective we turn to the simulated effects of reform on school budgets.

The solution to the model also yields optimal employment levels of our two personnel categories. Using these results we compute the effect of school reform (i.e., hiring the optimal number of instructional and noninstructional personnel) on spending, see Table 7.3.

Table 7.3

HOW SCHOOL REFORM AFFECTS SAMPLE SPENDING ON PERSONNEL

	Teachers	*Administrators*	*Total*
Expenditures (in millions)			
Status Quo (USD)	6,790	1,750	8,540
Post-Reform:			
Constant Cost	6,830	1,710	8,540
Constant Output	6,450	1,610	8,060

The first row of Table 7.3 summarizes current expenditure for the sample of school districts on teachers and administrators. If we hold budgets constant, and find the optimal employment of teachers and administrators based on the solution to the indirect distance function and use these to compute expenditures on our two personnel types, we see that relative to the status quo the Texas school districts in our sample would increase their spending on instructional personnel and decrease their spending on administrative personnel, see Table 7.3. We can also compute the cost savings that would arise if instead of holding the budget constant we hold output constant, but produce it efficiently.[8] In this case costs could be decreased by about 480 million dollars for the school districts in our sample, as shown in the third row of Table 7.3.

7.6 SUMMARY

In this chapter we show how DEA may be used to model and simulate educational reform. The particular reform we use here is a form of site-based management in which school districts are allowed to allocate their personnel budgets instead of having them determined by a central control. Our particular application is to a sample of Texas school districts operating in the 1997 school year. We find that there are potential gains of this type of reform or deregulation – about a 1% increase in output, or equivalently an aggregate decrease in costs of more than USD 400 million. These results are similar to earlier results for a smaller sample of Texas school districts from 1989.

We would like to emphasize that the tools we use – efficiency measures (distance functions) and DEA – are useful more generally in public sector applications. In fact, we would like to lobby for wider use of the model we used to simulate reform for the public sector in general. It allows for explicit inclusion of a budget constraint, which is typical in the public sector and otherwise imposes very minimal structure. It allows for multiple outputs or services and does not require output prices. As a benchmark, it seeks the greatest possible increase in outputs subject to satisfaction of the budget constraint, providing a natural performance measure – sorely needed in the public sector. When estimated using the linear programming techniques (DEA) as we did here, individual governmental units (in our case school districts) are compared to the best-practice frontier which is constructed from the data in the sample. All in all, this provides a modeling and analytical tool that we would argue is 'custom made' for the public sector.

APPENDIX

We begin with the notation to be used. Let

$x_f,\quad f=1,\ldots,F \in \Re_+^F$	*be a vector of fixed input quantities*
$x_v,\quad v=F+1,\ldots,N \in \Re_+^{N-F}$	*be a vector of variable input quantities*
$p_v \in \Re_{++}^{N-F}$	*be a vector of variable input prices*
$y \in \Re_+^M$	*be a vector of output quantities*
$c \in \Re+$	*be the total cost of budget available*

Next, we specify the best-practice frontiers we use for our two distance functions. The first best-practice technology (or production possibilities set) takes observed inputs (x) as given, and specifies all feasible outputs (y) that are technologically possible given those inputs and technology. In this most basic description of technology, we assume that administrators are not able to choose their personnel inputs; rather, these are taken as given – for example, by regulation. This is the technology we assume is faced by administrators under the current organizational structure, i.e. it is the *status quo* technology which we call $P(x)$.

Next we define the best-practice technology faced by administrators who face a budget constraint c, but are free to choose their variable inputs x_v (personnel) as long as they satisfy that budget constraint, $\Sigma_v p_v x_v \leq c$.[9] This is our model of best-practice technology under deregulation, which we call $IP(x_f, p_v/c)$.[10]

Thus, $IP(x_f, p_v/c)$ is the largest production possibility set allowing x_v to vary, but satisfy the budget constraint. These technologies may be related to each other. Specifically, if the variable input bundle x_v in $P(x)$ satisfies $p'_v x_v \leq c$, then we have

$$P(x) \subseteq IP(x_f p_v/c) \tag{A7.1}$$

In fact, one may think of $IP(x_f p_v/c)$ as the envelop of all $P(x)$ for which $p'_v x_v \leq c$.

Next we introduce our distance functions (efficiency measures). The direct output distance function is our measure of performance under the status quo technology and can be defined as

$$D_o(x_f x_v, y) = \min\ \{\theta : y/\theta \in P(x_f x_v)\}, \tag{A7.2}$$

Similarly, we may define a performance measure relative to the reform or deregulated best-practice technology. We call this the indirect output distance function and define it as

$$ID_o(x_f p_v/c, y) = \min\ \{\lambda : y/\lambda \in IP(x_f\, p_v/c)\}, \tag{A7.3}$$

The direct output distance function seeks the largest possible expansion of outputs feasible within the status quo best-practice technology $P(x)$. The indirect output distance function seeks the largest possible expansion of outputs feasible with the deregulated best-practice set $IP(x_f p_v/c)$, which allows for input choice.

Figure A7.1 illustrates the direct and indirect output distance functions for a typical school district that produces two outputs. The set $P(x)$, gives all possible combinations of the two outputs that can be produced under the status quo. The set $IP(x_f\, p_v/c)$ give all the possible combinations of two outputs that can be produced given the budget constraint c and variable input prices p_v faced by the school district. As long as x_v satisfies the

budget constraint, $P(x)$ will be a subset of $IP(x_f, p_v/c)$. Suppose that a particular school district has observed output bundle A, which they produce from their given input bundle x_A. The direct distance function tells us 'how far' that observed bundle is from the frontier of the direct technology, $P(x)$, holding the mix of outputs constant. This is equal to the ratio OA/OU, where U represents the maximum output feasible within $P(x)$, given the observed output mix and input bundle. The indirect output distance function tells us how far the observed output bundle is from the frontier of the indirect or budget-constrained technology, $IP(x_f, p_v/c)$. In this figure that is equal to OA/OT. Finally, we can think of the distance between maximum potential output under the status quo and the maximum potential output under reform as the net gain due to reform (after correcting for pre-existing technical inefficiency). In our figure that would be OU/OT.

Figure A7.1

$D_o(\cdot)$ AND $ID_o(\cdot)$

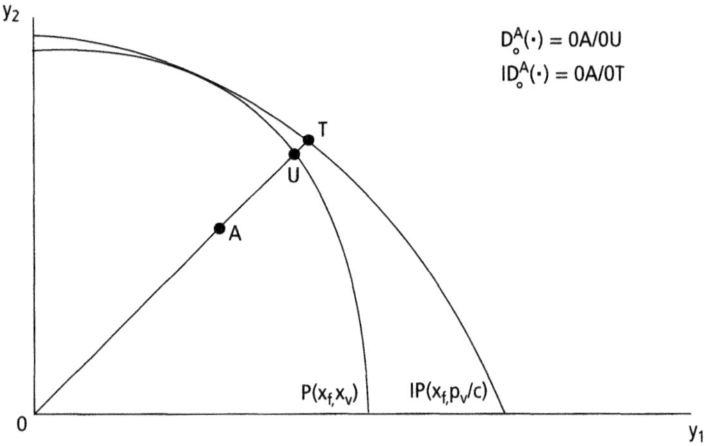

That is, the direct and indirect output distance functions take on values less than or equal to one as long as y is feasible. Values of one indicate that observed output is on the boundary of the respective best-practice technology. Equivalently, values of one indicate that the particular school district is efficient in the sense of Farrell.[11] The relationship between the status quo and reform technologies shows that allowing school districts to choose inputs subject to a budget constraint instead of facing a regulated or given input vector may increase their potential output.

This relationship is exactly what we exploit to simulate our version of 'deregulation'. Maximum potential output under a regulated regime in which school districts are restricted to using a fixed vector of inputs may be calculated as

$$D_o(x_f, x_v, y) \cdot y \tag{A7.4}$$

for any school district. That is, we are correcting the school's observed output for any technical ineffiency. In order to determine the maximum potential output achievable if school districts are allowed to choose any input bundle as long as it satisfies the budget constraint, we calculate

$$ID_o(x_f, p_v/c, y) \cdot y \tag{A7.5}$$

Any gains from this simulated deregulation (For expository purposes, assume that there is only one output) are calculated from

$$\begin{aligned} Gain &= (ID_o(x_f, p_v/c, y) \cdot y) / (D_o(x_f, x_v, y) \cdot y) \\ &= ID_o(x_f, p_v/c, y) / D_o(x_f, x_v, y) \end{aligned} \tag{A7.6}$$

We compute the direct and indirect output distance functions as solutions to linear programming problems. Specifically, for each school district k' = 1, ...,K we calculate

$$(D_o(x_{k'f}, x_{k'v}, y_{k'}))^{-1} = \max_{\theta, z} \theta \tag{A7.7}$$

subject to

$$\begin{aligned} \sum_{k=1}^{K} z_k y_m - \theta y_{k'm} &\geq 0, & m &= 1, \ldots, M \\ \sum_{k=1}^{K} z_k x_{kf} &\leq x_{k'f}, & f &= 1, \ldots, F \\ \sum_{k=1}^{K} z_k x_{kv} &\leq x_{k'v}, & v &= F+1, \ldots, N \\ z_k &\geq 0, & k &= 1, \ldots, I \end{aligned} \tag{A7.8}$$

and

$$(ID_o(x_{k'f}, x_{k'v}/c_{k'}, y_{k'}))^{-1} = \max_{\lambda, z, x_v} \lambda \tag{A7.9}$$

subject to

$$\begin{aligned} \sum_{k=1}^{K} z_k y_{im} - \lambda y_{k'm} &\geq 0, & m &= 1, \ldots, M \\ \sum_{k=1}^{K} z_k x_{kf} &\leq x_{k'f}, & f &= 1, \ldots, F \\ \sum_{k=1}^{K} z_k x_{kv} &\leq x_v, & v &= F+1, \ldots,] \\ z_k &\geq 0, & k &= 1, \ldots, I \\ \sum_{v=F+1}^{N} p_{k'v} x_v &\leq c_{k'} \end{aligned} \tag{A7.10}$$

The intensity vector z serves to construct convex combinations of the data to form the reference sets P(x) and $IP(x_f, p_v/c)$. The restriction that the intensity variables be nonnegative allows the technology to exhibit constant returns to scale.[12] These problems are solved for each school district in our sample.

Our output variables were based on the residuals of the following regression:

$$\begin{aligned} MATH97_{ig} = \alpha_g & + \beta_{1g} MATH95_{ig-2} + \beta_{2g} READ95_{ig-2} + \beta_{3g} SES_{ig} \\ & + \sum_{j=4}^{6} \beta_{jg} ETHNICITY_{jig} + \beta_{7g} XCOHORTN_{ig} \\ & + \beta_{8g} XSES_{ig} + \sum_{j=9}^{11} \beta_{jg} XETHNICITY_{jig} + \varepsilon_{ig} \end{aligned} \quad \text{(A7.11)}$$

where $MATH97_{ig}$ is the average Texas Assessment of Academic Skills TAAS mathematics score for school district i and grade level g in 1997, $MATH95_{ig-2}$ and $READ95_{ig-2}$ are the average TAAS scores in mathematics and reading for the same cohort two years previously, $ETHNICITY_{jg}$ is the fraction of the grade cohort that is BLACK, HISPANIC or ASIAN (respectively), SES_{ig} is the fraction of the grade cohort that is not receiving free or reduced-price lunches (the best available proxy for socio-economic status), $XCOHORTN_{ig}$ is percentage increase in cohort size between 1995 and 1997, $XSES_{ig}$ is the change in the share of students receiving free or reduced-price lunches, $XETHNICITY_{jig}$ is the change in the share of students in ethnic group j, and the estimated residual, ε_{ig}, represents the average value added to output by school district i in grade g, plus an error term.

NOTES

[1] They assume that school inefficiency has a half-normal distribution and use maximum likelihood techniques to estimate single-output frontier production functions.

[2] This modified version used is an indirect output distance function.

[3] This model is the indirect output distance function.

[4] These are modeled using the output distance function and cost indirect output distance function, respectively.

[5] Observations for which data were missing were deleted.

[6] We focus on mathematics because Bishop (1992) suggests that mathematics skills are disproportionately valued in the U.S. labor market.

[7] We note that our earlier study also used a different specification of inputs and outputs and 1988-89 data which reflected a somewhat more restrictive policy environment, see Grosskopf, Hayes, Taylor and Weber (1999a).

[8] This follows from our assumption of constant returns to scale. See Grosskopf, Hayes, Taylor and Weber (1999a) for details.

[9] In our simulation we construct a budget estimate from the observed input quantities and the average personnel prices in the metropolitan area. This implies that the current allocation of resources in feasible under the simulated deregulation.

[10] The input vector x may be partitioned into fixed and variable inputs x_f and x_v, respectively.

[11] In fact, the direct output distance function is the reciprocal of Farrell output increasing technical efficiency.

[12] Variable returns to scale may be imposed by adding the constraint that the sum of the intensity variables equal one.

REFERENCES

Barrow, M.M. 1991. Measuring Local Education Authority Performance: A Frontier Approach. *Economics of Education Review* 10 (1): 19-27.

Berger, M.C., and E.F. Toma. 1994. Variations in State Education Policies and Effects on Student Performance. *Journal of Public Policy Analysis and Management* 13 (Summer): 477-491.

Bessent, A.M., and E.W. Bessent. 1980. Determining the Comparative Efficiency of Schools through Data Envelopment Analysis. *Educational Administration Quarterly* 16 (Spring): 57-75.

Bessent, A.M., E.W. Bessent, J. Elam, and D. Long. 1984. Educational Productivity Council Employs Management Service Methods to Improve Educational Quality. *Interfaces* 14 (November/December): 1-8.

Bessent, A.M., W.W. Bessent, J. Kennington, and B. Reagan. 1982. An Application of Mathematical Programming to Assess Productivity in the Houston Independent School District. *Management Science* 28 (December): 1355-1367.

Betts, J.R. 1995. Does School Quality Matter? Evidence from the National Longitudinal Survey of Youth. Review of Economics and Statistics 77(2): 231-250.

Bishop, J. 1992. The Impact of Academic Competencies on Wages, Unemployment and Job Performance. *Carnegie-Rochester Conference Series on Public Policy* 37 (December): 127-194.

Bridge, R.G., C.M. Judd and P. Moock. 1979. *The Determinants of Educational Outcomes*. Cambridge Mass.: Ballinger Publishing Company.

Callan, S.J., and R.E. Santerre. 1990. The Production Characteristics of Local Public Education: A Multiple Product and Input Analysis. *Southern Economic Journal* 57 (2): 468-480.

Card, D., and A.B. Krueger. 1992a. School Quality and Black-White Relative Earnings: A Direct Assessment. *The Quarterly Journal of Economics* 57 (February): 151-200.

Card, D., and A.B. Krueger. 1992b. Does School Quality Matter? Returns to Education and the Characteristics of Public Schools in the United States. *Journal of Political Economy* 100 (February): 1-40.

Cooper, S.T., and E. Cohn. 1997. Estimation of a Frontier Production Function for the South Carolina Educational Process. *Economics of Education Review* 16(3): 313-327.

Deller, S.C., and E. Rudnicki. 1993. Production Efficiency in Elementary Education: The Case of Maine Public Schools. *Economics of Education Review* 12 (1): 45-57.

Eberts, R.W., and J.A. Stone. 1986. Teacher Unions and the Cost of Public Education. *Economic Inquiry* 24 (October): 631-643.

Eberts, R.W., and J.A. Stone. 1987. Teacher Unions and the Productivity of Public Schools. *Industrial and Labor Relations Review* 40 (April): 354-363.

Färe, R., S. Grosskopf, and W.L. Weber. 1989. Measuring School District Performance. *Public Finance Quarterly* 17 (October): 409-428.

Glass, J.C., D.G. McKillop, and G. O'Rourke. 1998. A Cost Indirect Evaluation of Productivity Change in UK Universities. *Journal of Productivity Analysis* 10 (2): 153-175.

Grosskopf, S., K.J. Hayes, L.L. Taylor and W.L. Weber. 1997. Budget-Constrained Frontier Measures of Fiscal Equality and Efficiency in Schooling. *Review of Economics and Statistics* 79 (1): 116-124.

Grosskopf, S., K.J. Hayes, L.L. Taylor and W.L. Weber. 1999a. Anticipating the Consequences of School Reform: A New Use of DEA. *Management Science* 45 (4): 608-620.

Grosskopf, S., K.J. Hayes, L.L. Taylor and W.L. Weber. 1999b. On the Determinants of School District Efficiency: Competition and Monitoring. manuscript.

Grosskopf, S. and C. Moutray. Evaluation of School Reform in Chicago Public Schools. *Economics of Education Review* (forthcoming).

Hanushek, E.A.1986. The Economics of Schooling: Production and Efficiency in Public Schools. *Journal of Economic Literature* 24 (September): 1141-1176.

Hanushek, E.A., S.G. Rivkin and L.L. Taylor. 1996. Aggregation and the Estimated Effects of School Resources. *Review of Economics and Statistics* 78 (4): 611-627.

Hanushek, E.A., and L.L. Taylor. 1990. Alternative Assessments of the Performance of Schools. *The Journal of Human Resources* 25 (2): 179-201.

Jimenez, E. 1986. The Structure of Educational Costs: Multiproduct Cost Functions for Primary and Secondary Schools in Latin America. *Economics of Education Review* 3: 25-39.

McCarty, T.A., and S. Yaisawarng. 1993. Technical Efficiency in New Jersey School Districts. In: Fried, H.O., C.A.K. Lovell, and S.S. Schmidt (eds.), *The Measurement of Productive Efficiency: Techniques and Applications*. New York: Oxford University Press.

Ray, S.C. 1991. Resource-Use Efficiency in Public Schools: A Study of Connecticut Schools. *Management Science* 37 (12): 1620-1628.

Ray, S.C., and K. Mukherjee. 1998. Quantity, Quality and Efficiency for a Partially Super-Additive Cost Function: Connecticut Public Schools Revisited. *Journal of Productivity Analysis* 10: 47-62.

Ruggiero, J. 1996. Efficiency of Educational Production: An Analysis of New York School Districts. *Review of Economics and Statistics* 78 (3): 499-509.

Wahlberg, H.J., and W.J. Fowler, Jr. 1987. Expenditure and Size Efficiencies of Public School Districts. *Educational Researcher* 16 (October): 5-13.

8 Economics of Scale and School Consolidation in Dutch Primary School Industry

Arnold H.Q.M. Merkies

8.1 Introduction

As with many other issues discussions on education have turned gradually from matters to money. In the Netherlands there were, at the beginning of the century, many religious schools. The objective of most of them was to preserve the religious views cherished by the founders of the schools and to pass them onto the next generation. This became increasingly possible from 1917 onwards when the Dutch Constitution was changed to acknowledge financial equality between private and public schools. At the same time, however, the government strengthened its grip on education by tying private schools to the same requirements as public schools. With the evaporation of religious convictions, the objectives of private schools gradually approached those held by public schools. This fortified the position of the government even more. The question arose whether the financial rules of the school system were still appropriate. More specifically a study was felt needed on the relation of the costs of a school to its scale and its performance. The consequences to the freedom of parents to choose their children's school, amply discussed elsewhere, see e.g. Boyd (1992), could not remain unseen, see e.g. Ritzen, Van Dommelen, and De Vijlder (1997).

A study of the performance of the Dutch schooling system was assigned to the Social and Cultural Planning Office (SCP) (Blank, Boef, Bronneman, Herweijer, Kuhry, and Schreurs 1990). It was a good choice as it fitted well in the new program of studies on the production of provisions in the public sector that started with Goudriaan, Van Tulder, Blank, Van Der Torre, and Kuhry (1989). Moreover, the SCP had completed several studies on planning of educational provisions. The study of Blank, Boef, Bronneman, Herweijer, Kuhry, and Schreurs (1990) lead, in 1991, to an assignment of the Dutch government to the SCP to support a revision of the system of closure norms for primary schools in order to realize a substantial school consolidation. The new system *Convenience*

and Accessibility (*T&B, Toerusting en bereikbaarheid*), Ministry of Education, Culture and Science 1991) was amended several times based on SCP calculations. In addition, the implementation of the new system was monitored by the SCP on a yearly basis (see also Blank 1993). A number of studies followed the initial study on school performance of Blank, Boef, Bronneman, Herweijer, Kuhry, and Schreurs (1990). Among these, Blank (1993) used the same approach as used by Färe and Primont (1995: 54-59) and Färe and Grosskopf (1994: 88-102) and fits within the methodological approach in this volume. In international literature on education several productivity studies have appeared since Coleman, Campbell, Hobson, McPartland, Mood, Weinfel and York (1966), Riew (1966), and Cohn (1968), for a survey see Hanushek (1989). However, the standard methodological approach of efficiency and productivity was only introduced in education by Jimenez (1986), Gyimah Brempong and Gyapong (1991 and 1992), and Callan and Santerre (1990). Applications to the field of education become more widespread, see e.g. Hanushek, and Jorgenson (1996), Grosskopf, Hayes, Taylor, and Weber (1997), Rao (1998) and Smet and Nonneman (1998) and the references therein.

In view of the purpose of this book, I will discuss only SCP experiences with primary education but within a broader context. I limit myself to the following questions:

1. are there economies of scale to be achieved in the Dutch primary schooling system?
2. is it possible to employ these economies if any, and what will be the effects on relevant policy variables such as the freedom of choice and transportation costs for parents and children?

The presentation starts in Section 8.2 with a survey of primary education in the Netherlands. Section 8.3 is devoted to the methodology that is used to study efficiency within the Dutch primary school system. In this section I will also discuss the model with which the effects were simulated of different rules for allowing schools to stay open. Of particular interest are the effects on the geographical dispersion of various kinds of schools and the (additional) transportation costs for parents and children, including costs due to the time of traveling. Section 8.4 describes the data underlying the computations. The empirical results, based on the work of Blank (1993), are given in Section 8.5. Conclusions and policy implications follow in Section 8.6.

8.2 Primary Education in the Netherlands

Education is universally accepted as one of the basic provisions that humans are entitled to. In the Netherlands this is acknowledged in the Constitution. Not all education in the field of *education permanente* is of equal concern to the government, but most important part – primary education – certainly is. Compulsory education has existed in the Netherlands since 1900, encompassing both public and private schools. It is not only a public responsibility but also private schools are allowed. The freedom to educate is founded in art. 23 par. 2 of the Constitution and was elaborated in the Primary Education Act (Lager onderwijswet) of 1920. This law boosted the start of many religious types of schools and dominated the scenery for many years after. The change came with the revolution in family planning in the late sixties and the decline in birth rates that followed. As a consequence, the number of pupils dropped significantly in the seventies and an increasing number of schools no longer met the required size-thresholds for existence laid down in the Act of 1920. In 1985 a new Act on primary education replaced the 1920 Act. The primary school and the nursery school were combined into a new style primary school or *basisschool*. Education is now compulsory from the age of 5 on. Children of 4 years old can and in fact almost generally do attend this new type of school.

At present around 12% of the population of the Netherlands of 15.8 million is aged between five and fourteen and therefore legally obliged to be in full-time education. A quarter of them has left primary schools, mostly to attend continued or special education. The others, together with some older ones and the children of age 4, make up the population of Dutch primary schools. In 1996 they numbered 1.5 million. Figure 8.1, constructed from Central Bureau of Statistics and SCP sources, shows the development of numbers of primary schools, pupils and school sizes from 1910 till 1996.

Figure 8.1

NUMBER OF SCHOOLS, PUPILS AND SCHOOL SIZE, THE NETHERLANDS, 1910-1994 (IN INDEXNUMBERS, 1910 = 100)[a]

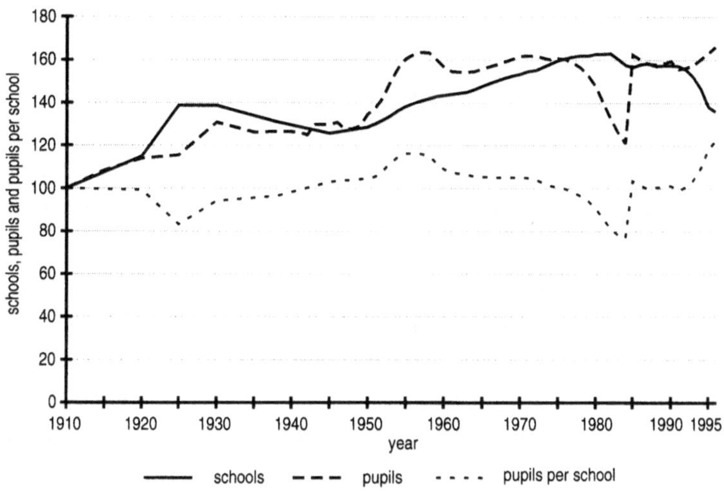

a In 1910 there were 5,700 schools and 900,000 pupils, indicating a mean school size of 169.

Source: Statistics Netherlands

Figure 8.1 should be handled with care. Neither the definition of a primary school nor of its inhabitants is constant over time. In the Primary Education Act of 1985 (*Wet op het basisonderwijs*) nursery and primary schools were combined (Daane and Hoeijmans 1985, 1986, 1988). From that date children could start school at age four, whereas before the 1985 Act children started school at the age of six years and eight months. Figure 8.1 shows how the average school size dropped from 1960 onwards. This was mainly due to the declining birth rate but it is remarkable that the number of schools still increased, suggesting a lenient attitude of the government towards the establishment of new schools. In 1985 the children from the nursery schools helped to bring the average school size suddenly back to its original level.

The merger of schools pushed the average school size upward but this was not sufficient. This raised the questions of whether the legal size thresholds still met the modern standards of efficiency. It was felt by many that increasing scales would lead to a reduction of costs which was said to create room for improvement of the quality of the education. It was acknowledged that a policy of increasing thresholds for existence would diminish the number of schools by merges or closures, having an impact on geographical dispersion and accessibility.

The law of 1920 did not recognize one single threshold for starting a school, but the threshold depended on the size of the municipality. These were divided in four classes depending on the number of inhabitants. Each class had its own minimum requirement for founding a school and its minimum for keeping it open (see Table A8.1 for an overview of these norms). Later, in 1993, a new law was issued where, in a typical Dutch way to reach for ultimate justice, each municipality obtained its own standard adorned with a whole casuistic of refinements (see Ministry of Education, Culture and Science 1991). The rule for starting a school now varies from 200 to 333 and for closing from 23 to 200. Exception rules are valid for schools for which there are no alternatives in the neighborhood.

In 1996 the average school size is 213 pupils. However, there is a considerable variation in school size. The distribution of schools over size in 1996 is given in Figure 8.2.

Figure 8.2

Distribution of Schools by Size, 1996

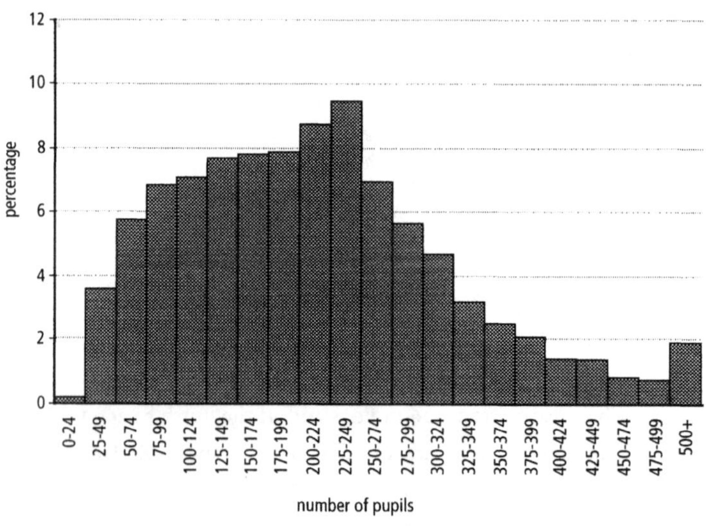

Source: Ministry of Education

Figure 8.2 shows that about 50% of the schools has between 150 and 300 pupils. About a third of the schools is smaller than 150 pupils, and about 10% has fewer than 75 pupils. On the other hand, there is a considerable number of large schools: about 20% of the schools has over 300 pupils.

Schools and pupils are not homogeneously spread. They may vary in size between regions, denominations and socio-economic classes. This is shown in the next paragraphs.

The size of schools varies over regions. Figure 8.3 gives an indication of the geographical variation of the school sizes by depicting of the average school size over regions in 1996.

Figure 8.3

REGIONAL DISTRIBUTION OF AVERAGE SCHOOL SIZE, THE NETHERLANDS, 1996

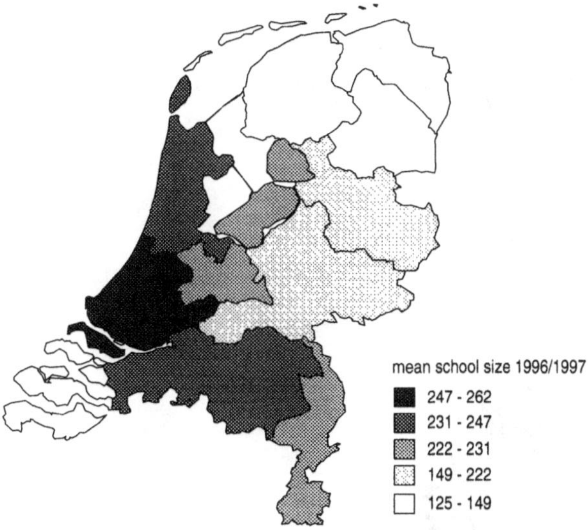

Source: Ministry *of* Education

The average size of a school varies from 126 in Friesland in the north to 260 in South Holland in the west. Many small schools are found in Drenthe and in Zeeland[1]. The South (Brabant and Limburg) with its predominant Catholic population shows a large average school size and dispersion over size is small in that part of the country.

The law of 1920 allowed groups of parents to start their own school. These were financed by public means if certain standards on quality and size were met. Many groups, especially Roman-Catholics and several Protestant denominations took that opportunity. The share of public schools dropped gradually from 58% in 1910 to 45% in 1930. At the same time, the share of Protestant schools rose from 19.5 % to 24% and that of Roman Catholic schools from 17% to 29%. The effect is even more marked in the

attendance of the pupils. To illustrate this the distribution of pupils over denominations is shown in figure 8.4.

Figure 8.4

Distribution of Pupils over Denomination, 1910-1996 (in percentages)

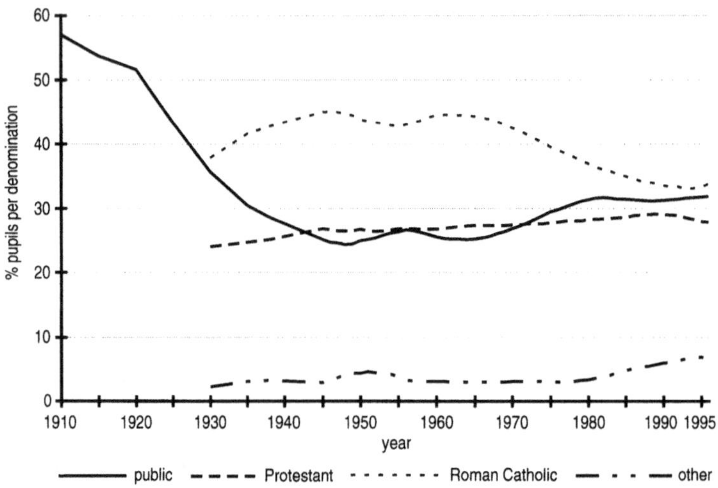

Source: Statistics Netherlands

Figure 8.4 shows that by 1930 public schools held only one third of the number of pupils. Roman Catholic and Protestant schools took the other part of the pupils with a small share of "others". It remained so for almost fifty years. It is remarkable that the share of denominational schools especially those of Protestant schools was hardly affected by the secularization in the Dutch society.[2]

The pupils' social background is used as an approximation of their initial level (see also Chapter 7). Hurn (1985: 144) shows that according to a large number of scientific papers, pupils' educational performance is strongly determined by their social background. The social background is identified here by three social groups. Each pupil is classified into one of three economic categories distinguished. The categories, referred to by a social-economic characteristic (SEC), are[3]:

1. pupils belonging to an ethnic minority group (SEC-1.90);
2. pupils whose parents have a low level of education but do not belong to an ethnic minority (SEC-1.25);
3. all other pupils (SEC-1.00).

These so-called Social Economic Class weights (SEC-weights) are then used as weights in allotting personnel to the schools.

Table 8.1 contains the distribution of pupils over denomination and SEC-class in the school year 1986/87.

Table 8.1

PUPILS PER DENOMINATION AND SEC-CLASS 1986/'87

	SEC 1 (in %)	SEC 1.25 (in %)	SEC 1.90 (in %)	total (x 1,000)
Public	48	38	14	450
Protestant	54	39	6	380
RC	50	42	7	500
Other	71	25	4	110
Total	52	39	9	1,440

Source: Blank, Boef, Bronneman, Herweijer, Kuhry, and Schreurs 1990

It is of interest to observe in Table 8.1 that public schools do take a larger proportion of the SEC-1.90 pupils than other schools. This could be expected as private schools have the right to refuse pupils with reference to their bylaws, whereas public schools do have to accept all pupils applying to them.

The discussion in this chapter moves gradually from matters to money. In 1995 the Dutch government spent 8.6 billion guilders (or approximately $ 4.3 billion) on primary education, which is almost 6,000 guilders per pupil or 1.2 million guilders per school. 70% of these costs are spent on personnel, a little less than 20% are costs of materials and the remaining 10% are costs of capital. In 1995 primary education required a little more than 75,000 teachers and education workers; personnel costs per teacher were approximately 80,000 guilders per year.

It is tempting to express the output in similar physical units. However, the measurement of output is sometimes called the Achilles heel of the economics of education, see Butel and Atkinson (1983: 51). Centuries of views on the purpose of education challenge the researcher here. For a range of definitions see Hanushek (1986 and 1989). Production in education was defined here based on the value-added concept. The value-added is represented by a number of variables. This includes the percentages of pupils in the SEC-classes as a proxy of initial skills and knowledge, and the outflow to higher sec-

ondary education as a proxy of final skills and knowledge. In Chapter 7 Grosskopf, Hayes, Taylor and Weber derive the value added from a separate analysis and add this variable to the efficiency analysis. Here we add the determinants of the value added variable directly to the efficiency analysis

8.3 Methodology

8.3.1 *Regression Analysis*

To know whether there are economies of scale to attain in the Dutch schooling system Blank (1993) selected from data sets of the Ministry of Education, Culture and Science a consistent set of 1,784 primary schools for which total costs and the number of pupils were given. In Figure 8.5 the number of pupils and the average costs per pupil are depicted for these schools.

Figure 8.5

Average Cost per Pupil per School, 1986/'87 (in guilders)

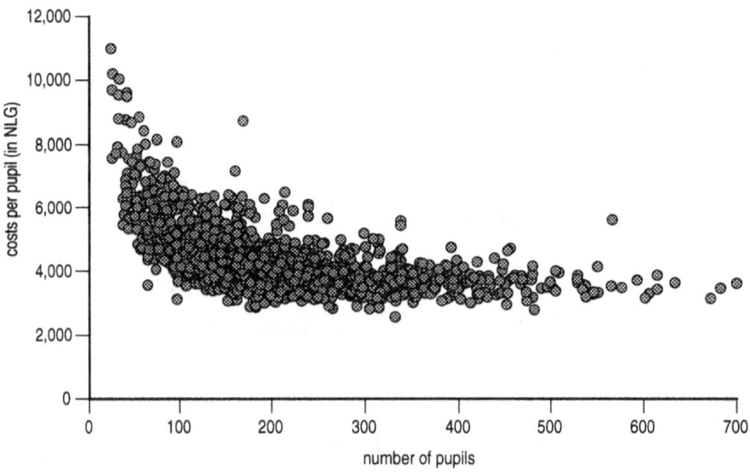

Source: Ministry of Education

Figure 8.5 shows that average costs decrease when the number of pupils increases. If all data points in Figure 8.5 had been on the same curve it would have been simple to find a function that is representative for Dutch primary schools, the so-called average cost curve. However, there are deviations from this curve, caused by external circumstance,

inefficiencies, and measurement errors. In this volume, most studies are concerned with the inefficiencies, i.e. they are concerned with finding the minimum cost-curve (the best practice). However since this chapter focuses on the effects of school size on costs, the average cost curve is more appropriate than the minimum cost curve, since changing the scale of production is not likely to change the managerial inefficiencies. A best practice approach (stochastic frontier analysis, SFA) as described by Lovell in Section 2.4 is conducted here as well. The regression results hardly differ between the two techniques, except for the intercept. Therefore no further attention is paid to the SFA estimates.

The relation of costs to production can be described by using a mathematical function, which can be estimated by using regression methods. In the regression analysis the costs are related to the number of pupils in the various SEC-categories, the expected fraction of pupils moving on to higher advanced education, and the prices of the various inputs (labor, materials, capital). The translog function proposed by Christensen, Jorgenson, and Lau. (1973) is employed here.[4] Indications of scale effects can be obtained from the estimated coefficients by calculating the cost flexibility as defined by Caves, Christensen, and Swanson(1981). The formulae of the cost function employed here and the corresponding global characteristics such as cost flexibility and marginal costs are given in the Appendix A8.1 (see also Blank 1993).

8.3.2 *Simulation*

If there are economies of scale, the government can take advantage of them by requiring minimum numbers of pupils per school. The question arises how such minima should be chosen: one single relatively high threshold for the existence of schools or particular minimum conditions for each school separately? To identify a proper standard, it helps to study the effects of proposed norms by simulation. Such a simulation requires the formalization of "law proposals" with closure norms and reactions of parents when the school of their children is closed. The results of the simulations can be evaluated according to rules of equity, justice, cost and personnel. A study of this kind was performed by Merkies, Steijn, and De Vos (1985) on behalf of the Council of Boards of Protestant Schools to inform them on the possible effects of some government plan to require higher thresholds for the existence of secondary schools. The idea was followed by the SCP to analyze the government plans with respect to primary and secondary schools in the beginning of the nineties (see Blank, Boef, Bronneman, Herweijer, Kuhry, and Schreurs 1990). The procedure of the simulation for primary education is summarized in Figure 8.6.

Figure 8.6

The Simulation Model

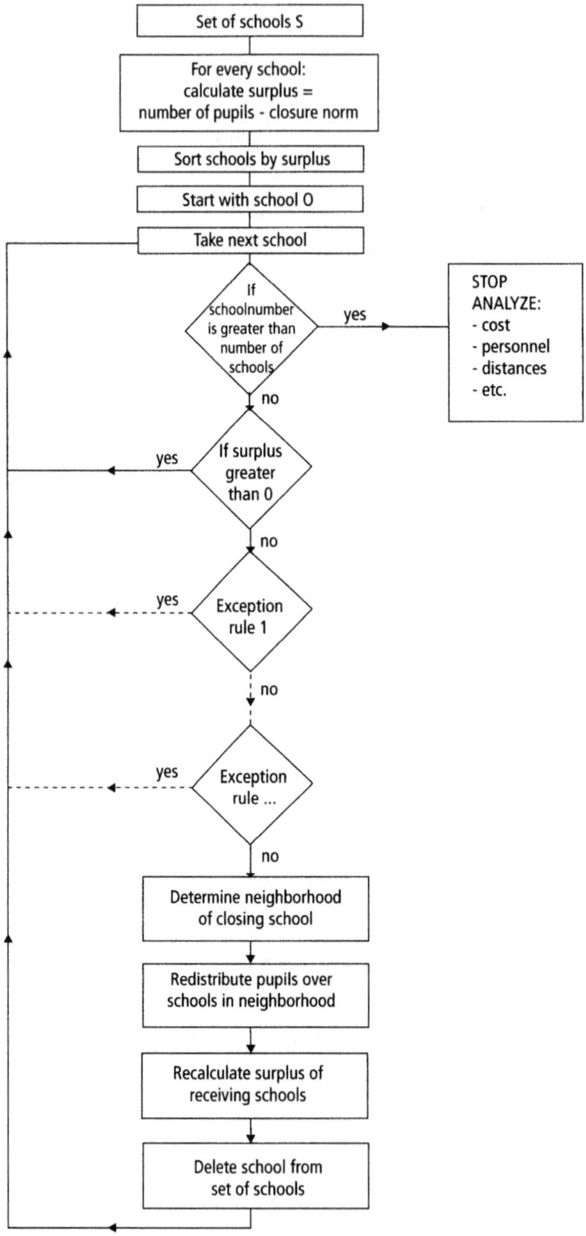

Source: Blank (1993)

The simulation model works as follows. The starting point is formed by the closure norms. The complete list of schools is run through in succession. Whenever a school does not comply with the closure norm, it is 'closed down' in the simulation. Before it is closed, however, it is first verified whether a number of preconditions are satisfied. An example of such a precondition is that an alternative for pupils must be present within a reasonable distance of the school to be closed down.[5] If preconditions are not satisfied, the school may stay open through an exception rule. When a school is closed down in the simulation, the pupils from that school are distributed among other schools in the neighborhood. This redistribution is dependent on the distance to and the denomination of the alternative schools and the preferences of parents of pupils. Once all schools are run through in this manner and the pupils of 'closed' schools redistributed among other schools, the new situation is assessed.

This simulation procedure is followed for seven different sets of closure norms and exception rules. The number of pupils required for staying open (closure norm) and the exception rules differ between the simulation alternatives. From enquiries among parents it is known that Dutch parents base their school choice mainly on denomination and geographical distance to the school. Therefore the simulation is also repeated for different alternatives that reflect strong denominational rules or rules without taking denominations into account.

The simulation provides information on the number and dispersion of schools that remain in the various situations, together with the reduction of costs that can be obtained. These cost reductions alleviate the financial burden for the government but there is an increased burden of transportation costs for the pupils and their parents, since they are, on average, faced with greater distances towards the school. Blank (1993) measured transportation costs, and included the additional travel time needed. He also estimated the loss of utility of parents who could no longer find the school of their denominational preference by comparing the simulation results with those obtained with alternatives where parents stick to their conviction at all times. The difference in transportation costs between the two simulations yields an upper bound to the utility loss of the parents. The loss of capital stock (school buildings) was neglected.

8.4 Data

In the empirical work, two separate data sets were employed. The first data set comprising 1,784 primary schools in the Netherlands in the year 1986/87 was available for cross-sectional analysis, i.e. for the estimation of the cost function. This data set is

referred to as the SCP analysis data set.[6] The cross-section contains a great deal of data about production, resources used and other characteristics of individual schools. The simulation study was based on a second data set covering all 8,410 primary schools in 1990/91. This data set contained information on numbers of pupils and their SEC weights and the denominations and locations of all schools present. However, no detailed characteristics of the schools are available there.

In the cross-sectional data set, the number of pupils is used as a quantitative output, the fraction of pupils with the different SEC weights were used as proxies for initial skills and knowledge, and the fraction of pupils proceeding to advanced secondary education are used as a proxy for the final skills and knowledge. The resources used refer to the budgets for personnel and material allocated to schools or municipalities during a given period. This data also contains the number of squared meters of floor space, used in the quantity index in the derivation of the price of capital inputs (housing). The main characteristics of the SCP analysis data set are given in Table 8.2.

Table 8.2

Synopsis of the SCP Dataset, 1986/'87

	mean	*minimum*	*maximum*
Use of resources[a]			
labor costs (x NLG. 1,000)	540	120	2,330
material costs (x NLG. 1,000)	140	20	410
variable costs (x NLG. 1,000)	680	160	2,680
capital costs (x NLG. 1,000)	130	10	690
total costs (x NLG. 1,000)	810	190	3,180
Input prices			
labor (per fte)	62,200	42,300	95,400
capital (per m^2)	108	25	296
material (per weighted pupil and weighted m^2)	247	85	480
Production and other characteristics			
pupils (L)	200	24	700
expected fraction of outflow to secondary education (Q1)	0.31	0.00	0.98
fraction of pupils SEC-1.00 (L1.00%)	0.53	0.00	1.00
fraction of pupils SEC-1.25 (L1.25%)	0.41	0.00	1.00
fraction of pupils SEC-1.90 (L1.90%)	0.06	0.00	0.92

a 1 NLG ≈ USD 0.5.

Source: SCP analysis data set

Table 8.2 shows that school size ranges from 24 to 700 pupils with an average of 200. The average costs per school in 1986/87 was 810,000 Dutch guilders, 540,000 guilders of which is spent on labor. The average costs per unit of labor is about 62,000 guilders.

The average school "produces" about 200 pupils. Over 50% of the pupils have SEC-score 1.00 and 6% of pupils have SEC-score 1.90; the remaining 41% of the pupils have SEC-score 1.25. There are schools which deal exclusively with pupils from ethnic minorities or with parents with a low level of education (a minimum value of 0% of SEC-1.00 pupils), although these are not all from the weakest (the SEC-1.90) group as the maximum number of pupils per school in this category is only 92%. On the other hand, there are also schools with no pupils at all from the two higher SEC groups (a maximum value of 100% SEC-1.00 pupils).

8.5 Empirical Results

8.5.1 *Regression Results*

The core estimation results of the regression study are given in Table A8.2 in the Appendix (see also Blank 1993). They provide an appropriate representation of the cost structure of primary schools in the Netherlands in 1986/87. The results show that the costs depend mainly upon the number of pupils, i.e. on the volume of production. First we deal with the scale effects (Section 8.5.2), since this chapter focuses on that aspect of the production structure. Some other regression results are discussed in Section 8.5.3. Finally, the simulation results are discussed in Section 8.5.4.

8.5.2 *Economies of Scale*

The cost flexibility measures the percentage cost increase for an increase of production by 1% and is therefore a measure for the economies of scale. A value for the cost flexibility of less than 1 indicates that there are economies of scale: the costs increase less than proportional to the output. Alternatively, if the cost flexibility exceeds 1 there are diseconomies of scale. The measure of cost flexibility depends on the level of output at which the increase occurs. For example, if the level of production is indicated by μ, $\mu = 2$ means that the production is twice that of the average school (of about 200 pupils), while the remaining factors, such as composition of the pupil population (according to SEC-class and outflow to higher secondary education) are kept constant at their respec-

tive means. The cost flexibility as calculated from the regression results in Table A8.2 is presented in Table 8.3.

Table 8.3

Cost Flexibility 1986/'87

production (= μ)	*cost flexibility*
0.5	0.69
1	0.83
1.5	0.91
2	0.96
2.5	1.01

Source: Blank (1993)

The cost flexibility of a school of average size (μ=1) is 0.83. This means that in a school of average size each additional percentage of pupil attendance raises cost by 0.83%. Smaller schools with say 100 pupils (μ = 0.5) have a lower cost flexibility (0.69) but usually higher average costs (see Figure 8.7). Large schools of 500 pupils (μ = 2.5) have a much higher cost flexibility (1.01) and lower average costs, leading to much higher marginal costs (10,000 guilders).

This is also reflected in Figure 8.7, where the relation between the average school costs and school size is drawn at the average values of the exogenous variables, such as the distribution of pupils over SEC categories. This figure can be taken as an approximation of the data shown in Figure 8.5.

Figure 8.7

AVERAGE COSTS, 1986/'87

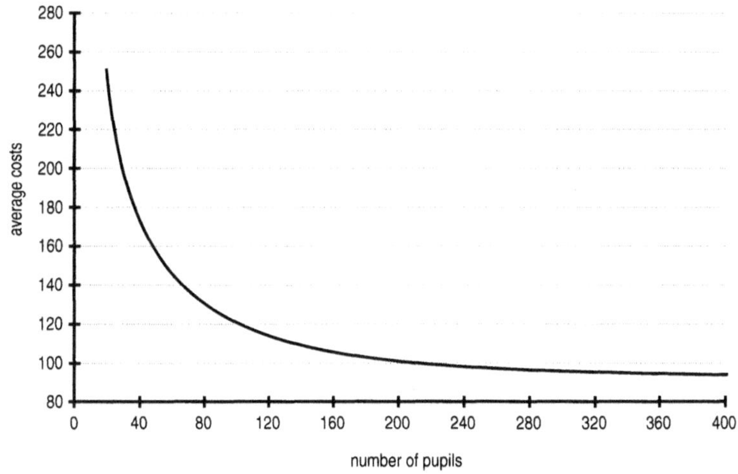

Source: Blank (1993)

Table 8.3 and Figure 8.7 indicate that considerable economies of scale can be acquired by small schools. These benefits dissipate as schools get larger. From the average school (200 pupils) onwards the average costs remain virtually constant. For schools with more than twice the average number of pupils there are no more economies of scale. The optimal size is around 450 pupils. Whether this is also optimal from an educational point of view falls outside the framework of this study.

8.5.3 *Some Other Regression Results*

The regression results indicate that, apart from the scale of production, costs are heavily affected by changes in prices of labor due to the large cost share of labor in the total costs. Changes in prices of materials and housing have relatively small impact on total costs. Marginal costs, measuring the extra costs of educating one additional pupil, are about 3,500 guilders (≈ $2,100). They increase as schools become larger.

Higher attendance from high education backgrounds (SEC-1.00) diminishes costs significantly; a significant additional cost effect occurs for the ethnic minority groups (SEC-1,90). A shift from SEC 1.25 pupils to SEC 1.00 pupils diminishes the input share of labor. This is not surprising since the labor-budget for SEC 1.25 pupils is higher than that for 1.00 pupils. The question is whether the higher budgets are justified by the

deficiencies of the pupils from the SEC-1.25 and SEC-1.90-categories. Blank (1993) examined the related issue whether schools in areas with exceptionally low living standards deviate from other schools.[7] A formal statistical test (see Test 1 in table A8.3) shows there is no evidence that those have a systematically different cost structure.

A higher performance of a school in the sense of a higher throughput to advanced secondary education, incurs only slightly more costs. It requires a higher labor input but fewer materials. A formal statistical test shows, however, that higher costs and better performance do cohere (see Test 2 in Table A8.3 of the Appendix). A higher performance incurs more costs in schools with a high proportion of SEC-1.00 pupils compared with schools with a higher share of SEC-1.25 pupils, but not in schools with high share of SEC-1.90 pupils. Apparently it is more difficult to improve performance in schools with pupils with high initial skills.

School boards are responsible for the school and hence for the purchase of materials. Boards may also govern more than one school. Such schools, for short 'branch schools', may profit from economies of scale, the more so as regulations exist requiring that supernumerary staff in 'branch schools' are obliged to fill vacant positions and to replace other teachers who are absent through illness. However, a formal statistical test of this conjecture (Test 3a and 3b in Table A8.3) did not provide sufficient evidence for the existence of higher productivity of 'branch schools'.

8.5.4 *Simulation Results*

The results of the simulation study are discussed below using the summary of the seven simulation alternatives in table 8.4.

The *reference situation* describes the actual situation in the school year 1990/91. In the school year 1990/91 the costs of the 8,410 existing primary schools amounted to 6,130 million guilders. The cost of the parents and children to cover the distance to school were estimated at 310 million guilders.[8]

With strict application of the *original norm* valid until 1991 (see Table A8.1) 670 schools should have been closed which would have reduced costs of the schools by 110 million guilders. Some 3% of the children will then need to find another school. This adds 20 million guilders to the transportation costs already born by the parents.

Table 8.4

Summary of Simulation Results 1990/'91

		remaining schools	closures	costs[a] (x mln gld)	transportation costs (x mln gld)[a]	relocated pupils(in %)
0	reference	8,410	-	6,130	310	-
1	original norm	7,750	670	6,020	330	2.8
2	weighted	7,700	710	6,020	330	4.7
3	simple	7,430	980	5,970	340	3.4
4	law proposal Min. of Education	7,170	1,240	5,950	340	8.2
5	strong denominations	7,160	1,250	5,950	340	8.3
6	no denominations	7,020	1,390	5,920	340	9.0
7	maximum	3,580	4,840	5,610	460	45.0

a 1 guilder ≈ USD 0.5.

Source: Blank (1993)

Given that schools with a large proportion of pupils from ethnic minority groups and of parents with a low level of education (SEC-classes 1.25 and 1.90) require higher funding per pupil the government might consider using a norm with a *weighted* number of pupils. Schools which should have been closed according to the previous alternative, may now stay open if they have enough pupils of the higher SEC categories. It is expected that 40 additional schools will be closed, leaving a remainder 7,700 schools. There will be a greater number of closures of schools with a high share of SEC-1.00 pupils. About 5% of the pupils will go to another school; only 9% of them are SEC-1.90 pupils and 59% are children with SEC-1.00, whereas their representation in the pupils' population are 11% and 54% respectively.

There is a challenge to *simplify* the rules by choosing a system with one single norm for all schools. This alternative is simulated with a minimum requirement of 75 pupils for each school. This would be a cheaper alternative than a strict application of the original norm (simulation 2) but it would require an additional 310 closures. Which would be be more of the exceptionally located schools. The additional transportation costs are a modest 10 million guilders extra compared to the original norms.

The complicated set of rules proposed by the Ministry of Education in 1991, see *Toerusting en bereikbaarheid* (*Convenience and Accessibility*), (Ministry of Education and Science,

1991) lead to a strong reduction in costs that the simple alternative: they lead to a considerable number of additional closures. The evaluation of the 260 additional school closures requires more careful scrutiny than on the basis of costs alone.

Additional insight is given by alternatives 5 and 6 concerning denomination. Both are based on the proposal for closure norms of the Ministry of Education, Culture and Science simulated in alternative 4. In alternative 5 the parents' *denominational preferences* are followed strictly. Alternative 6 ignores denomination completely. The results show that the strong denomination alternative hardly differs from the law proposal of the Ministry of Education, Culture and Science. It is not surprising that foregoing denomination (simulation 6) would lead to a smaller number of schools. However, also this alternative hardly reduces costs compared with alternative 4. The additional transportation costs of alternative 5 compared to alternative 6 are small. It is explained in Section 8.3.2 that these costs can be seen as an upper bound on the utility loss of parents that arise from not being able to choose the preferred school.

The last alternative, allowing each municipality one school, but requiring at least 250 pupils for each additional school, is called the *maximum alternative*. It would lead to a battle field of closures. Almost 60% of the schools would cease to exist, 45% of the pupils would be left to find a new school but the reduction in costs would be limited to 520 million guilders; a relative low 8.5% of costs. This indicates that there is no ample room for cost reduction in this field. In addition, a part of this cost reduction is offset by a substantial increase in the transportation costs of 140 million guilders (+35%).

8.6 CONCLUSIONS, POLICY IMPLICATIONS AND FURTHER RESEARCH

The foregoing describes the production structure of primary education in the Netherlands and simulation of the impact of different closure norms on the costs of primary schools. This section summarizes and formulates a number of policy recommendations.

A short survey was given of the development of the Dutch schooling system through the 20th century. It was shown that the government allowed and supported financially the founding of schools of several denominations, which led to an abundance of primary schools. These could no longer be sustained when admissions declined and government funds became scarce. The result was that a multitude of schools operated below optimal scale. One of the most striking results of the study was the magnitude of returns to scale that could be obtained at the end of the century. Policy recommendations all circle

around the way how to gain the benefits of these scale economies. Whatever the alternatives offered the optimal choice will always depend on a number of effects involved such as costs of schools, simplicity of the laws, transportation costs of pupils, number of schools remaining, attention for individuals, time of implementation, quality of services etc.

The study showed that the positive scale effects decrease as school size increases. Unrestrained increases therefore make little sense. Extreme closure norms do not produce large savings. More specifically, a reduction of costs by closing schools will be limited to approximately 200 million guilders. A further reduction would lead to a significant increase of the transportation costs of the parents.

This will also be the case if a law is envisaged with a simple general norm for school size. Of course, it is always attractive to keep laws as simple as possible so that they are easily understood by everyone and are cheaper to implement. However, a general single norm will lead to an unjustified distribution of transportation costs, this time not borne by a large group parents (8.2%) but by only those who happen to live at great distances from schools (3.4 % of all pupils). On the other hand, a plea for simplicity is supported by dynamics. Complex laws need more time to be implemented and starting conditions may have changed by the time laws become effective. Time is a factor of importance also for other reasons. In the short run, savings will be small due to redundancy compensations and capital losses caused by unused buildings.

The attraction of the use of socio- economic characteristics (SEC-weights) to create a norm is that the burden of closures is less likely to be born by schools with a large proportion of pupils from ethic minority groups and of parents with a low level of education. However attentuon should be paid to the educational arguments here. It would be wise in any case to try to incorporate educational and sociological studies in studies of the present kind as the lack of information on output performance is rather restrictive.

Policy recommendations on school consolidation are implicit in the foregoing. Another recommendation is to seek ways of using economies of scale in costs of materials or capital e.g. by creating common stocks of materials or shared buildings. Another is that common management may help: 'branch' schools are found to be cheaper although the difference was statistically not significant. Efficiency of schools, not specifically studied here, seems to prone to be improved by incentives e.g. by introducing lump sum funding to reduce costs of labor by creating substitution possibilities with material inputs. See for example the previous chapter.

Ample recommendations can be given for further research. Most obvious, of course, is to start an efficiency study with methodologies described in Chapter 2 of this book. Almost equally challenging are studies with improved output figures, using better quality indicators as is done in Chapter 7. Also the use of actual cost rather than budgets set by the government may improve results.

An important observation in the context of this book is that the results from the (simulation) model described in this chapter are used in actual policy making. The Minister of Education, Culture and Science based his plans for new closure norms (T&B) on the results provided by this model. Moreover, the progress of the implementation of the T&B closure norms was monitored on a yearly basis. Each year the Ministry of Education, Culture and Science was provided with an overview of the results already attained, and a revised prediction of the gains at the end of the implementation period. This project shows that efficiency and productivity measurement can be a very useful tool in policy making and implementation.

APPENDIX

A8.1 MINIMUM SIZE STANDARDS

The standards were adapted several times. Table A8.1 shows the standards as they were in 1986/87, the number of schools in each class and the number of schools in each class that failed the requirements.

Table A8.1 shows that the threshold for staying open varies form 50 pupils in small municipalities to 125 pupils for the largest group. Only a few schools did not meet those standards. The threshold for starting a school is higher and varies form 80 to 200 pupils per school. In addition, the table indicates that almost half the schools are situated in the smallest municipalities.

Table A8.1

NUMBER OF PUPILS REQUIRED FOR STARTING AND MAINTAINING SCHOOLS IN 1986/'87, AFTER SIZE OF MUNICIPALITY WITH NUMBERS OF SCHOOLS BELOW THE NORMS

	Size of municipality (x 1,000 inhabitants)			
	< 25	25- 49	50-99	>100
number of pupils required for starting a school	80	120	160	200
number of pupils required for staying open	50	75	100	125
number of schools	4,060	1,640	1,270	1,440
schools below threshold for staying open	9	11	15	18

Source: Blank, Boef, Bronneman, Herweijer, Kuhry, and Schreurs (1990)

A8.2 REGRESSION EQUATION AND RESULTS

Regression Equation

The regression equation to estimate the cost function of primary schools used is (see also Blank 1993):

$$\begin{aligned} \ln(C) = {} & a_0 + b_0\ln(L) + b_1\ln(L1.00\%) + b_3\ln(L1.90\%) + b_4\ln(Q1) \\ & + .5b_{00}\ln^2(L) + b_{01}\ln(L)\ln(L1.00\%) + b_{03}\ln(L)\ln(L1.90\%) + \ln(L)b_4\ln(Q1) \\ & + .5b_{11}\ln^2(L1.00\%) + b_{13}\ln(L1.00\%)\ln(L1.90\%) + b_{14}\ln(L1.00\%)\ln(Q1) \\ & + .5b_{33}\ln^2(L1.90\%) + b_{34}\ln(L1.90\%)\ln(Q1) + .5b_{44}\ln^2(Q1) \\ & + c_1\ln(WL) + c_2\ln(WM) + c_3\ln(WH) \\ & + .5c_{11}\ln^2(WL) + c_{12}\ln(WL)\ln(WM) + c_{13}\ln(WL)\ln(WH) \\ & + .5c_{22}\ln^2(WM) + c_{23}\ln(WM)\ln(WH) + .5c_{33}\ln^2(WH) \\ & + e_{01}\ln(L)\ln(WL) + e_{02}\ln(L)\ln(WM) + e_{03}\ln(L)\ln(WH) \\ & + e_{11}\ln(L1.00\%)\ln(WL) + e_{12}\ln(L1.00\%)\ln(WM) + e_{13}\ln(L1.00\%)\ln(WH) \\ & + e_{31}\ln(L1.90\%)\ln(WL) + e_{32}\ln(L1.90\%)\ln(WM) + e_{32}\ln(L1.90\%)\ln(WH) \\ & + e_{41}\ln(Q1)\ln(WL) + e_{42}\ln(Q1)\ln(WM) + e_{43}\ln(Q1)\ln(WH) \end{aligned} \tag{A8.1}$$

where

C	= total costs of a school
L	= number of pupils
L1.00%	= proportion of pupils SEC-1.00
L1.90%	= proportion of pupils SEC-1.90
Q1	= expected proportion outflow of pupils to higher secondary education
WL	= price of labor
WM	= price of material
WH	= price of housing

From (A8.1) a number of global characteristics of the production structure can be derived, e.g.

Cost flexibility according Caves, Christensen, and Swanson(1981) gives:

$$v'(L) = \frac{\partial \ln(C)}{\partial \ln(L)} = b_0 + b_{00}\ln(L) \tag{A8.2}$$

Marginal costs for a school with L pupils are equal to:

$$mc(L) = \frac{\partial C}{\partial L} = \frac{C}{L}\frac{\partial \ln(C)}{\partial \ln(L)} \tag{A8.3}$$

Note that the costs and the number of pupils in the function are centered around their mean. Accordingly, (A8.3) represents marginal costs standardized for the average school for which average costs are unity. For an arbitrary school (A8.3) must be multiplied by the average variable costs per pupil to generate the marginal costs hence.

Average costs for a school with L pupils are equal to:

$$avc(L) = \frac{C}{L} \tag{A8.4}$$

Estimation Results

The maximum likelihood estimates of the parameters of equation (A8.1) are given in Table A8.2.

Table A8.2

ESTIMATION RESULTS FOR THE COST FUNCTION OF PRIMARY SCHOOLS 1986/87

	Parameter	Estimate	t-value
constant	a_0	0.01	2.3
L	b_0	0.83	111.1
L1.00%	b_1	-0.10	14.9
L1.90%	b_3	0.04	12.7
Q1	b_4	0.01	1.5
L x L	b_{00}	0.20	13.4
L x L1.00%	b_{01}	-0.01	1.2
L x L1.90%	b_{03}	0.00	0.7
L x Q1	b_{04}	0.00	0.5
L1.00% x L1.00%	b_{11}	-0.03	6.7
L1.00% x L1.90%	b_{13}	-0.01	4.6
L1.00% x Q1	b_{14}	0.01	2.2
L1.90% x L1.90%	b_{33}	0.01	10.5
L1.90% x Q1	b_{34}	0.00	1.9
Q1 x Q1	b_{44}	0.00	0.3
WL	c_1	0.71	24.5
WM	c_2	0.13	4.9
WH	c_3	0.11	19.4
WL x WL	c_{11}	0.04	0.2
WL x WM	c_{12}	-0.14	1.0
WL x WH	c_{13}	-0.05	1.4
WM x WM	c_{22}	0.15	2.4
WM x WH	c_{23}	-0.05	2.1
WH x WH	c_{33}	0.02	2.1
L x WL	e_{01}	0.00	0.1
L x WM	e_{02}	-0.05	1.8
L x WH	e_{03}	-0.03	4.1
L1.00% x WL	e_{11}	-0.04	1.2
L1.00% x WM	e_{12}	-0.02	0.9
L1.00% x WH	e_{13}	-0.01	1.0
L1.90% x WL	e_{31}	-0.03	3.4
L1.90% x WM	e_{32}	0.00	0.7
L1.90% x WH	e_{33}	0.00	0.2
Q1 x WL	e_{41}	0.01	0.4
Q1 x WM	e_{42}	-0.02	1.0
Q1 x WH	e_{43}	0.01	1.4
R^2		0.96	

a *Variables standardized at their means and in logarithms.*

Source: Blank (1993)

A number of hypotheses were tested formally. The test-results are given in Table A8.3.

Table A8.3

TESTS OF HYPOTHESES

null-hypothesis (H_0)	likelihood ratio	degrees of freedom	$\chi^2_{(0,05,j)}$	reject H_0 [a]
1 There is no relation between a school's being located in a so-called minimum-income area and its use of resources	33.9	26	38.9	-
2 There is no relation between the percentage outflow to higher secondary education and the resources used	19.0	7	14.1	+
3a There is no relation between the size of the school board and total costs	37.1	26	38.9	-
3b There is no relation between thesize of the school board and the number of personnel	8.4	26	3.8	+

a + = reject H_0 - = do not reject H_0.

Source: Blank (1993)

NOTES

[1] In Drenthe several very large schools overcompensate this somewhat but in Zeeland the average school size is the second lowest in the country.

[2] Recently, new denominational at schools such as Islamic, appear on the scene.

[3] The term between brackets is the customary designation for these groups. The fgure (1.90, 1.25 or 1.00) refers to the weight assigned to a given pupil in determining the budget for personnel. Thus, a pupil from an ethnic minority group is entitled to 1.90 times as much personnel as a pupil in the SEC-1.00 category.

[4] The usual economic properties such as symmetry and homogeneity are imposed.

[5] The model for secondary education (Merkies, Steijn, and De Vos 1985) also contained a layer for the decision of the school board. Closures of secondary school could sometimes be avoided by mergers or cooperation with other "types" of secondary education. As this type of behavior does not play any role in primary education this layer has been deleted from the model.

[6] The SCP analysis data set was derived from information from the Ministry of Education, Culture and Science, the RION Institute for Educational Research and the market research bureau NSS Marktonderzoek BV.

[7] Areas with a low living standard were identified as areas where minimum incomes prevail.

[8] These costs include both the actual costs incurred and the time of the parents and children spent on traveling.

References

Blank, J.L.T. 1993. *Kosten van kennis: een empirisch onderzoek naar de produktiestructuur van het basisonderwijs in Nederland.* Rijswijk/Den Haag: Sociaal en Cultureel Planbureau/VUGA (Sociale en Culturele Studie 17).

Blank, J.L.T., S. Boef-van der Meulen, H.M. Bronneman-Helmers, L.J. Herweijer, B. Kuhry, and R.A.H. Schreurs. 1990. *School en schaal.* Rijswijk/Den Haag: Sociaal en Cultureel Planbureau/VUGA (Cahier 81).

Boyd, W.L. 1992. The Power of Paradigms: Reconceptualizing Educational Policy and Management. *Education Administration Quarterly* 28: 504-528.

Butel J.H., and G.B.J. Atkinson. 1983. Secondary School Size and Costs. *Educational Studies* 1983(3):151-157.

Callan, S.J., and R.E. Santerre. 1990. The Production Characteristics of Local Public Education: A Multiple Product and Input Analysis. *Southern Economic Journal* 57 (2): 468-480.

Caves, D.W., L.R. Christensen, and J.A. Swanson. 1981. Productivity Growth, Scale Economies and Capacity Utilization in U.S. Railroads, 1955-1974. *American Economic Review* 71 (5): 994-1002.

Christensen, L.R., D.W. Jorgenson, and L.J. Lau. 1973. Transcendental Logarithmic Production Frontiers. *The Review of Economics and Statistics* 55: 28-45.

Cohn, E. 1968. Economies of Scale in Iowa High School Operations. *Journal of Human Resources* 3 (4): 422-434.

Coleman, J.S., E.Q. Campbell, C.J. Hobson, J. McPartland, A.M. Mood, F. Weinfeld, and R.L. York. 1966. *Equality of Educational Opportunity.* Washington D.C: Government Printing Office.

Daane, A.M.M., and B.G.A. Hoeijmans. 1985. *Overgangswet op het basisonderwijs.* Zwolle: Tjeenk Willink.

Daane, A.M.M., and B.G.A. Hoeijmans. 1986. *Wet op het basisonderwijs, Overgangswet Wet op het basisonderwijs.* Zwolle: Tjeenk Willink.

Daane, A.M.M., and B.G.A. Hoeijmans. 1988. *Wet op het basisonderwijs.* Zwolle: Tjeenk Willink.

Färe, R., and S. Grosskopf. 1994. *Cost and Revenue Constrained Production.* New York: Springer.

Färe, R., and D. Primont. 1995. *Multi-output Production and Duality: Theory and Applications.* Dordrecht: Kluwer Academic Publishers.

Goudriaan R., F. van Tulder, J. Blank, A. van der Torre, and Bob Kuhry. 1989. *Doelmatig dienstverlenen. Een onderzoek naar de produktiestructuur van vier voorzieningen in de kwartaire sector.* Rijswijk/Alphen aan den Rijn: Sociaal en Cultureel Planbureau/Samsom (Sociale en Culturele Studie 11).

Grosskopf, S., K.J. Hayes, L.L. Taylor, and W.L. Weber. 1997. Budget-constrained Frontier Measures of Fiscal Equality and Efficiency in Schooling. *Review of Economics and Statistics* 79 (1): 116-124.

Gyimah-Brempong, K., and A.O. Gyapong. 1991. Characteristics of Education Production Functions: An Application of Canonical Regression Analysis. *Economics of Education Review* 10 (1): 7-17.

Gyimah-Brempong, K., and A.O. Gyapong. 1992. Elasticities of Factor Substitution in the Production of Education. *Economics of Education Review* 11 (3): 205-217.

Hanushek, E.A. 1986. The Economics of Schooling: Production and Efficiency in Public Schools. *Journal of Economic Literature* 24 (3): 1141-1177.

Hanushek, E.A. 1989. The Impact of Differential Expenditures on School Performance. *Educational Researcher* 18: 45-51, 62.

Hanushek, E.A., and D.W. Jorgenson (eds.). 1996. *Improving America's Schools: The Role of Incentives*. Washington DC: National Academic Press.

Hurn, C.J. 1985. *The Limits and Possibilities of Schooling: An Introduction to the Sociology of Education*, 2nd Edition. Newton Mass: Allyn and Bacon Inc.

Jimenez E. 1986. The Structure of Educational Costs: Multiproduct Cost Functions for Primary and Secondary Schools in Latin America. *Economics of Education Review* 5 (1): 25-39.

Merkies, A.H.Q.M., I.J. Steijn, and A.F. de Vos. 1985. *Gevolgen van de hef-vo operatie: structuur en kosten van het voortgezet onderwijs 1985-1995 onder diverse opheffingsnormen voor scholen*. Amsterdam: Vrije Universiteit.

Ministry of Education, Culture and Science. 1991. *Toerusting en bereikbaarheid van basisscholen*. Zoetermeer: Ministerie van Onderwijs en Wetenschappen.

Rao, P.D. 1998. *Economics of Primary Education*. New Delhi: Vedams Books International.

Riew, J. 1966. Economies of Scale in High School Operation. *Review of Economics and Statistics* 48: 280-288.

Ritzen, J.M.M., J. van Dommelen, and F.J. de Vijlder. 1997. School Finance and School Choice in the Netherlands. *Economics of Educational Review* 16 (3): 329-335.

Smet, M., and W. Nonneman. 1998. Economies of Scale and Scope in Flemish Secondary Schools. *Applied Economics* 30 (9): 1251-1258.

Statistics Netherlands. 1994. *Vijfennegentig jaren statistiek in tijdreeksen, 1899-1994*. Den Haag/Heerlen: Sdu/CBS.

Part Four:
Law Enforcement

9 Cost Efficiency in Florida Prisons

Kwabena Gyimah-Brempong[1]

9.1 Introduction

This chapter investigates short run cost (in)efficiency in the provision of prison services in the state of Florida. Specifically, it investigates short run cost inefficiency in the provision of custodial and security services in the state's prison system by estimating a *best practice* short run variable cost function with a panel data from forty five major adult prisons for the 1995-98 period and uses the estimates from the cost function to calculate measures of inefficiency.[2] I compare the estimates of cost inefficiency in Florida prisons to estimates of cost inefficiency from other studies that use either public or private sector data to see if cost inefficiency in Florida prisons is out of the ordinary. Given the controversy surrounding the privatization of prisons, I use the estimated cost inefficiencies from the sample to *speculate* on the relative efficiency of private prisons in Florida.

In this chapter, I measure the output of prisons as consisting of custodial and security services.[3] Prisons are charged with multiple objectives, including custodial services, specific deterrence, education, and rehabilitation of inmates, among others. Some of these services cannot be easily measured (e.g. rehabilitation). Efficiency in the provision of such multiple services may therefore not be adequately captured by cost efficiency. However, given that one cannot obtain data on some of the qualitative aspects of prison production, this chapter limits itself to the study of cost (in)efficiency while acknowledging the limitations of such an approach. I use a cost function approach because I believe that the custodial and security output of prisons are outside the control of prison management and hence, are exogenous. Prison management is charged with the production of these exogenously determined outputs as efficiently as possible, hence cost minimization is the appropriate framework for investigating (in)efficiency in prisons (see also Section 2.2.1). Besides the exogeneity of output, the cost function approach allows me to model prisons as multi-product firms.

The concern with crime and its effects on society has led many state governments as well as the Federal government to impose mandatory minimum prison sentences on guilty offenders in recent years. In addition, the minimum sentences themselves have been increasing. For example, many states have adopted versions of California's "three strikes and you're out" laws under which an offender's prison term increases with the number of previous offenses even though the offense for which he/she is being sentenced may be a misdemeanor.[4] Increased incarceration for longer periods implies the need for increased prison capacity, hence states have expanded prison beds at unprecedented rates. For example, between 1988 and 1998, Florida doubled its prison capacity as well as the inmate population to 66,280 from 33,681. This increase in the prison population cannot be attributed to Florida's rapid population growth alone. The incarceration rate per 100,000 in Florida increased from 274 in 1988 to 440 in 1998 – a 60% increase during the 11 year period. The average percentage of sentence served by all offenders increased from about 34% in 1991 to about 73% in 1998.[5]

Accompanying the increased and longer periods of incarceration has been the predictable increase in the cost of providing prison services. In Florida, the average inmate cost per day (for both public and private prisons) increased from USD 42.56 per day in 1992 to USD 49.44 per day in 1998; for custodial services, the increase was from USD 33.87 to USD 38.67 per day. The increasing cost of operating prisons as well as the general unwillingness of the populace to pay higher taxes to support the provision of public services in general and prisons in particular has led states to search for ways to increase efficiency in the provision of prison services. Among the suggested solutions to the cost problem are privatization of state prisons, building larger state prisons to take advantage of economies of scale, reducing the *quality of services* by denying inmates certain amenities such as access to exercise machines, and establishing prison enterprises to defray the cost of running the prisons. Privatization of prisons will increase efficiency in the Florida prison system if private prisons are more cost efficient than those managed by the public sector.

To achieve efficiency in the provision of prison services, there is the need to understand the structure of production and cost in the provision of such services. Unfortunately, there is very little empirical work on the production or cost structure of prison production. The few investigations into prison efficiency have focused almost exclusively on cost differentials between publicly run prisons and private prisons. The quantity of resources devoted to the provision of prison services as well as the number of people in the prison system in the US warrant careful analyses and empirical work in this area in order to guide policy. If prison services are being provided above the lowest possible cost, then prison managers can take measures to move towards the cost frontier as a way

of decreasing the cost of production. On the other hand, if prisons are operating at the lowest possible cost, then society will either have to pay more for the provision of prison services or produce fewer of these services. If private prisons are more cost efficient than those managed by the Florida Department of Corrections (FDOC), then policy makers could reduce the cost of running the state's prison system through privatization of prisons. While privatization of prisons may entail some legal issues, I only concern myself with cost differences in this chapter.

Whereas there have been a few efforts to estimate cost functions for prisons, this study, as far I know, is the only one that employs the *best practice* cost function and *panel data* to investigate cost efficiency in a state prison system. Most estimates of economic efficiency have relied on cross sectional data. To improve the quality of estimates, this chapter uses panel data and panel estimation methodology as one of the specifications to estimate the short run cost inefficiency in prison production. The contribution of this chapter to the literature lies in using several specifications, including panel estimation, of a *best practice* short run cost function to estimate cost (in)efficiency in the production of prison services in Florida. The chapter also uses the estimated cost to speculate on the prison privatization debate.

The most comprehensive work on estimating the cost and production functions for prisons are the two studies by Schmidt and Witte (1984), and Witte et al. (1979), and the study of Federal and California prisons by Trumbull and Witte (1981-82).[6] These studies use data from large Federal as well as state prisons in California, Texas, New York, and North Carolina to model prison production and cost functions. These studies find that, in general, the estimated prison cost functions are consistent with cost minimization. They also find that prison cost is significantly influenced by the *quality* of inputs as well as service *conditions* such as how hardened the inmates are, prison overcrowding, and the gender and age composition of inmates. They conclude that it is not possible to make any generalizations about economies of scale in the production of prison services. While some prisons exhibited decreasing average cost throughout the output ranges, others exhibited the usual U shaped curve while still others exhibited increasing cost throughout their production ranges.

Block and Ulen (1979) uses data from California and a Cobb Douglas specification to estimate prison production functions. Using time series data for individual prisons, they concluded that the optimal prison size in California is about 5700 inmates, a number that will imply substantial economies of scale. A long run prison cost function they estimate based on cross sectional data of city and county jails indicates the existence of constant returns to scale in California prisons. None of the studies reviewed above cal-

culate indices of efficiency in prison production. Ganley and Cubbin (1992) uses a Data Envelopment Analysis (DEA) model to study efficiency in 33 UK prisons. Based on this non stochastic frontier function, they found that 13 of the prisons they studied were technically inefficient relative to the efficient prisons in the sample, and, on average, the operating cost of the inefficient prisons could be reduced by 12%.

Outside the studies cited above, there have been few empirical studies on the estimation of prison production or cost functions. Gyimah-Brempong (1989) estimates a cost function for police production in Florida municipalities and find socioeconomic characteristics of communities to be important non purchased inputs and argues that cost functions for the provision of public goods should include such socioeconomic characteristics of communities. I incorporate this argument in this chapter by including some characteristics of prisons in the cost function. The studies mentioned above were not interested in investigating cost efficiency; they merely estimated "representative" cost functions that are not necessarily the most economically efficient way of producing the output. Second, none of the papers estimate a *best practice* cost function, hence their approaches are not suitable for the estimation of efficiency. I do so in this chapter.

Filippini and Prioni (1994) investigated whether regulation is the source of inefficiency in the Swiss bus industry. They concluded that regulation does not seem to contribute to inefficiency in the Swiss bus industry. Burns and Weyman-Jones (1996) investigated inefficiency in the distribution of electricity in the United Kingdom and found a significant degree of inefficiency among firms. Moreover, they find that inefficiency estimates depend upon the specification of the distribution of the (in)efficiency parameter. I partially follow this approach by estimating different specifications of the short run cost function and calculate inefficiency in prison production from the different models.

There has not been any serious econometric study on the privatization of prisons despite the attention and controversy the subject has generated in recent years.[7] There is, however, a growing body of work that compares the efficiency of private prisons to those of public prisons.[8] A number of studies argue that private prisons are more efficient than public prisons (e.g. Logan 1987, Sellers 1993, Lanza-Kaduce, Parker, and Thomas 1999, among others). Another group of researchers contend that privatization does not confer any efficiency gains in the production of prison services and that it may lead to reductions in the *quality* of prison services (Avio 1993, DiIulio 1988, Edwards 1996). Hart, Shleifer, and Vishny (1997) note that the relative efficiency of private prisons depends on the nature of the contract between the state and the private prison. Where rights to make changes are given to the private company, there are strong incentives to improve quality of service as well as decrease the cost of production. On the other hand,

if prison contracts require the governments to retain control over the facility and any changes in the facility or its operation has to be renegotiated, the quality of service is likely to decrease with cost. I do not estimate an econometric model of prison privatization given the small number of privately managed prisons in my sample. I, however, use the estimates of cost inefficiency in this study to *speculate* on the relative efficiency of private prisons in Florida.

The rest of the chapter is organized as follows. Section 9.2, following this section, provides a brief description of Florida prisons. Section 9.3 introduces the short run cost function. Section 9.4 discusses the data used to estimate the model while Section 9.5 presents and discusses the estimates of the various specifications of the *best practice* short run cost function as well as the estimates of cost inefficiency in the provision of prison services. It also uses the estimates to speculate on differences in cost inefficiency between private and public prisons in Florida. Section 9.6 summarizes and concludes the chapter.

9.2 Brief Description of Florida Prisons

As of June 30 1998, Florida had 66,300 persons under some form of incarceration in its penal system in about 131 correctional facilities. The 131 correctional facilities consist of 60 major correctional institutions, 26 work camps, 11 treatment centers, 29 work release centers, and 5 road camp prisons. The 60 correctional institutions house about 83% or 55,000 of the total Florida penal population while work camps account for another 10.4% or 6,900. All other facilities combined account for less than 10% of people in the Florida penal system. Of the 60 correctional institutions, 55 are male institutions, 4 are female institutions while 1 houses both males and females. Seven of the institutions are Reception Centers while another 8 are youthful correctional facilities. Seven of the correctional institutions are maximum security prisons, another 36 are closed custody prisons, 13 are medium custody prisons, and the rest are minimum security prisons.[9]

Besides its size, the Florida prison population has grown rapidly. Between 1989 and 1998, it grew from 38,000 to 66,300 a 74% growth in a decade. Part of this rapid growth is due to mandatory sentences imposed by the legislature as a means to fight crime. For example, in 1998, fully 42% of all inmates in Florida prisons were sentenced for a crime for which there was a mandatory prison sentence. In the 1980s, the state went on an accelerated prison construction program to accommodate the rapidly increasing prison population. The result is that a large number of new prison facilities were opened for operations in the late 1980s and 1990s.

The Florida inmate population is predominantly male and black. Of the total inmate population of 66,300, 62,800 or 95% is male while only 5% is female. Blacks account for a total of 36,700 or 55% of the inmate population while whites account for 43%. A relatively large proportion of the Florida inmate population has prior conviction records. In 1998, 50% of the inmate population had prior conviction records. This ratio rose from about 40% in 1989 and peaked at 57% in 1994. Of the number of inmates in the Florida prison system in 1998, 21% are characterized as habitual offenders. This indicates that the rate of recidivism is relatively high among Florida prison releasees. Inmate escapes from Florida prisons are relatively low, having dropped from 1.3 per 1,000 inmates in 1993/94 to about .2 per 1000 in 1998. The average inmate in Florida's prison system is relatively young with a mean age of 34 years and 57% of the prison population is 34 years or younger.

In the 1997/98 fiscal year, Florida spent a total of USD 1.2 billion on the state's prison system, a 16% increase from the amount of USD 1 billion in 1990. Part of this rapid increase is due to the rapid increase in the prison population. The average daily cost per prisoner in 1999 was USD 51 for the state's public prisons while it was USD 49 for the private prisons, a 25% increase from the average daily cost of USD 40 in 1990.[10] The average daily cost ranged from a low of USD 34 to a high of USD 102. Prisons' share of the state budget increased from 8% in 1990 to 9% in 1997/98 fiscal year. Generally, female prisons, youthful offender prisons, and specialty prisons (such as mental prisons) tend to be more expensive to operate than male adult prisons which tend to be the least expensive to operate. Of the average daily cost of USD 50, about 53% was spent on security, 19% on health care, 10% on energy and supplies, and another 7% on food. There is a slight variation in the composition of the average daily cost across prisons.

Faced with escalating costs of building and operating prisons and an unwillingness to increase taxes to finance these operations, the Florida legislature has been searching for a means to decrease the cost of operating prisons. Some legislators have seen the answer in privatization of prisons. In 1993, the Florida legislature passed legislation that authorized the Florida Correctional Privatization Commission (FCPC) to "enter into contracts for the designing, acquiring, financing, leasing, constructing, and operating" of state prisons (Florida Statutes, §95157.03). The legislation also required that the quality of service and program provided by these private prisons be "at least equal to those provided by the Department" (of Corrections) and to provide these services at a cost savings of at least 7% over the cost of operating comparable facilities by the Florida Department of Corrections (FDOC). As a result of this legislation, FCPC has moved aggressively to privatize some Florida prisons. As of June 30 1998, 5 medium security prisons (3 adult prisons and 2 youthful offender prisons) were managed by two for-prof-

it private prison corporations.[11] These prisons are generally new and modern. For example all the five prisons being managed by the private sector were opened in 1995 or later. The private prisons are also slightly different in terms of inmate characteristics and restrictions imposed in management.[12] It is possible that the cost structure of the private prisons are different from the cost structure of prisons run by the FDOC.

9.3 Methodology

There are several problems with modeling the cost of production of public services especially the provision of prison services. For one thing, it is not clear what output prisons produce – custodial services, rehabilitation, education, or the end product safety? If safety is the output of prisons, how is safety measured? There are also differences in the levels of security required to safely confine different categories of offenders. In this chapter, I concentrate on the output that is directly produced by prisons (see also Section 9.4). I measure the direct output of a prison as the number of prisoners per year it holds in custody and the level of security of the prison. While concentration on a few dimensions of a prison's output could lead to biased estimates, introducing variables that cannot be measured could introduce another set of biases. The second issue is whether prison administrators minimize cost. Although prison administrators may behave like bureaucrats and strive to maximize their budgets rather than minimize cost, pressure from state comptrollers and citizens at large, to minimize cost will be strong enough to force them to behave as if they minimize cost.[13] The final problem associated with analyzing prisons cost is determining the unit of analysis. If the entire prison system is the unit of analysis, then modeling will be different from the situation where the individual prison is the unit of analysis. Schmidt and Witte (1984) argue convincingly that the unit of analysis is the individual prison. I follow their argument and adopt the individual prison as the unit of analysis and treat individual prisons as multi-product firms.

My working assumption is that the prison manager is provided with a fixed prison capacity and a given number of prison inmates and charged with choosing variable inputs to minimize the cost of housing these inmates at a given level of security. This implies that a prison is seen as a multi-product firm producing custodial services and security. Prisons may not be able to produce the given output at the minimum cost, hence it is possible that the actual cost of production in a prison systematically exceeds the minimum cost of producing the given output. Cost inefficiency could result from the fact that a producer cannot produce the maximum output that is possible given a set of inputs (produce a given output level with the minimum amount of inputs) a con-

dition referred to as production inefficiency. Cost inefficiency could arise if the prison is productively efficient (produce maximum output with a given set of inputs) but does not use the input combination that minimizes production cost. This is referred to as allocative inefficiency. The cost inefficiency I calculate here is a combination of productive and allocative inefficiency. The degree to which actual cost of production exceeds the minimum cost of production is the measure of cost inefficiency estimated in this study.[14] Actual cost of production will depend, in part, on the degree to which a prison is inefficient; the more inefficient a prison is, the more its cost will exceed the minimum cost of production, all things equal. Any deviation of production cost from minimum cost of production will consist of a systematic component due to inefficiency and the usual white noise component. Minimizing the cost of producing the prison output subject to the prison production technology, and given input prices, the resulting cost function depends on actual output, input price vector, and the production inefficiency parameter.[15] The dependence of the cost function on the inefficiency parameter makes this cost a stochastic one.

Because I treat prison capacity and other capital inputs as fixed, the *best practice* cost function I estimate is a short run variable cost function. Any short run cost function depends, in part, on the quantities of fixed inputs (capital in this case) available to the manager as well as on output and input price.

The prison manager is assumed to combine two purchased variable inputs – labor and energy – with the fixed prison capacity to efficiently produce prison services. The cost function will also depend on the environment within which the prison management operates. So short run cost depends on inmate population, the level of security, the wage rate, the price of energy, the ratio of black inmates to black prison officers, health care personnel per prisoner, the age of a prison, and prison capacity.

Prison population (POP) and the level of security (S) are indicators of output and therefore expected to be positively correlated with the cost of production. In accordance with any well behaved cost function, I expect the coefficient of the first order relative wage (W/P_e) to be positive in the short run variable cost function. CROWD, defined as the ratio of prison population to designed capacity of the prison, is my proxy for fixed capital. Its coefficient in the variable cost function is expected to be negative since prison managers can substitute it for some variable inputs. Three environmental variables – health care personnel per prisoner (HP), age of the prison (AGE), and the ratio of black inmates to black prison personnel (BLK) – are included in the cost function as non purchased inputs (Gyimah-Brempong 1989, Schmidt and Witte 1984, Trumbull and Witte 1981-82). I expect the coefficients of HP and AGE to be positive since a large HP is an

indicator of health problems among inmates and old prisons imply high maintenance cost. If a concentration of blacks in a prison increases "cultural harmony" among inmates, this may reduce tensions, hence make the prison more manageable. This is likely to decrease the cost of production of custodial services in prisons. The reverse will be the case if the concentration of blacks in a prison increases tensions. The coefficient of BLK, therefore cannot be signed *a priori*. I use the estimates from the *best practice cost* function to estimate cost (in)efficiency as shown by Jondrow, Lovell, Meterov, and Schmidt (1982), Battese and Coelli (1988), and Lovell (Chapter 2 of *this volume*).

9.4 Data

The dependent variable, variable cost is measured as the annual total operating cost, excluding health, education, and vocational training cost for each prison. It excludes all capital cost including construction cost, debt repayment, and other capital purchases. The short run total variable cost is a function of the prices of two purchased inputs – labor and energy and two outputs. The wage rate of labor (W) is measured as the average annual security personnel wage of a full time equivalent employee in a prison. The wage rate was calculated as the total annual security wage bill divided by the total annual full time equivalent of security workers. The wages and salaries of prison personnel in Florida are set by the state legislature and can therefore be considered as exogenously determined. Though set by legislative fiat, the wages vary spatially to reflect cost of living and quality of life differences. In 1998, the average wage differential between officers in the urbanized southern part of the state and the rural northern parts was about USD 3,700, a significant difference. The wage data did not include fringe benefits. I adjusted the wages upwards by 31% to reflect the fringe benefits component.[16] The short run cost function has the *quantity* of fixed capital as an argument. I proxy the quantity of fixed capital by the ratio of inmate population to the designed capacity of the prison (CROWD). Capacity, used to calculate overcrowding, is measured as the design capacity of the prison defined as the number of inmates a prison was designed to carry.[17] I proxy the price of energy by the price of electricity (P_e). Security level (S) is indexed from 1 to 7 with 7 being maximum level security and 1 being minimum security prisons.

I measure output as the total number of prisoners per year equivalent kept by a prison. I measured CROWD as the ratio of prison population to the designed capacity of the prison. BLK is measured as the ratio of black inmates in a prison to the ratio of black employees in that prison. This variable is intended to measure the racial balance of black prison population relative to that of black prison professionals. HP is measured as

the ratio of full time equivalent health personnel in the prison to the full time equivalent resident population of the prison. Finally, I measured the age of the physical facility of prisons (AGE) as current year minus the year the facility began operations. Although the Florida legislature has mandated that prison guards in high cost urban areas be paid wages that are higher than those of their counterparts in rural areas, the wage differentials are not adjusted to reflect *annual changes* in the cost of living across the state. This implies that real wages of prison guards will differ from nominal wages not only through time but also across space if there are annual geographical changes in the relative cost of living in Florida.[18] It is therefore necessary to adjust all nominal variables for the changing cost of living across Florida and through time. All nominal variables were converted to 1996 constant USD and to Florida's average cost of living using the US CPI and Florida CPI (FCPC) respectively to account for differences in the cost of living through time and across Florida.[19] Data for FCPC and the price of electricity were obtained from *Florida Statistical Abstract*, (Gainesville, University of Florida, various years). The CPI data were obtained from the Department of Commerce, Bureau of Economic Analysis.

With the exception of FCPC, CPI, and the price of electricity, all data used to estimate the cost function were obtained from the files of FDOC. The data is a four year panel of 45 major adult residential correctional institutions and the data are annual data from 1995 to 1998. Excluded are all reception centers, special facilities such as rehabilitation centers and hospitals, all juvenile facilities, and all community correctional institutions. Reception centers, juvenile correctional facilities, and community correctional facilities produce a different mix of services compared to those produced by adult prisons. The cost structure of these facilities seem to be different from those of adult facilities. I therefore excluded these institutions in order to make the product mix of prisons in the sample more comparable. Not all prisons had usable observations for all four years, hence I have an unbalanced panel of 168 observations.[20] The maximum number of observations for each prison is four while the minimum number of observations for each prison is two.

A summary of statistics of the data is presented in Table 9.1. The summary statistics show that there are large variations in the characteristics and cost in Florida prisons. Prisons range from small size with average population of about 200 to large with an average population of 1,900. The sample mean prison population is 1,110 while about 1.9% of inmates use illegal drugs in prison. The difference in size is also reflected in the range of total cost. The average variable cost for custodial and security services is USD 1.9 million with a standard error of USD 64 million. This large standard error reflects the extreme range in the size and security levels of prisons operated in the state of Florida.

Table 9.1

SUMMARY STATISTICS OF DATA

	Mean[a]	Standard Deviation	Min.	Max.
prison population	1,110	370	210	1,900
wage rate (in USD)	32,500	2,900	26,900	48,200
price of electricity (in USD/KWH)	0.06	0.01	0.01	0.13
security level	4.7	1.0	2.0	7.0
ratio of inmates to capacity	1.6	0.5	0.3	3.4
health care personnal per prisoner	0.04	0.01	0.00	0.07
total costs (in USD million)	1.9	64.3	0.9	54.2
age of prison	17.7	16.8	2.0	108.0
capacity	710	240	150	1,810
ratio of black inmates to black employees	5.9	3.9	4.7	6.7

a *These are unweighted means.*

The variance in the wage rate across prisons is much smaller than that of prison size and total variable cost. This may, in part, reflect the fact that salary scale of corrections officers are controlled by the legislature. Of the 45 prisons covered, three house females only, while the rest house males only. Two of the prison facilities, both medium security prisons, are operated by for-profit private corporations while the rest are operated by the Florida Department of Corrections. About 56% of the inmate population is black while only 11% of corrections professionals is black. About 17% of inmates in the sample are between 18 and 24 years old. Although Florida prisons seem to be operating at 60% above design capacity, this unweighted ratio may reflect the influence of smaller and older institutions.

9.5 RESULTS

9.5.1 *Coefficient Estimates*

The efficiency component in a stochastic best practice analysis can be specified in various ways. I refer to them as exponential, panel, half normal and truncated normal. For a technical discussion I refer the reader to Coelli (1995), Jondrow, Lovell, Meterov, and Schmidt (1982) and Chapter 2 of this volume. The **best practice** short run cost function is estimated with data from 45 major adult institutions over the 1995-1998 period.

Coefficient estimates of the cost function are presented in Table 9.2. Columns 2 and 3 present the coefficient estimates for the exponential specification, columns 4 and 5 the estimates for the panel specification, columns 6 and 7 the estimates for the half normal specification, while columns 8 and 9 present the estimates for the truncated normal specification.

Table 9.2

COEFFICIENT ESTIMATES OF THE STOCHASTIC FRONTIER COST FUNCTION

Variable	*Exponential*		*Panel*		*Half Normal*		*Truncated Normal*	
	coefficient	*t-value*	*coefficient*	*t-value*	*coefficient*	*t-value*	*coefficient*	*t-value*
Constant	-2.47	1.09	-4.42	2.73	-4.42	1.75	-4.48	1.40
W/P_c	0.64	13.50	0.69	13.24	0.63	10.87	0.63	10.87
Q	0.80	9.77	0.67	5.40	0.79	10.96	0.79	10.96
S	1.43	3.39	0.92	4.14	1.49	3.35	1.42	3.17
QxS	-0.31	2.57	-0.32	3.72	-0.31	2.49	-0.31	2.48
Q^2	-0.09	2.02	-0.17	2.83	-0.17	1.71	-0.17	1.69
W^2/P_c	-0.01	2.01	-0.01	1.60	-0.01	1.99	-0.01	1.99
BLK	-1.42	2.78	-1.36	4.12	-1.36	2.59	-1.36	2.56
HP	0.09	5.50	0.11	8.14	0.10	5.80	0.10	5.59
AGE	0.06	1.71	0.45	2.02	0.07	1.71	0.07	1.70
CROWD	-0.08	1.88	-0.09	2.60	-0.08	1.88	-0.08	1.88
σ_u^2/σ_v^2	-		0.34	2.96	-		-	
σ_v^2	0.12	6.20	0.01	8.96	-		-	
θ	8.31	7.85	-		-		-	
σ_u/σ_v	-		-		0.40	0.32	0.40	0.32
$\sqrt{\sigma_u^2 + \sigma_v^2}$	-		-		0.13	2.25	0.13	2.25
μ/σ_u	-		-		-		0.00	0.05
N	168		168		168		168	
Log L	113		122		108		108	
Pseudo R^2	0.86		0.91		0.82		0.82	
σ_u^2	0.02		0.01		0.00		0.02	
σ_v^2	0.00		0.00		0.00		0.00	

In general, the parameters of the *best practice* cost function are precisely estimated and the regression statistics indicate that the *best practice* cost function fits the data relatively well. I could not reject the null hypothesis that all slope coefficients are jointly equal to zero at

any reasonable confidence level in all specifications of the cost function. The calculated pseudo R^2 also indicates that a relatively large proportion of the variance in the dependent variable is explained by the *best practice* cost function that I used to estimate the model.[21]

The estimates indicate that the cost function is well behaved in the sense that it is increasing in outputs and input prices in the range of operation. They are also consistent with the postulates of cost minimization. The elasticity of cost with respect to security is greater than unity indicating that cost rises faster than security levels. The coefficients of the quadratic output terms are negative and statistically significant in all specifications. This is an indication that total cost is increasing at a decreasing rate with respect to output. The coefficient of the quadratic wage term is negative and statistically significant in all specifications. The coefficient of the interaction term between the two output indices is negative and statistically significant in all the specifications. There are indications that the combination of large size and high levels of security may moderate the growth of operating costs in Florida prisons.

The coefficients of HP and AGE are positive and highly significant in all specifications of the *best practice* cost function, indicating that health problems of inmates and age of the physical structure of prisons increases the cost of prison services. This is in accord with prior expectations. The coefficient of BLK is negative and significant at $\alpha = .05$ or better in all specifications. There are indications that increasing the relative population of blacks in Florida prisons is associated with decreasing cost. Given the large proportion of the Florida prison inmate population that is black, the negative coefficient estimate is reasonable. An increase in BLK may result in less racial tensions among inmates and hence lower the cost of custodial services. The coefficient of CROWD is negative and significant at $\alpha = .05$ in all specifications. The negative coefficient of CROWD indicates that there is some capacity slack in the Florida prison system. This may be especially true in the face of rapid expansion of prison capacity in Florida during the last decade. It also indicates that prison managers substitute fixed capacity for some variable inputs in order to decrease short run operating cost. The coefficient estimates obtained in this chapter are consistent with those estimated by Schmidt and Witte (1984). They are also consistent with the estimates obtained by Ganley and Cubbin (1992) as well as those obtained by Block and Ulen (1979).

9.5.2 *Cost Inefficiency*

I use the estimates from the cost function to estimate cost inefficiency in Florid prisons. I estimate average cost inefficiency from the various specifications of the cost

function as well as provide prison-specific inefficiency for all 45 prisons using the panel estimates in this subsection. I employ the methodology suggested by Battese and Coelli (1988) and Jondrow, Lovell, Meterov, and Schmidt (1982) in my calculations. Point estimates of the averages of cost inefficiency, as well as their 90% confidence limits, for all prisons and all years are presented in Table 9.3. In addition, I present the minimum and maximum values of the inefficiency indices for all prisons and all years for each specification as well as observation numbers (in parentheses) of the prison so ranked. Table 9.3 indicates that Florida prisons exhibit significant cost inefficiency in the short run. The average cost inefficiency ranges from 12.1% to about 15.7%. I am unable to indicate which portion of this cost inefficiency is due to technical inefficiency and which is due to allocative inefficiency.

Table 9.3

Indices of (In)efficiency[a]

	Exponential	*Panel*	*Half normal*	*Truncated normal*
Mean	0.16	0.14	0.12	0.12
Minimum	0.05	0.05	0.07	0.07
Maximum	0.49	0.37	0.22	0.22
Lower	0.12	0.09	0.04	0.04
Upper	0.27	0.21	0.21	0.21

a *Proportion by which actual cost exceeds the cost frontier.*

The calculated confidence bands for the inefficiency estimates are relatively wide. This, of course, is one of the characteristics of inefficiency estimates obtained from *frontier* functions (Horrace and Schmidt 1996). There is a wide variation in the degree of inefficiency across prisons for specific years. These estimated cost inefficiencies suggest that Florida prisons are not operating at the cost frontier either because of technical inefficiency and/or allocative inefficiency. This implies that more prison services can be produced at the same cost or the same quantity and quality of service can be produced at a lower cost, all things equal.

In addition to differences in average inefficiency indices across specifications, prison specific cost inefficiency vary widely across specifications. The range of cost inefficiency is narrowest for the half normal specification while it is widest for the exponential specification. In general, all specifications tend to agree on clusters of prisons with high, low, or medium inefficiency indices. However, the magnitude of the inefficiency score

for specific prisons differ. For example, all specifications rank observation 125 as the most efficient prison. This observation has an inefficiency index of .05 in the exponential specification and an inefficiency index of .07 in the half normal specification. While all specifications rank observation 149 as the most inefficient, the exponential specification gives it an inefficiency index of .49 while the truncated normal gives it an inefficiency index of .22. The fact that all specifications agree on the ranking should engender some confidence in the estimates.

The discussion of average prison inefficiency in this section has been based on estimates of inefficiency for all prisons for all years. No attempt was made to estimate average inefficiency index for each prison. This implies that a prison that operated at unusually high level of (in)efficiency in a particular year but reverted to its usual pattern will have a larger influence in the estimated average inefficiency than is warranted. This problem is compounded if that particular prison is not represented in the sample in all four years. I therefore present average prison-specific inefficiency indices, based on the panel specification and calculated over the periods of observation for each prison. One should be cautioned, however, that the thinness of these panel could affect the precision of the estimates.

Table 9.4

Prison-specific Average Inefficiency[a]

Prison	Inefficiency (%)	Prison	Inefficiency (%)	Prison	Inefficiency (%)
1	4.3	16	3.2	31	15.8
2	8.6	17	4.1	32	8.3
3	9.2	18	3.6	33	13.8
4	4.3	19	16.5	34	33.0
5	13.4	20	9.6	35	8.6
6	14.7	21	18.6	36	6.5
7	4.5	22	7.8	37	19.6
8	12.1	23	15.7	38	7.1
9	8.3	24	6.4	39	11.2
10	5.3	25	17.3	40	7.8
11	9.0	26	3.9	41	2.5
12	13.7	27	9.9	42	17.8
13	3.9	28	0.5	43	12.7
14	0.5	29	9.8	44	14.0
15	9.7	30	2.9	45	12.2

a Proportion by which cost exceeds the cost frontier.

Estimates of precision-specific inefficiency indices are presented in Table 9.4. The mean inefficiency index for all 45 prisons is 11%, indicating that the average cost of production in Florida prisons is about 11% above the cost frontier. This figure is an unweighted average. The most efficient prison in the sample is prison 41, with an inefficiency index of 2%. Prison 41 is a large, male, medium security prison in the northern rural part of the state. This prison is relatively new, having been opened only in the 1990s. The least efficient prison is prison 34 with an inefficiency index of 33%. This prison is a large, old maximum security prison located in the northeast part of the state. The high degree of inefficiency estimated for this prison may reflect the possibility that maintenance required by the aging infrastructure adds to operating cost. The estimates of cost inefficiency presented here are similar to those estimated by Ganley and Cubbin (1992) who find that the average index of inefficiency in UK prisons is about 12%. These figures suggest that the estimate of cost inefficiency in Florida prisons are not different from those estimated by earlier researchers.

9.5.3 Prison Inefficiency and Privatization

Florida currently has five private prisons in its penal system and more prisons are under privatization considerations. The reason for privatization is the perceived cost efficiency of private prisons relative to those managed by the state. Are private prisons more cost efficient than state-run prisons? If they are, then shifting the inmates from the state to private prisons will improve the overall cost efficiency of the state prison system. There are two private prisons (prison numbers 15 and 25) in my sample. I use information on these two prisons to *speculate* on the relative efficiency of private prisons by comparing the average inefficiency index of the state run prisons to that of the private prisons.

A straight forward way of testing efficiency differences between private prisons and prisons managed by FDOC would be to estimate separate cost functions for the two sets of prisons and compare the cost structures. However, with only two private prisons in the sample, this approach is not possible. An alternative, indirect strategy is to estimate the cost function with and without the two private prisons in the sample and compare the estimates from the two samples. I follow this strategy to investigate whether private prisons are more cost efficient than those managed by the FDOC. I refer to the sample of publicly run prisons as the public sample for lack of a better terminology.

The estimates of the cost functions using the public prison sample are presented in Table A9.1 in the Appendix. Generally the coefficients of the cost function are as expect-

ed and they are precisely estimated. The coefficient estimates from the public sample are remarkably close to those estimated from the full sample. In particular the coefficients of relative wage, outputs and variance composition are very close to those presented in Table 9.2. I cannot reject the null hypothesis that the two sets of estimates (Tables 9.2 and A9.1) are equal.[22] The equality of the two sets of cost estimates may be an indication that the cost structure in private prisons is not different from those managed by the FDOC.

The estimated mean inefficiency indices calculated from the public prison sample are presented in the Appendix (Table A9.2). These estimates are similar to those estimated from the full sample and may indicate that inefficiency in private prisons are not different from those managed by the public sector. I compare the mean inefficiency index of public prisons calculated from the panel specification to an average inefficiency index of the two private prisons calculated from the panel specification. The average inefficiency index estimated for these two prisons is .135 compared to the estimated mean inefficiency index of .11 calculated from the public sample estimates. I find no statistically significant difference in inefficiency between the private prisons and those managed by the state.[23]

Parametric statistical tests are not reliable when the sample size is small as in this case. I therefore use another technique, known as non parametric kernel density estimation, to further explore whether the density distribution of the inefficiency indices of the two private prisons lie outside the bounds of those managed by FDOC.[24] These estimates, with the two private prisons evaluated as if they were public prisons, indicate that this is not the case. This shows again that the private prisons could be treated as part of the public prisons since their cost structure is not different from those of public prisons. Although not definitive, these tests seem to suggest that private prisons are no more cost efficient than those managed by the Florida Department of Corrections.

I conclude from this exercise that private prisons are not more cost efficient than public prisons in Florida. However, given the limitations of the data, this statement is only *speculative*. If the speculation that private prisons are no more cost efficient than those managed by the state is true, then it means that privatization of state prisons is unlikely to lead to any significant cost reductions. Privatization decisions may await further investigation of the relative cost efficiency of privately managed prisons. Of course, it is possible that the objective of privatization is not cost reduction but some other objective such as education, rehabilitation, or vocational training and it is possible that private prisons provide these services more efficiently than state agencies.

Why are there no cost advantages of private prisons in Florida? Hart, Shleifer, and Vishny (1997) argue that when a contract requires renegotiation for major changes to be made, private prisons have no incentive to make changes that will improve quality of service and decrease cost. Although contracts between private prisons and the FCPC give the private prisons a lot of managerial discretion not allowed FDOC prison managers, the contract requires that major changes to facilities and operational guidelines be renegotiated. In addition, FDOC maintains staff in the private prisons to supervise and monitor activities to ensure that guidelines are followed. This restrictive nature of the contract may explain why there is no evidence of efficiency advantage in private prisons relative to publicly run prisons in Florida. Possibly, it may be too early to know whether privatization of prisons confers any advantages in efficiency.

Cost (in)efficiency may not reflect all the differences between private and public prisons. It is possible that private prisons provide a wider range of services at a higher quality level than FDOC. The evidence does not support the idea that the private prisons provide a wider range of services than those provided by FDOC. It is, however, possible that the *quality* of services private prisons provide is superior to those provided by the public sector. Lanza-Kaduce, Parker, and Thomas (1999) argue that recidivism among releasees from two private prisons in Florida is much lower than that of a comparable group from prisons managed by the state. However, they are careful to note that the relatively small sample and the short period of the study (one year after release) makes it impossible to make any definitive statement about the relative efficiency of private versus public prisons in this regard. If privatization of prisons in Florida is intended to achieve this *qualitative* dimension of prison output rather than to achieve cost efficiency then it may be that private prisons are more efficient in this regard. It is not possible to show differences in efficiency on quality with the cost data from just two private prisons. However, given that private prisons *choose* to house only prisoners who are easy to control, lower recidivism rates among their releasees relative to those from the state run prisons may not imply more efficient operations; it may only reflect differences in product mix. Given the available data, a detailed analysis of this subject will have to await future research.

9.6 Conclusion

This chapter uses an unbalanced panel data from 1995 to 1998 and a *best practice* variable cost function to investigate cost inefficiency in the production of custodial and security services in Florida prisons. Using different specifications of the distributions of the inefficient parameter of a *best practice* cost function, I find that there is significant cost inefficiency in the provision of prison services in the short run in Florida prisons. The average inefficiency index estimated is about 11%. This cost inefficiency cannot be apportioned between technical and allocative inefficiency. I also find that prison cost inefficiency varies widely across prisons. The existence of substantial cost inefficiency in prison production implies that Florida prisons can provide more services without substantially increasing cost by improving cost efficiency. Though based on very sparse information on private prisons, there is no evidence that private prisons are more cost efficient than those run by the Florida Department of Corrections. If this is the case, privatization of prisons will not confer any cost advantages to the state.

The results of this chapter, although interesting, should be treated as exploratory rather than definitive for various reasons. First, the assumption that prison managers minimize cost is tenuous, at best. Second, I am unable to control for the regulatory environment within which prisons are operated. Third, my measure of prison output is not comprehensive enough to capture all aspects of prison production. Given that prisons produce a myriad of services, it may be desirable to estimate a multi product cost function that includes all the services produced by prisons and control for all inputs instead of the cost function I estimated which accounts for only a partial list of the output produced by prisons. In spite of these weaknesses, I hope that this chapter will encourage more research on the topic.

APPENDIX

A9.1 STOCHASTIC FRONTIER COST FUNCTION

In this appendix, I briefly sketch out how I arrived at the general cost function (1). I approach it in a very non-technical way. Readers who are interested in the formal exposition of this approach should refer to Chapter 2 of the book.

My working assumption is that the prison manager is provided with a given prison capacity and a given number of prison inmates and charged with choosing variable inputs to minimize the cost of housing these inmates at a given level of security. This implies that a prison is seen as a multi-product firm producing custodial services and security. I assume that the underlying prison production relation is given as:

$$F(Qe^{\gamma}, X) = 0 \qquad \gamma \geq 0 \tag{A9.1}$$

where Q is a vector of actual outputs, X is a vector inputs including fixed inputs and γ is an inefficiency parameter. For an inefficient prison, the actual output produced is less than the maximum output Qe^{γ} that can be produced given technology and the input vector. γ can be interpreted as output inefficiency as in Kumbhakhar (1996).

The actual cost of production is given as:

$$C = WX'\beta e^{\delta} \geq 0 \tag{A9.2}$$

where W is a vector of input prices, X is a vector of inputs and δ is addition to cost due to allocative inefficiency. I note that when δ is equal to zero, the cost of production will be minimized if there is no *productive* inefficiency. Minimizing the cost of production subject to the prison production relation in (1) and market determined input prices, the resulting cost function depends on actual output (Qe^{γ}), input price vector (W), fixed capital (K), and the allocative inefficiency parameter (δ). One can combine the two inefficiency parameters to obtain a composite cost inefficiency index (u) which a combination of productive and allocative inefficiency. The general form of the cost function can be written as:

$$C = c(Q, K, W, u) \qquad u \geq 0 \tag{A9.3}$$

The dependence of the cost function on the inefficiency parameter makes this cost a stochastic one.

I specify the general form short run cost function as:

$$VC = C(Q, K, W) + \varepsilon \qquad \varepsilon = v + u,\ u \geq 0 \tag{A9.4}$$

where *VC* is total variable cost, K is fixed capital, Q is a vector of outputs, *W* is a vector of input prices, ε is a composite error term, *u* is the inefficiency component of the error term that reflect cost inefficiency, while *v* is white noise. Any well behaved total cost function is non decreasing in outputs and input prices. I therefore expect the variable cost function to be increasing in Q and W.[25]

The prison manager is assumed to combine two purchased variable inputs – labor and energy – with the fixed prison capacity to efficiently produce prison services. The cost function will also depend on the environment within which the prison management operates. I approximate the short run variable cost function by the Generalized Cobb-Douglas cost function.[26]

$$\ln(VC/P_e) = \beta_0 + a_1 \ln POP + \beta_1 \ln(VC/P_e) + \alpha_2 \ln S + .5\alpha_{11}(\ln POP)^2 + .5\cdot\beta_{11}(\ln(W/P_e))^2 + \delta_1 BLK + \delta_2 HP + \delta_3 AGE + \delta_4 CROWD + \varepsilon \quad (A9.5)$$

where POP is inmate population, S is the level of security, W is the wage rate, P_e is the price of energy, BLK is the ratio of black inmates to black prison officers, HP is health care personnel per prisoner, AGE is the age of a prison, CROWD indexes prison capacity, α, β, γ, δ, are coefficients to be estimated, and all other variables are as defined above. Since any cost function is homogenous of degree one in input prices, I avoid perfect collinearity in estimation by using the price of energy as the nummeraire. I note that this cost function reduces to the simple deterministic Cobb-Douglas variety if all the second order coefficients are equal to zero and u_i is zero for all observations.

A9.2 Estimation Results for Public Prisons

Table A9.1

Coefficient Estimates of the Stochastic Frontier Cost Function, Public Prisons

	Exponential		Panel		Half Normal		Truncated Normal	
	coefficient	*t-value*[a]	*coefficient*	*t-value*[a]	*coefficient*	*t-value*[a]	*coefficient*	*t-value*[a]
Constant	-2.38	1.11	-8.54	2.82	-4.09	1.46	-4.09	1.33
W/P_e	0.72	13.39	0.70	11.62	0.68	13.63	0.68	13.63
Q	0.71	9.69	0.84	4.42	0.92	12.96	0.92	12.89
S	0.86	3.21	1.32	3.03	1.23	3.16	1.23	3.12
QxS	-0.39	2.76	-0.60	2.80	-0.37	2.78	-0.37	2.78
Q^2	-0.31	5.93	-0.19	2.76	-0.09	1.83	-0.09	1.82
W^2/P_e	-0.01	2.02	-0.01	1.62	-0.01	2.27	-0.01	2.27
BLK	-0.60	3.15	-1.24	1.61	-1.45	2.79	-1.45	2.79
HP	0.09	5.72	0.11	6.29	0.09	5.32	0.09	5.10
AGE	0.07	1.99	0.02	1.29	0.09	1.83	0.09	1.82
CROWD	-0.08	2.33	-0.10	2.89	-0.06	2.13	-0.06	2.13
σ_u^2/σ_v^2	-		4.00	1.79	-		-	
σ_v^2	0.12	5.58	0.01	7.92	-		-	
θ	6.93	7.68	-		-		-	
σ_u/σ_v	-		-		0.51	0.68	0.51	0.09
$\sqrt{(\sigma_u^2+\sigma_v^2)}$	-		-		0.14	1.74	0.14	1.70
μ/σ_u	-		-		-		0.00	0.01
N	161		161		161		161	
Log L	116		119		101		101	
Pseudo R^2	0.86		0.87		0.82		0.82	
σ_u^2	0.00		0.01		0.02		0.02	
σ_v^2	0.02		0.03		0.00		0.00	

a *Absolute value of asymptotic 't' statistics.*

Table A9.2

INDICES OF (IN)EFFICIENCY, PUBLIC PRISONS[a]

	Exponential	Panel	Half Normal	Truncated Normal
Mean	0.15	0.15	0.13	0.13
Minimum	0.02	0.02	0.03	0.03
Maximum	0.68	0.48	0.22	0.22
Lower	0.05	0.05	0.04	0.04
Upper	0.27	0.26	0.25	0.25

a Proportion by which actual cost exceeds the cost frontier.

Table A9.3

PRISON-SPECIFIC AVERAGE INEFFICIENCY, PUBLIC PRISONS[a]

Prison	Inefficiency (%)	Prison	Inefficiency (%)	Prison	Inefficiency (%)
1	3.2	16	24.9	31	44.9
2	10.9	17	3.8	32	5.1
3	8.9	18	4.1	33	12.4
4	7.7	19	4.0	34	48.4
5	10.7	20	10.3	35	7.4
6	8.6	21	9.9	36	7.6
7	10.1	22	10.9	37	26.3
8	7.7	23	6.8	38	5.9
9	9.8	24	11.7	39	13.5
10	6.7	25	NA	40	5.0
11	8.4	26	6.3	41	12.2
12	21.5	27	13.1	42	14.8
13	6.8	28	1.1	43	18.1
14	3.0	29	8.8	44	11.3
15	NA	30	4.3	45	13.6

a Proportion by which cost exceeds the cost frontier.

Notes

[1] Earlier versions of the chapter were presented at the annual meetings of the Midwest Economics Association in Chicago, IL and the Southern Economics Association in Baltimore, MD. I thank Eve Cantral and Julie Harrington of the Florida Department of Corrections for data collection assistance and Colleen Davis for able research assistance. Bill Frances provided helpful comments on an earlier draft. The usual disclaimer applies here.

[2] In technical language, this *best practice* cost function is referred to as *frontier* cost function. See Chapter 2 of this book for a technical analysis of frontier functions.

[3] See a discussion of the problems associated with measuring prison output in Section 9.3 below.

[4] A man was recently sentenced to 26 years imprisonment in California for stealing four chocolate chip cookies under such laws. Source: Associate Press Newswire Service, March 15, 1998 as reported in *The Oracle* (University of South Florida Campus Newspaper) of March 16, 1998.

[5] See the *Annual Reports*, of Florida Department of Corrections.

[6] Avio (1998) provides an excellent survey of the economics of prisons.

[7] Prison privatization as used here implies that the government contracts with a private firm to run *all* aspects of the prison for the state. It does not refer to contracting out for specific services such as food, medicines or transportation.

[8] See Avio (1998) for an excellent review of the literature on prison privatization.

[9] The Florida Department of Corrections indexes prisons security levels as follows: Level 1 is minimum custody, Levels 2 & 3 are medium custody, Level 4, 5, & 6 are closed custody, and Level 7 is maximum custody.

[10] The cost of operating private prisons do include some administrative and book keeping services provided by the state prison system.

[11] The Wackenhut Corporation manages two adult medium security prisons (South Bay and Moore Haven Correctional facilities) while the Corrections Corporation of America manages two adult prisons (Bay and Gadsden Correctional institutions) and one youthful offender institution (Lake City Correctional Facility).

[12] The private prisons are allowed to be selective in the choice of inmates and are allowed to send inmates with behavioral problems back to the state sector. They also have air-conditioned dormitories and are allowed to purchase TV sets and recreational equipment for inmates use: items that are prohibited by law in the state sector.

[13] Inspection of individual prison budgets indicate that some prisons spent less money than was allocated to them for operating expenses in some years. This may be an indication of cost minimizing behavior.

[14] For a technical exposition of this approach to calculating cost efficiency, see Jondrow, Lovell, Meterov, and Schmidt (1982), Coelli (1995), Cornwall and Schmidt (1996), and Chapter 2 of this volume.

[15] A very simple and non technical derivation of the cost function is provided in the Appendix. Those interested in more involved discussion of frontier functions should consult Chapter 2 of this book.

[16] The 31% figure was provided to us by the staff of FDOC as the appropriate rate for fringe benefits.

[17] This measurement is different from the official measure of capacity set by legislative fiat and revised upwards according as the legislature feels desirable.

[18] In a fast growing state like Florida, the relative cost of living changes very drastically on an annual basis. For example in 1995, the most expensive city to live in was Ft. Lauderdale; in 1997, that "honor" went to Gainesville. Wages of prison guards are not adjusted for changes in the relative cost of living.

[19] The Florida CPI is designed to measure variations in the cost of living across the state. Its base is a population weighted average cost of living for the state of Florida.

[20] The number of major adult institutions included in the sample (45 out of 50) and the number of years covered for each prison (minimum of 2 and maximum of 4) were dictated by the availability and completeness of data. Florida did not keep detailed budget data for individual prisons until recently.

[21] A likelihood ratio test rejects the simple Cobb-Douglas specification of the cost function at $\alpha = .01$. The calculated χ^2 statistics are 38.89, 48.29, 37.89, and 37.88 for the exponential, panel, half normal, and truncated normal respectively.

[22] A Hausman **m** test to test the equality of the two sets of estimates could not reject the hypothesis that the two sets of estimates are equal at $\alpha = .01$. He calculated χ^2 statistics are 1.821, 1.032, 0.8921, and 0.8629 for the exponential, panel, half normal and truncated normal specifications respectively.

[23] A test of equality between the mean inefficiency of prisons managed by private corporations and those managed by the state produced a "t"statistic of -.2628.

[24] Given the sample size for private prisons, it appears there is no test that can allow any strong statements to be made about the relative efficiency of private versus public prisons.

[25] It is reasonable to expect the cost of housing prisoners in maximum security prisons to be higher than in a minimum security prison.

[26] Because of small variation in input prices across prisons as well as the small sample size, a full translog specification of the cost function could not be estimated because of high degree of collinearity.

References

Avio, K. 1998. Economics of Prison. *European Journal of Law and Economics* 6: 143-175.

Avio, K. 1993. Remuneration Regimes for Private Prisons. *International Review of Law and Economics* 13: 35-45.

Battese, G., and T. Coelli. 1988. Prediction of Firm Level Efficiencies with a Generalized Frontier Production Function and Panel Data. *Journal of Econometrics* 38: 387-399.

Block, M., and T. Ulen. 1979. Cost Functions for Correctional Institutions. In: Gray, C. (ed.), *The Cost of Crime*. Beverly Hills CA: Sage Publications 187-209.

Burns, P., and T.G. Weyman-Jones. 1996. Cost Functions and Efficiency in Electricity Distribution: A Stochastic Frontier Approach. *Bulletin of Economic Research* 48 (1): 41-64.

Coelli, T. 1995. Recent Developments in Frontier Modelling and Efficiency Measurement. *Australian Journal of Agricultural Economics* 39 (3): 219-245.

Cornwall, C. and P. Schmidt. 1996. Production Frontiers and Efficiency Measurement. In: Matyas, L., and P. Severste (eds.), *The Econometrics of Panel Data: A Handbook of the Theory with Applications*. 2nd Edition. Dordrecht, Netherlands: Kluwer Academic Press.

DiIulio, J. 1988. What is Wrong with Private Prisons? *Public Interest* 92: 66-83.

Edwards, G. 1996. Public Crime, Private Punishment: Prison Privatization in Queensland. *International Journal of Social Economics* 23: 391-408.

Filippini, M., and P. Prioni. 1994. Is Scale and Cost Inefficiency in the Swiss Bus Industry a Regulatory Problem? Evidence from a Stochastic Cost Frontier Approach. *Journal of the Economics of Business* 1 (2): 219-231.

Florida Department of Corrections. *Annual Reports* various years. Tallahassee: Florida.

Ganley, J. and J. Cubbin. 1992. *Public Sector Efficiency Measurement: Application of Data Envelopment Analysis*. Amsterdam: North Holland.

Gyimah-Brempong, K. 1989. Production of Public Safety: Are Socioeconomic Characteristics of Local Communities Important Factors? *Journal of Applied Econometrics* 4: 57-71.

Hart, O., A. Shleifer, and R. Vishny. 1997. The Proper Scope of Government: Theory and Application to Prisons. *Quarterly Journal of Economics* 112 (4): 1127-1161.

Horrace, W.C., and P. Schmidt. 1996. Confidence Estimates for Efficiency Estimates from Stochastic Frontier Models. *Journal of Productivity Analysis* 7: 257-282.

Jondrow, J., C.A.K. Lovell, I.S. Meterov, and P. Schmidt. 1982. On the Estimation of Technical Inefficiency in the Stochastic Frontier Production Model. *Journal of Econometrics* 19: 233-238.

Kumbhakar, S.C. 1996. Efficiency Measures of Multiple Output and Multiple Inputs. *Journal of Productivity Analysis* 7: 225-255.

Lanza-Kaduce, L., K.F. Parker, and C.W. Thomas. 1999. A Comparative Recidivism Analysis of Releasees from Private and Public Prisons. *Crime and Delinquency* 45 (1): 28-47.

Logan, C. 1987. The Propriety of Proprietary Prisons. *Federal Probation* 51: 35-40.

Schmidt, P., and A.D. Witte. 1984. *An Economic Analysis of Crime and Justice: Theory, Methods, and Applications.* Orlando, FL: Academic Press.

Sellers, M. 1993. Private and Public Prisons: A Comparison of Costs, Programs and Facilities. *International Journal of Offender Therapy and Comparative Criminology* 33: 241-256.

Trumbull, W. and A.D. Witte. 1981-1982. Determinants of the Costs of Operating Large-Scale Prisons with Implications for the Cost Correctional Standards. *Law and Society Review* 16: 115-137.

Witte, A.D. et al. 1979. *Empirical Estimations of Correctional Cost Functions.* Report to the National Institute of Law Enforcement and Criminal Justice. Washington, D.C.

10 The Revenue Approach to Dutch Police Departments

Frank P. van Tulder

10.1 Introduction

As in many countries, law enforcement is high on the Dutch political agenda. Till some years ago crime figures showed steadily rises. Funds to police and the judicial system were extended rapidly. The volume of inputs of the police increased with almost 50% from 1980 up to 1997. Prison population rose by a factor of three in the same period. This expansion took place in a period in which there was general agreement about the need to curb public expenditure. Efficiency in the public sector became an issue. Declining solution rates and increasing backlogs within the judicial system increased doubts about the efficiency of the law enforcement system. Plans to restructure Dutch police has been in existence for a long time, but now the time was ready for it. A major restructuring operation took place, which had improved efficiency as one of its main goals.

Knowledge about production structure and determinants of productivity and efficiency can help to shape policy measures in this field. Such knowledge is often lacking or sketchy. The analysis presented in this chapter intends to improve the knowledge about the production structure of Dutch police.[1] The focus will be on general characteristics of the production structure, with special attention paid to economics of scale. The model used tries to fit, as well as possible, the institutional structure of Dutch police and pays attention to the various types of outputs of the police.[2] An analysis of the consequences of this restructuring will be part of the analysis presented.

The content of the rest of this chapter is as follows. Section 10.2 sketches the tasks, the institutional structure and the allocation of funds of the Dutch police. Section 10.3 presents some important elements of the international economic literature on the production structure of the police and elaborates on some methodological points. Section

10.4 introduces the approach chosen here. Section 10.5 gives the empirical estimation of the production structure and some characteristics of it. It also shows results of some simulations on the effects of reallocating and restructuring operations on productivity. Finally, Section 10.6 gives summary and conclusions.

10.2 DUTCH POLICE: TASKS, ORGANIZATION AND FUNDING

Dutch police has many tasks. According to the law its task is to maintain the legal order by acts and to give assistance to those who need this. In practice, an important part of the police work, in the Netherlands as elsewhere, has to do with crime and law enforcement in all its aspects: helping victims, recording crimes, solving, arresting, sending official reports to the public prosecutor, and so on. Another important part of the work has to do with traffic: dealing with accidents and processing minor traffic offences. A third part is related to maintaining public order and a fourth and softest part relates to assistance in a wide variety of circumstances.

Until 1993/1994 there were two types of police forces patrolling in the Netherlands.[3] The *municipal police* controlled 148 urban municipalities, each having their own police department. These municipal police departments ranged in size from approximately 40 to 5,000 labor years. The average department had an input of approximately 180 labor years. In addition there was a police force (Rijkspolitie) for the rural areas, divided into 17 districts. Table 10.1 presents a rough indication of the growth of Dutch police inputs and outputs in the period 1980-1997 (in indices). Because the analysis in the rest of this chapter is based on data from 1983 and 1986, some special attention is paid to the level of inputs in the year 1986.

The volume of police inputs rose from 1980 to 1997 by 48%. The fastest growth took place after 1990. In terms of labor years the growth is more modest: 17%. In the eighties, growth of labor input was influenced by a shortening of working hours, for which is corrected. The output of police, constructed as a weighted index of the numbers of recorded and solved crimes, processed minor offenses, traffic accidents and (estimated) cases of assistance, grew with 36%, mainly at the end of the period.[4] In 1986 approximately two thirds of police inputs was allocated to the municipal police departments, which are all included in the analysis. The average costs per labor year are somewhat higher with the municipal police than for the other police units.

Table 10.1

DUTCH POLICE: INPUTS AND OUTPUTS 1980-1997

	Volume of inputs	Labor years (x 1,000)[e]	Output index[f]
1980 (index, 1980 = 100)	100	100	100
1985 (index, 1980 = 100)	110	106	112
1990 (index, 1980 = 100)	119	106	116
1997 (index, 1980 = 100)	148	117	136
Municipal police - 1986 (x million NLG)	2,070[a]	25.8	
Rural police - 1986 (x million NLG)[b]	970[c]	13.1	
Central police departments - 1986 (x million NLG)[c, d]	120	1.5	
Total police - 1986	3,150	40.4	

a Including part of housing costs (for municipal police).
b Total of a division serving in rural areas (divided in 17 districts) and a number of special nationwide services, like the water police and a traffic controling division.
c Excluding housing costs.
d Includes some central specialized services, like the criminal investigation division and police schools.
e Corrected for shortening of labor hours.
f Weighted average of numbers of recorded crimes, solved crimes, processed minor offenses, traffic accidents and (estimated) cases of assistance (see Goudriaan, Van Tulder, Blank, Van der Torre, and Kuhry 1989: 150).

As already said, a mayor restructuring operation of Dutch police took place between the years 1993-1994. The two different forces with their departments merged into 25 regional police departments, together covering the whole country. The new regional departments have an average size of approximately 1,500 labor years. This restructuring operation had major consequences, both in terms of organization and scale of production. An important aim of the restructuring was to improve the efficiency of the police by abolishing the patchwork of many small departments beside each other and creating regionally homogenous new ones. Later in this chapter we will elaborate on the rationale behind this view.

The management structure of Dutch police was complicated and is even more so after the restructuring. In the first place, there is the so called triangular management college, consisting of the police chief, the mayor of the city and the chief regional public prosecutor. The mayor is head of the police department in its municipality and responsible for maintaining public order, the public prosecutor for maintaining law enforcement. There may be conflicts between these aims. After reorganization regional police departments have to do business with many mayors. The mayor of the main munici-

pality in the region is the head of regional police department. Conflicting interests and the diversity of parties involved here may give problems. A further complication is that the regional police partition does not fit the democratic regional partition, so control by democratically elected bodies is difficult. These management and control aspects are important and may have consequences for efficiency.[5] However, they will here not be dealt with explicitly.

Allocation of funds to Dutch police departments is mainly in hands of the central government. The allocation of central funds to individual police departments is determined by general formulas based mainly on area and population characteristics. In the last decade, regional differences in crime rates also played an indirect role in these formulas, as did the allocation of separate funds for building and IT-activities. Municipal authorities sometimes give some extra funds to the local police, but these funds are relatively small and mostly meant for special purposes (e.g. new uniforms). For our analysis it is important to note that the budget allocated to departments is mainly lump sum and fixed (given) for the department.

10.3 Methodology

10.3.1 *Existing Literature*

The police have been a frequent object of study in the 'economics of crime' literature, starting with Becker (1968) and with the mainstream in the seventies. See Pyle (1983) for an overview. Nearly all authors define the production in terms of direct output (D-output), not in terms of consequences (C-output). Output measurement is often limited to crime solutions. There is much more data available in this area than for the "softer" forms of policing, say in traffic policing or assisting the public. The emphasis on combating crime fits in the tradition of the 'economics of crime', in which most of this research takes place.

Early studies in this field were inspired by production function literature. However, Ehrlich (1973) and other pioneers in this area (e.g. Carr-Hill and Stern 1973, 1979) diverge from the traditional production function approach in several ways. In the first place, they use the solution rate rather than the absolute number of crimes as the output indicator. This indicator is a measure of the probability of being caught which is a central variable in 'economics of crime' literature. Secondly, the number of recorded crimes is seen as one of the explanatory variables, as it were a type of input alongside the traditional inputs such as labor etc. The argument is that a higher number of crimes

with given police inputs will result in a lower solution rate. We will return to the issue of the role of recorded crimes in productivity analyzes later (Section 10.3.2).

A pressing problem, especially for all authors using production functions, is the heterogeneous nature of police output. A production function can only handle one type of output. Even in analyzes restricted to solving crimes activities this is a problem: aggregating murders and petty thefts is unsatisfactory. Disaggregation of outputs brings however new problems: weighting the various types into one measure of outputs or incorporating various types of outputs in the productivity analysis. Votey and Phillips (1972) solve this problem by introducing two simplifications: (1) the police production process can be divided into different parts which each produce one type of output (technically speaking: these parts are non-joint) and (2) the fraction of police inputs devoted to different production processes is constant across the observation units. Other authors (e.g. Walzer 1972) construct one measure of output by weighting different types of output by estimates of needed inputs, based on time keeping research.

Later studies reflect the new theoretical developments in the production literature (see Chapter 2). Cost functions and indirect approaches of production analysis are introduced. This makes it possible to deal with more types of output. In a simple cost function approach (Goudriaan,Van Tulder, Blank, Van der Torre, and Kuhry 1989), the question is which costs are essential to produce outputs in a predetermined or fixed quantity. Because in the real world some police outputs are certainly not predetermined, some authors have tried to relax this assumption.

Darrough and Heineke (1978, 1979) introduce a 'value maximization' model, inspired by profit maximization models in the market sector. In this, they formulate and test the hypothesis that police departments maximize the added value: the difference between value of outputs and costs. Outputs are defined as number of solved crimes of different types. The 'prices' of different outputs in the sphere of property crimes are based on the average values of stolen goods. The value maximization hypothesis cannot be maintained when it is tested, however.

Some authors have tried to measure C-output of police directly or indirectly. Popp and Sebold (1972) formulate a quasi cost function in which unrecovered losses of crime play a role. This a sort of measure of the inverse of the C-output of police.[6] Chapman, Hirsch, and Sonenblum (1975) link in a cross section analysis population characteristics to an 'normal' level of crime. Using this relation in a time series they estimate the preventive effect of police activities by the residuals over time. Their approach is very bold, somewhat old fashioned now and methodologically not really convincing.

Nevertheless, it must be remembered that the total reduction of crime achieved by police activities is really the holy grail in this field. Gyimah-Brempong (1989) and Finney (1997) replace the output measures in terms of solutions by a 'security indicator', the inverse of the number of crimes. This indicator is related not only to the inputs of police but also to the socioeconomic characteristics of the regions involved. Thus, the difference with an approach in which the crime level is linked directly to regional characteristics and the police inputs (Swimmer 1974, Van Tulder 1992) is small. These analyzes give however little or no information about the technical properties of the police production structure. In our analysis we will return to the measurement of prevention.

10.3.2 *The Role of Recorded Crimes*

It is curious to note that there are at least four theoretical approaches to the role of the recorded crimes in production analyzes of the police. All four can be found in literature. Recorded crimes can be seen as:

1. an *input*, like labor and raw materials. According to this line of thought it is essential that no crimes can be solved when there are no crimes committed and recorded (the 'fisheries argument'). Examples are: Ehrlich (1973) and Votey and Phillips (1972). The observation in police practice that with an important number of crimes (especially in the sphere of violence) suspects are immediately known gives ground for this argument[7];
2. a *throughput*, in other words, neither an input or an output. Darrough and Heineke (1978: 262) rightly state that it is questionable whether the interaction between number of recorded and solved crimes exists on the actual level experienced in practice. They conclude that the studies of Ehrlich and Votey and Phillips do not give a firm affirmation in this respect;
3. an *output* with *positive* value attached to it (a 'good'). In practice, the recording of crimes is certainly a task of police, so victims often require it as proof for their insurance company. In practice one sees more general value attached to activities in this field, e.g. to produce statistical information to policy makers, et cetera. Therefore Goudriaan, Van Tulder, Blank, Van der Torre, and Kuhry (1989) take this approach;
4. an *output* with *negative* value attached to it (a 'bad'). This approach concentrates on the effective output of police. Every crime not committed is seen to be the direct esult of crime prevention work carried out by the police. This approach is found in the work of Chapman, Hirsch, and Sonenblum (1975), mentioned earlier.

These different approaches will be referred to later in this chapter.

10.3.3 *Theoretical Framework of this Analysis*

We have attempted to formulate a model of police production structure which reflects the heterogeneous nature of police output on the one hand and the institutional structure and the actual decision making process at the other. Police departments are faced with a fixed budget, which is exogenously determined (see Section 10.2). With this budget and the available technology, they produce different types of output. The quantity of some outputs can be seen as fixed.[8] The number of reported crimes (excluding victimless crimes), traffic accidents and cases of assistance) fall into this category.[9] With other outputs, like the solving of crimes and the processing of minor offences, the produced quantities depend on the efforts and priorities of the police departments themselves, so are endogenous in the model. The departments implicitly attach values, in economic terms 'shadow prices', to different types of endogenous outputs. In this analysis, the term 'revenue' is therefore synonymous with 'weighted endogenous outputs'.

This allows us to adopt the revenue function approach in production analysis literature (see Section 2.2). This approach is well established in economic literature (McFadden 1978, Lau 1978, Cornes 1992), but application in this field is rather new. The revenue function used here relates the revenue to the other variables describing the police department, i.e. the budget and the fixed outputs. When revenue is maximized there is a dual relation between the characteristics of the revenue function and the characteristics of the production structure. Therefore the revenue function can be used to draw conclusions about features of the production structure, like scale elasticities, marginal costs et cetera.[10] An advantage of a revenue function approach is that there is no need to discriminate between exogenous inputs and outputs in advance. This is important in light of the discussion in Section 10.3.2 about the unclear role of recorded crimes in this respect.

Police departments are, then, regarded as producers which, given their fixed budget and given the fixed quantity of some outputs, maximize the value of their endogenous outputs (the 'revenues') according to a scheme of shadow prices (their value system). Essential in the analysis are the shadow prices which the police attaches to different types of output. Direct market prices are of course not available here. Darrough and Heineke (1978) use the average value of stolen goods as a measure for the shadow price of a solution of different types of crimes. This approach has at least two problems: (1) it can only be used for property crimes, so is only partial and (2) it is doubtful whether the valuation by the police corresponds with these values of stolen goods (see e.g. Pyle and Deadman 1980).

Anderson (1974) points out that the functioning of the police ought to be in accordance with the rest of law enforcement. The output of the police is intermediate within the 'chain' of law enforcement activities. What is the value of solutions of certain type of crime and bringing suspects to the Public Prosecutor, when the Public Prosecutor decides not to pursue the suspects because of other priorities? This means that the value the police attaches (or at least ought to attach) to the solution of different types of crime and the processing of offenses is consistent with the way these are handled in the rest of the judicial system. As data on the prosecution and sanctioning process are available, part of the analysis is a test whether there is a certain type of consistency in valuing different types of crimes and offenses throughout the law enforcement system. Institutionally, this consistency is pursued by the influence of the Public Prosecutor on the police (see Section 10.2). Thus one tries to minimize the 'loss' within the law enforcement chain.[11] Research on the Dutch criminal investigation process shows that both the seriousness of crimes as the expected readiness of the Public Prosecutor to prosecute the case are relevant criteria to set priorities. These priorities may be dependent on local circumstances and characteristics of the offenders.

A standard assumption in the framework of a revenue function is that the producers are 'price-takers', i.e. prices are fixed for them. But the shadow prices defined here, i.e. the sanction levels are influenced by and thus endogenous for the law enforcement system. Nevertheless, in our concept these prosecution and sanction levels are primarily the result of how serious the crime is perceived to be at local level and are fixed conceptually before the outputs are produced. In practice, standard criteria are often used to set priorities in the criminal investigation process during a longer period.

There are regional differences within homogenous group of crimes at both prosecution and sanction levels as well as differences in the "relative value"of crimes. These differences are used in the analysis, but it should be noted that these variations may be due to differences in the quality of the underlying criminal investigation activities of the police. E.g. the information given by different police departments to the Public Prosecutor may differ in quality, thus making convictions more or less easy. In that sense the value attached to the outputs may reflect both a quality aspect and a real price aspect. We generalize the model in question by assuming that a fixed part of the differences in sanction levels is due to differences in quality and the other remaining part is really a shadow price aspect. In the following, shadow prices of the endogenous outputs are based on the sanction levels, measured by the average *severity of punishment* in cases in which punishment took place. Punishment can be a settlement with the Public Prosecutor or a conviction by a judge.[12] Appendix 10.1 presents the formal representation and the type of tests of the plausibility of the model.

10.3.4 *The Definition of Outputs*

The functions of Dutch police as described in law (see Section 10.2) are defined in terms of *tasks* or direct outputs. No mention is made of the ultimate *aims* or consequences to pursue. Thus in literature a wide variety of these, often in vague terms, can be found. In this analysis we define a number of types of direct outputs without further mentioning consequences.

Four types of fixed outputs are used in the empirical analysis:

1. the number of recorded serious crimes, i.e. the aggregate of violent crimes and breaking and entering offenses;
2. the number of other recorded crimes, excluding those types which are dependent almost or totally on own police tracing activities. This means that cases of drunk-driving and drug trafficking are excluded[13];
3. the number of traffic accidents with personal injury. These more serious accidents are always handled by the police. Systematic data on the handling of less serious accidents are not available[14];
4. the number of cases of general assistance, including public order maintenance activities. This includes a wide variety of activities, e.g. mediation between quarreling neighbors, pacifying pub fights, delivery of first aid, collecting returning lost property. Data on these "softer" activities are scarce. Two population surveys are used to approximate the size of these activities. This means that the assistance of companies and institutions is not registered. Requests for this type of assistance are indeed exogenous for the police. In reality, police can choose to a certain extent whether or not to react to these requests. Such empirical data that exists reveal that there are not very large differences between police departments in this respect. [15]

Four types of endogenous outputs are used in the empirical analysis:

1. solutions of serious crimes, i.e. violent crimes, breaking and entering offenses and drug trafficking;
2. solutions (approximately equals recordings) of cases of drunk driving;
3. solutions of other types of crime;
4. number of processed minor offenses (mostly traffic offenses).

The differentiation of types of crime is rather basic, but further disaggregation proved not to be fruitful with the data at hand.

10.4 Data

The data for analysis relate to the 148 municipal police departments operating in the larger Dutch municipalities in 1983 and 1986. Table 10.2 presents some descriptive statistics. The shadow prices are based on the average severity of punishment with the various types of endogenous outputs (see Section 10.3.3).

Table 10.2

Descriptive Statistics Departments of Police 1983 and 1986 (N = 274)

	Symbol	Mean	Standard Deviation	Minimum	Maximum
Fixed outputs (quantities)					
Recorded serious crimes	q_{r1}	2,480	7,650	62	88,900
Recorded other crimes	q_{r2}	3,030	5,280	130	47,900
Traffic accidents	q_{r3}	1,490	2,520	140	23,800
Other cases of assistance	q_{r4}	6,010	9,730	328	77,100
Endogenous outputs (quant.)					
Solved serious crimes	q_{m1}	490	997	18	7,800
Solved cases of drunk driving	q_{m2}	184	393	5	3,600
Solved other crimes	q_{m3}	707	1,300	30	9,720
Processed minor offenses (x 1,000)	q_{m4}	14.7	38.2	0.6	394
Shadow prices (NLG)					
Solved serious crimes	p_{m1}	3,480	2,850	611	23,000
Solved cases of drunk driving	p_{m2}	1,280	212	837	2,510
Solved other crimes	p_{m3}	898	545	282	4,260
Processed minor offenses	p_{m4}	70	11	48	105
Revenue (million NLG)[a]	P	4.5	13.3	0.2	137
Revenue share q_{m1} (in %)[a]	P_1	43	17	5	85
Revenue share q_{m2} (in %)[a]	P_2	9	6	1	34
Revenue share q_{m3} (in %)[a]	P_3	19	10	3	67
Revenue share q_{m4} (in %)[a]	P_4	28	13	2	71
Budget (prices of 1986, x million NLG)	B	12.6	31.4	2.9	260.0
Patrolling area (x 1,000 ha.)		6.7	7.2	0.6	46.6
Population size (x 1,000 inhabitants)		63.3	85.6	12.0	687.4

a With shadow prices according to conditional severity of punishment (x billion NLG; 1 day prison sentence = NLG 50).

Table 10.2 shows that the minor offenses and cases of assistance are the most important parts in the production of the Dutch police departments. Cases of driving under the influence are relatively minor in numbers. Differences between minimum and maximum values and standard deviations are large. The shadow price variables have large variances too. It is important to note that there is also large variation in revenue shares of the various outputs. Further analysis shows that the distribution of all variables is skew. E.g. the budget of the four largest police departments took nearly 40% of the aggregate budget. Appendix A10.2 presents some additional descriptive results.

10.5 Empirical Results

10.5.1 *The Estimated Revenue Function*

As said, the assumption is that the shadow prices used by the police are measured by the mean severity of punishment. In testing, this hypothesis can be maintained and gives better results than some possible alternative measures.[16] Of course, the results give no certainty that police departments actually maximize revenue according to shadow prices in this scheme. It is fair to say that further testing does not show really convincing results. More throughgoing analysis shows that the effects of varying shadow prices on the output mix is actually very limited. The elasticities measuring the effects of 1% (shadow) price rise on output mix are mostly not significantly different from zero.[17] This may be due to the characteristics of the production structure: exogenous factors and rigidities in allocation may be more important. Another possible explanation is that the real shadow prices of police departments do have not much to do with the prosecution and sanction levels or at least, not in the way which is assumed here. Nevertheless the model is not rejected statistically and is therefore the base for further analysis here.

The explanation that exogenous factors are important is supported by the results of the revenue function. Table 10.3 presents the main results, in terms of elasticities for the average municipal police department. A complete version of the estimated revenue function is presented in Appendix 10.2.

The revenue function shows that shadow prices of serious crimes solutions have the most effect on total revenue (elasticity of 49%), with shadow prices of minor offenses at second place (elasticity of 26%). It is noteworthy that none of the fixed (exogenous) outputs has a significant negative contribution to police revenue. On the contrary: both types of recorded crimes clearly contribute positively to the revenue. This sheds new

light on the discussion in Section 10.3.2, and suggests that it is more appropriate to consider the recorded crimes as an input rather than an output in police production.

Table 10.3

Revenue Function Municipal Police Departments, 1983 and 1986 (N = 274)a

	Coefficient	*T-value*
Dummy 1983	0.09	2.6
Patrolling area	0.03	1.5
Shadow price solved serious crimes (P_{m1})	0.49	70.3
Shadow price solved cases of drunk driving (P_{m2})	0.07	24.0
Shadow price solved other crimes (P_{m3})	0.18	46.3
Shadow price minor offenses (P_{m4})	0.26	36.8
Recorded serious crimes (q_{r1})	0.28	4.8
Recorded other crimes (q_{r2})	0.35	4.8
Traffic accidents (q_{r3})	0.01	0.2
Cases of assistance (q_{r4})	-0.08	1.3
Budget (B)	0.48	4.3
R^2 revenue function	0.95	
R^2 share solved serious crimes	0.66	
R^2 share solved cases of drunk driving	0.48	
R^2 share solved other crimes	0.62	
R^2 share processed minor offences	0.35	

a *Variables scaled around their means.*

The other types of fixed production play no important role. This may be because they do not only consume inputs, but may also deliver revenue. For instance, the handling of traffic accidents may sometimes be linked to the detection of cases of drink driving. Another possible explanation is that the allocation of police inputs to different types of output is rigid. In that case, fixed parts of inputs are allocated to traffic, crime, et cetera, then the production process can be split in different parts.[18] Further testing suggests, however, that traffic accidents can not be left out the revenue function altogether.[19] With the exogenous output fixed an 1% larger police budget means 0.48% more revenue for the average department. Of course, this relatively modest effect is linked to the relative importance of the number of recorded crimes.

10.5.2 *Scale Effects and Optimal Scale*

The estimated revenue function can be used to derive various characteristics of the production structure of Dutch police. Table 10.4 presents the relation between scale elasticities and marginal revenue according to scale of departments. We distinguish two types of scale elasticities, dependent on the interpretation of the number of recorded crimes, traffic accidents and cases of assistance as fixed outputs or as inputs (see Appendix A10.1). The first definition implies that the police budget is the only input and that we are interested in the relation between this budget and the total quantities of endogenous and exogenous outputs produced. With the second definition the elasticity gives the relation between all inputs together (budget, recorded crimes, traffic accidents and cases of assistance) and the produced endogenous outputs, in terms of revenue. The marginal revenue here is the extra revenue (in 'guilders of conditional punishment') of 1,000 guilders extra budget allocated to the police.

Table 10.4

SCALE ELASTICITIES AND MARGINAL REVENUE MUNICIPAL POLICE DEPARTMENTS, 1983 AND 1986

Size of department	*Scale elasticity* v_1		*Scale elasticity* v_2[b]		*Marginal revenue per 1,000 guilders*[c]
	elasticity	T-value	*elasticity*	T-value	
Small[a]	1.62	1.7	1.13	2.3	81
Small to medium[a]	1.36	1.5	1.08	1.9	92
Medium to large[a]	0.90	0.7	0.97	0.6	104
Large[a]	0.84	2.2	0.88	1.6	208
Average department	1.12	1.5	1.05	1.7	138
Median department	1.54	1.5	1.10	2.1	77

a Small departments: up to 75 labor years; small medium departments: 75 up to 125 labor years; medium large departments: 125 up to 200 labor years; large departments: 200 and more labor years.

b With both exogenous outputs and shadow prices of endogenous outputs fixed at mean level of class. The T-values of the difference with 1 are given.

c With both exogenous outputs and shadow prices of endogenous outputs fixed at mean level of class.

The results show the existence of scale effects with Dutch police departments. Strong economies of scale, i.e. advantages of a larger scale, are found with the second definition with small and medium-sized departments. With the first definition large departments show diseconomies of scale. Scale effects, according to the first definition, are stronger than those according to the second. This is understandable, because according to the

first definition, scale economies imply a growth of exogenous outputs relative to the budget growth, which stimulates the growth in endogenous outputs, whereas the second definition implies in itself that the exogenous output is constant relative to the budget.

The last column of Table 10.4 shows that the marginal revenues increase with the size of departments. This may seem to contradict the scale elasticities. The results can however be attributed to two effects. Firstly to differences in output mix. The number of recorded serious crimes and the related output with relatively high shadow prices is more important with larger departments. Secondly, the marginal revenues reflect differences in shadow prices, i.e. the conditional severity of punishment, between smaller and larger departments. These are by definition not incorporated in (technical) elasticities of scale, but certainly play a role in marginal revenues. The median department scores relatively low because of a relative low shadow price (severity of punishment) of solutions of other crimes.

Figure 10.1 illustrates the scale effects graphically. The horizontal axis shows the budget (volume of inputs), the vertical axis the output per unit of input. Both are normalized at the value of the average department. The exogenous output per unit of input and the patrolling area are kept fixed at the value of the average department.

The average revenue increases with the size of the department up to a certain level and decreases thereafter. The optimal scale for producing the mean exogenous output mix is around a budget of 18 million guilders or 260 labor years. With this output mix the optima for the four types of endogenous outputs differ only slightly: for minor offenses it is somewhat higher and for the other outputs, somewhat lower. The average costs per unit of output are of course the inverse of the mean output per unit of input in the figure. In many Dutch studies of the production structure of public sector producers (Goudriaan,Van Tulder, Blank, Van der Torre, and Kuhry 1989, Blank 1993, Van Tulder 1994) such a U shaped cost curve is found.

10.5.3 *Sensitivity Analysis of the Optimal Scale*

Three earlier studies of productivity Dutch police estimated the optimal scale of production. (Goudriaan, Van Tulder, Blank, Van der Torre, and Kuhry 1989, Goudriaan and Van Tulder 1993, Van Tulder 1993). They were based on (roughly) the same data and the only difference is in the applied methodology. It is interesting to note that these differences had consequences for conclusions about optimal scale. The first mentioned

study was based on a statistical cost function approach and found a optimal scale of 12 million guilders versus 18 million now. This is due in part, to the fact that we differentiate now between serious and other crimes, which was not the case in the earlier study. The influence of this differentiation can be traced by comparing the results with Van Tulder (1993). He used also a revenue function there, without distinguishing between serious and other crimes, and found an optimal scale of 16 million guilders. Another part of the difference (12 versus 16 million) apparently has more to do with the difference in approach chosen (cost versus revenue function).

Figure 10.1

DEPARTMENTS OF MUNICIPAL POLICE: VOLUME OF INPUTS AND MEAN OUTPUT PER UNIT OF INPUT[a, b]

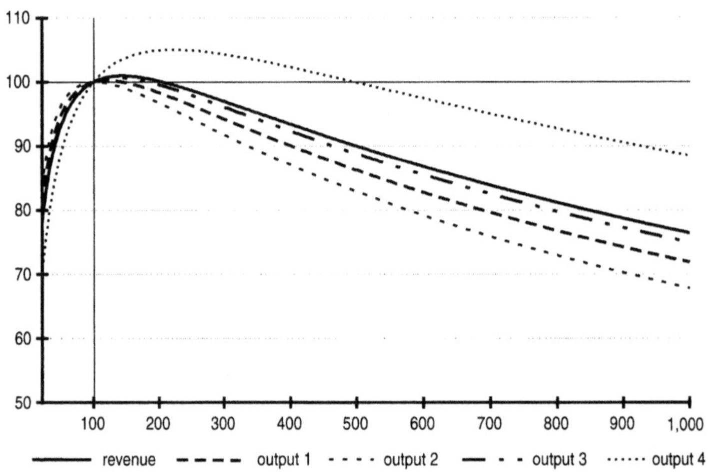

a *Average department in years 1983-1986 normalized at 100.*

b *Revenue: outputs of 4 types weighted by shadow prices,*

output 1: solutions of serious crimes,

output 2: solutions of drunk driving,

output 3: solutions of other crimes,

output 4: processed minor offences.

Goudriaan and Van Tulder (1993) used a non parametric approach (see Section 2.4). This analysis, in which serious and other crimes were distinguished between, derived an optimal scale between 4 and 14 million.

These comparisons show that the chosen methodology and differentiation of outputs can have important effects on the derived characteristics of the production structure, in this case, scale effects. This necessitates caution in interpreting the results. On the other hand, all the studies mentioned, for all their differences in methodology, agree on one important aspect: there is an optimal scale in Dutch police departments and this is clearly above the minimum size and far below the maximum size of the departments in the Netherlands in 1986.

Even within the methodology used here different optimal scales can be found, dependent on the output mix involved. With the average output mix of small departments the optimal scale is with a budget of approximately 11 million guilders. With the output mix of large departments it is clearly more: 25 million. This has to do with the relative weight of serious crimes in the output mix of larger departments. Further analysis shows that doubling the weight of serious crimes (compared to the average mix of exogenous outputs) means, all things being equal, an increase of optimal scale from 18 to 27 million. This in agreement with the general idea of policy makers that the scale of operation with fighting more serious crimes needs to be larger than by less serious crimes or cases of assistance.

10.5.4 *Prevention*

Until now, the preventive activities of the police were only taken into account in the analysis indirectly. Firstly, by the solutions of crimes and the possible deterrent effect they have on criminals. Secondly, by the activities in the sphere of assistance and public order maintenance which were one of the exogenous outputs. Police can try to reduce crime and increase safety in other ways, e.g. by patrolling in areas which are popular with burglars or by giving advice to the public about security precautions and so on. These activities use inputs, but are not fully accounted for in the analysis presented thus far. Maybe police departments attach value not only to solutions of crimes, but also to prevention of crimes in other ways. That means that they consider crimes as a 'bad' (see Section 10.3.2) and prevented crimes as a 'good'.

We tried to test this hypothesis as follows. Suppose every municipality has a sort of 'normal' crime level, dependent on autonomous demographic, social and characteristics of the population. Police influence the number of crimes in two ways: by solving crimes, which is already in the analysis and by other means, which are not fully covered in the analysis thus far. This second type of output (crimes prevented in other ways than by solutions) is measured now by the difference between the actual level of crimes and the

'normal' level. This normal level is based on an analysis relating the municipal crime level with autonomous background factors, which is corrected for differences in solution rates. It is beyond the reach of this chapter to describe this part of analysis exactly. More details can be found in Van Tulder (1994, 1995). Of course, it is improbable that the difference between real and 'normal' levels of crime can be ascribed totally to activities of the police. This approach is only valid insofar these differences (residuals) are not influenced to a great extent by autonomous background factors which are not in the model. A large number of factors is incorporated into the model of the normal crime level.[20] Since we cannot be certain that all the important factors are included, this part of the analysis is rather tentative.

Because the number of recorded crimes is endogenous and no longer exogenous, the estimated 'normal' crime levels of serious and other crimes are used instead of the number of recorded crimes as exogenous outputs. Thus a revenue function is estimated with the two types of prevention added. Therefore we have to assume that there is a constant ratio between the shadow price of a prevented crime and that of a solved crime of the same type. The resulting revenue function is roughly comparable with that presented before (see Table 10.3), and indeed indicates significant positive shadow prices of a prevented serious and other crime. So police see prevented crimes as a 'good'. The estimates suggest however that police value these prevented crimes rather low: 12% of the shadow price of a solution with serious crimes and 16% with other crimes. Of course, the problematic measurement of prevention may lead to a bias in the estimates here.

10.5.5 *Some Simulations*

The relations can be used to estimate the effects of various policy measures in organizing and allocating funds to Dutch police. Two policy measures are the starting-point here. They have a clear relation to discussions and policy plans which have played or still play a role in Dutch politics (see also Section 10.2).

1. the relative allocation of funds over small and large departments or over rural and urban areas is often subject of debate. In 1988, the government was advised to allocate funds over departments more in accordance with the 'demand' of police services by the public. So this advise is based on an equity argument (more equality in services) rather than an efficiency argument (more services). This reallocation means, in practice, a reallocation from rural to city areas and from smaller to larger departments. It is simulated here;
2. Section 10.2 already sketched the main features of the restructuring of Dutch police around 1994. This had large consequences for organization and scale of activities.

One of the main arguments here was an efficiency argument: larger departments would be capable to do the job more efficiently. It is interesting to see the results of a simulation of this restructuring. To that end, we assume that the new police departments are really integrated units, whose production structure can be compared with the old departments of the same size. Insofar that is not the case, the effects of the restructuring will be smaller;

3. in reality the restructuring was combined with a reallocation of funds, which was to a certain extent based on the ideas mentioned under 1. Added to which, not only he municipal police departments were involved, but also the police which controlled the rural areas. In the third simulation these changes are combined to estimate an over all effect of the policies pursued in the first half of the nineties regarding police organization and allocation.

Table 10.5 presents the results, in terms of the consequences on total aggregate revenue and various types of endogenous outputs.

Table 10.5

RESULTS OF SIMULATION OF EFFECTS ON OUTPUT OF REALLOCATION AND REORGANIZATION OF DUTCH POLICE (PERCENTAGES)

	revenue	*Endogenous output*[a]			
		q_{m1}	q_{m2}	q_{m3}	q_{m4}
1. reallocation	1.6	0.8	2.3	0.6	1.2
2. restructuring - municipal police	-8.4	-10.6	-15.5	-6.8	-1.0
3. restructuring/reallocation total police	-4.6	-10.8	-10.9	-7.4	2.7

a *Type 1: solutions of serious crimes; type 2: solutions of driving under the influence; type 3: solutions of other crimes; type 4: minor offenses.*

The reallocation of funds (variant 1) from smaller to larger departments has a positive effect on all types of output. The effect on total revenue is relatively greater because of the redistribution of output to larger departments with their higher shadow prices. This measure, which is mainly based on equity arguments, also has a positive effect on the productivity of Dutch police.

The simulated result of the restructuring (variant 2) is totally different. It has a negative effect on nearly all types of output. It is curious to note that this is effect is particularly strong on the solution of serious crimes. The efficiency argument, i.e. that larger units are more productive, is in practice often linked with the fight against serious and organ-

ized crime. The results presented here give no ground for this reasoning, to say the least. The result is explained by the diseconomies of scale which are found in our analysis.[21]

The combined effect of measures in the sphere of restructuring and reallocation (variant 3) are negative too. It is not surprising they are generally somewhere in between the effects of the two separate measures. The effect is modified somewhat because of the incorporation of the former rural police in the simulations.

10.6 Summary and Conclusions

This chapter presents a model of police production structure which reflects the heterogeneous nature of police output at the one hand and the institutional background and the actual decision making process at the other hand. Police departments have to deal with a fixed budget, exogenously determined. With this budget and the available technology, they have to produce different types of output. The quantity of some outputs is seen as fixed. The number of reported crimes (excluding victimless crimes), traffic accidents and cases of assistance are in this category. With other outputs, like the solving of crimes and the processing of minor offences, the size of output depends on the efforts and priorities of the police departments themselves, so are endogenous in the model. The department implicitly attaches value, a shadow price, to different types of endogenous outputs. This applies the revenue function approach to Dutch police. This approach is found in economic literature, but application in this field is rather new.

On what can the shadow prices be based? It is reasonable to assume that the value the police attaches (or at least ought to attach) to the solution of different types of crime and the processing of offenses depends on the way these are handled in the rest of the judicial system. Data on the prosecution and sanctioning process are available. In testing the assumption that shadow prices are based on the conditional severity of punishment, i.e. the average severity of punishment which can be maintained. This shadow price scheme is used in estimating the revenue function.

The revenue function shows that solutions of serious crimes have the most effect on total revenue with minor offenses at second place. None of the fixed (exogenous) outputs has a significant negative contribution to police revenue. On the contrary: both types of recorded crimes contribute clearly positively to the revenue. It suggests that it is more appropriate to see the recorded crimes as an input rather than an output in

police production. With the exogenous output fixed an 1% bigger police budget means 0.48% more revenue for the average department.

The average revenue increases with the size of the department up to a certain level and decreases thereafter. The optimal scale for producing the mean exogenous output mix is around a budget of 18 million guilders or 260 labor years. Further analysis shows that doubling the weight of serious crimes (compared to the average mix of exogenous outputs) means, all shings being equal, an increase of optimal scale from 18 to 27 million. This in agreement with the general idea of policy makers that the scale of the operation to fight more serious crime needs to be larger than that which deals with less serious crime or with cases of assistance.

The idea of these policy makers is also that the major restructuring of the Dutch police, which finished in 1994, must be beneficial for productivity. The analysis in this chapter is not in accordance with this perception. This is because of the diseconomies of scale, which have in the analysis a very negative effect on the productivity of the large regional police departments which are created.

Of course, one can say that the new departments are not like the large departments which existed before restructuring, as the analysis assumes.[22] Maybe everything will be better now. Large departments are broken down into smaller units, but this was also the case in the pre-restructured departments[23] as well. This maybe so, but one once tried to decentralize the old large departments too. We know from economic analysis that in general large units have an inherent tendency to bureaucracy and an overload of overheads. So from this economic view point doubts about the efficiency effects of the whole restructuring process remain. Maybe an economic analysis comparing the 25 new regional departments and indicating the more efficient ones may learn more about overcoming the negative tendencies of creating large units.

Appendix

A10.1 The Revenue Function

The revenue function used is of translog structure. This is a second order approximation of any revenue function, so of any production technology (Christensen, Jorgenson, and Lau 1971, 1973, Lau 1978: 196). It is a rather flexible form, although it has its restrictions. Guilkey, Lovell, and Sickles (1983) conclude that it is an adequate representation, as long as the partial substitution elasticities do not differ much from 1. It is rather restrictive in testing separability assumptions.

The revenue function is defined as :

$$\begin{aligned}\ln(\Pi(p_m,q_r,x)) &\equiv \ln(\textstyle\sum_i p_{mi}q_{mi}) \\ &= \alpha_o+\textstyle\sum_i\alpha_i\ln(p_{mi})+\sum_i\sum_j\frac{1}{2}\alpha_{ij}\ln(p_{mi})\ln(p_{mj}) \\ &+\textstyle\sum_i\beta_i\ln(q_{ri})+\sum_i\sum_j\frac{1}{2}\beta_{ij}\ln(q_{ri})\ln(q_{rj}) \\ &+\textstyle\sum_i\gamma_i\ln(x_i)+\sum_i\sum_j\frac{1}{2}\gamma_{ij}\ln(x_i)\ln(x_j) \\ &+\textstyle\sum_i\sum_j\delta_i\ln(p_{mi})\ln(q_{rj}) \\ &+\textstyle\sum_i\sum_j\varepsilon_i\ln(p_{mi})\ln(x_j) \\ &+\textstyle\sum_i\sum_j\zeta_i\ln(q_{ri})\ln(x_j)\end{aligned} \tag{A10.1}$$

with: Π - total revenue
p_m - vector of (shadow) prices p_{mi} of endogenous outputs
q_m - vector of quantitities q_{mi} of endogenous outputs
q_r - vector of quantitites q_{ri} of exogenous outputs
x - vector of (the volume of) inputs

The usual restrictions of symmetry and homogenity are:

$$\begin{aligned}\alpha_{ij} &= \alpha_{ij} \\ \textstyle\sum_i\alpha_i &= 1 \\ \textstyle\sum_j\alpha_{ij} &= \textstyle\sum_j\alpha_{ji}=0 \\ \textstyle\sum_i\delta_{ij} &= 0 \\ \textstyle\sum_i\varepsilon_{ij} &= 0\end{aligned} \tag{A10.2}$$

The kth element of the endogenous vector of outputs q_m is with maximization of revenue (Cornes 1992: 115):

$$\begin{aligned}s_k &= \frac{P_{mk}Q_{mk}}{\Pi} \\ &= \frac{\partial\ln(\Pi(p_m,q_r,x))}{\partial\ln(p_{mk})} \\ &= \alpha_k+\textstyle\sum_j\alpha_{jk}\ln(p_{mj})+\sum_j\delta_{jk}\ln(q_{rj})+\sum_j\varepsilon_{jk}\ln(x_j)\end{aligned} \tag{A10.3}$$

The quality indicator is introduced in this way. A difference in prices between producers of 1% is assumed to be partly, i.e. for $(1-k_j)$% an indication for differences in quality of the outputs, while the remaining part, i.e.

k_i%, is assumed to be a real difference in (shadow) prices. In formal sense there is a redefinition of the output size and the prices per unit of endogenous output i leading to q^*_{mi} and p^*_{mi} respectively via:

$$\begin{aligned} q^*_{mi} &= p_{mi}^{(1-k_i)} q_{mi} \\ p^*_{mi} &= p_{mi}^{k_i} \end{aligned} \tag{A10.4}$$

Note that the total revenue and the shares of the different endogenous outputs remain unchanged. The parameters k_i only at the right hand side of the equations and can be estimated together woth the other parameters. Logically the values of these quality-parameters must lie between 0 and 1.

The resulting revenue function gives, among other things, information about the scale effects in the production process. These can be derived (compare Caves, Christensen, and Swanson 1981) as follows. Suppose that the transformation function for the production process is F, with the same notation as earlier:

$$F(\ldots q_{mi} \ldots q_{ri} \ldots x_k \ldots) = 0 \tag{A10.5}$$

The total differential of F is:

$$\begin{aligned} &\sum F_{q_{mi}} d\ln(q_{mi}) + \sum F_{q_{ri}} d\ln(q_{ri}) + \sum F_{x_i} d\ln(x_i) = 0 \\ &\text{with} \quad F_{q_{mi}} = \frac{\partial F}{\partial \ln(q_{mi})} \\ &\qquad\quad F_{q_{ri}} = \frac{\partial F}{\partial \ln(q_{ri})} \\ &\qquad\quad F_{x_i} = \frac{\partial F}{\partial \ln(x_i)} \end{aligned} \tag{A10.6}$$

The scale elasticity is defined as the increase in production, with output mix unchanged, as a result of an increase in all inputs with 1%. Thus the scale elasticity v_1 is:

$$\begin{aligned} v_1 &= -\frac{\sum F_{x_i}}{\sum F_{q_{mi}} + \sum F_{q_{ri}}} \\ &= -\frac{\sum F_{x_i}}{\sum F_{q_{mi}}} \cdot \frac{\sum F_{q_{mi}}}{\sum F_{q_{mi}} + \sum F_{q_{ri}}} \\ &= -\frac{\sum F_{x_i}}{\sum F_{q_{mi}}} \Big/ \left(1 + \frac{\sum F_{q_{ri}}}{\sum F_{q_{mi}}} \right) \end{aligned} \tag{A10.7}$$

Thus:

$$\begin{aligned} \frac{\partial \ln(\Pi)}{\partial \ln(x_i)} &= -\frac{F_{x_i}}{\sum F_{q_{mi}}} \\ \frac{\partial \ln(\Pi)}{\partial \ln(q_{ri})} &= -\frac{F_{q_{ri}}}{\sum F_{q_{mi}}} \end{aligned} \tag{A10.8}$$

And so:

$$v_1 = \sum_i \frac{\partial \ln(\Pi)}{\partial \ln(x_i)} \cdot \left(1 - \sum_j \frac{\partial \ln(\Pi)}{\partial \ln(q_{rj})} \right)^{-1} \tag{A10.9}$$

Sometimes another interpretation is possible. The exogenous outputs q_{ri} are rather quasi inputs. In that case the scale elasticity v_2 relates to the consequences of an increase of all quasi and other inputs for the outputs:

$$v_2 = \Sigma_i \frac{\partial \ln(\Pi)}{\partial \ln(x_i)} + \Sigma_j \frac{\partial \ln(\Pi)}{\partial \ln(q_{rj})} \tag{A10.10}$$

Note that:

$$\begin{aligned} &v_1 = 1 \Rightarrow v_2 = 1 \\ &v_1 > 1 \Rightarrow v_2 > 1 \\ &v_1 < 1 \Rightarrow v_2 < 1 \end{aligned} \tag{A10.11}$$

So the domain of economies and diseconomies of scale and the optimal size are the same in both approaches.

A10.2 EMPIRICAL RESULTS

Table A10.1 presents some descriptives which are relevant for the interpretation of the results. Because scale of production is an important aspect of the analysis, the departments are divided in four classes according to scale. Firstly, the number of departments in various classes is presented. Departments which are in the analysis in both years are counted twice. Secondly, the table presents the number of inhabitants and the patrolling area, both related to the budget available (per million guilders). Thirdly, the solution rates of serious and other crimes. Note that these rates are slightly different from the usual definition: in the denominator only those recorded crimes are counted whose recordings are not solely dependent on police activities, defined as the fixed outputs q_{r1} en q_{r2}. The descriptives related to inhabitants and area are indices which are normalized at the mean level of all departments together.

Table A10.1

DESCRIPTIVES: INHABITANTS AND AREA PER MILLION GUILDERS BUDGET AND SOLUTION RATES, 1983 AND 1986 (N = 274)

	Number of departments	*Inhabitants per million NLG*	*Area per million NLG*	*Solution % serious crimes*	*Solution % other crimes*
Small departments[a]	148	145	304	27	22
Small to medium departments[a]	59	147	130	25	22
Medium to large departments[a]	31	138	142	26	20
Large departments[a]	36	72	30	17	25
All departments (index)		100	100		
All departments (level)	274	5,033	535	20	23

a *Small departments: up to 75 labor years; small medium departments: 75 up to 125 labor years; medium large departments: 125 up to 200 labor years; large departments: 200 and more labor years.*
The descriptives related to inhabitants and area are normalized at the mean level of all departments.

Most departments are in the class 'small', which is related to the skew distribution of size. An important measure in allocating funds to departments is the number of inhabitants. Thus the partial correlation between number of inhabitants and budget is high. The relation is, however, not linear. The budget per capita shows an inverted U shaped relation with size. This is caused partly by a fixed base sum in the budget and partly because the budget per inhabitant increases with the number of inhabitants. The budget per capita is lowest for small to medium departments with approximately 40,000 inhabitants.

The area per million guilders of budget decreases with size of the department. The lower population density of small departments means on the average larger distances to cross for patrolling units. For the analysis it is important to note that there is no, or only a very indirect, linkage of the allocation of funds to the departments and the output measures used in the analysis. So the relations found are generally not the result of institutional arrangements, but rather of properties of the production process. The solution rates of serious and other crimes show only slight differences. It is interesting to note the differences in pattern according to size. With serious crimes, the solution rate of large departments is lower than those of small ones. With other crimes it is the reverse.

Table A10.2 presents the complete estimated revenue function. The variables involved are in natural logarithms.

Table A10.2

REVENUE FUNCTION MUNICIPAL POLICE DEPARTMENTS, 1983 AND 1986 (N = 274)[a]

	Coefficient	*T-value*	*Variable*	*Coefficient*	*T-value*
Constant	−0.21	5.2	$q_{r4}{\cdot}q_{r4}$	−0.02	0.2
Dummy 1983	0.09	2.6	B·B	0.29	1.8
Patrolling area	0.03	1.5	$p_{m1}{\cdot}q_{r1}$	0.09	7.3
p_{m1}	0.49	70.3	$p_{m1}{\cdot}q_{r2}$	−0.01	0.5
p_{m2}	0.07	24.0	$p_{m1}{\cdot}q_{r3}$	−0.01	0.6
p_{m3}	0.18	46.3	$p_{m1}{\cdot}q_{r4}$	0.01	0.6
p_{m4}	0.26	36.8	$p_{m2}{\cdot}q_{r1}$	−0.03	5.7
q_{r1}	0.28	4.8	$p_{m2}{\cdot}q_{r2}$	−0.02	3.6
q_{r2}	0.35	4.8	$p_{m2}{\cdot}q_{r3}$	0.00	0.8
q_{r3}	0.01	0.2	$p_{m2}{\cdot}q_{r4}$	0.00	0.8
q_{r4}	−0.08	1.3	$p_{m3}{\cdot}q_{r1}$	−0.03	4.0
B	0.48	4.3	$p_{m3}{\cdot}q_{r2}$	0.06	6.5
$p_{m1}{\cdot}p_{m1}$	0.22	12.8	$p_{m3}{\cdot}q_{r3}$	0.00	0.1
$p_{m1}{\cdot}p_{m2}$	−0.04	7.4	$p_{m3}{\cdot}q_{r4}$	0.00	0.6
$p_{m1}{\cdot}p_{m3}$	−0.09	10.8	$p_{m4}{\cdot}q_{r1}$	−0.03	2.4
$p_{m1}{\cdot}p_{m4}$	−0.09	6.8	$p_{m4}{\cdot}q_{r2}$	−0.03	1.7
$p_{m2}{\cdot}p_{m2}$	0.08	3.4	$p_{m4}{\cdot}q_{r3}$	0.00	0.3
$p_{m2}{\cdot}p_{m3}$	−0.01	2.4	$p_{m4}{\cdot}q_{r4}$	−0.01	0.6
$p_{m2}{\cdot}p_{m4}$	−0.03	1.3	$p_{m1}{\cdot}B$	−0.09	4.6
$p_{m3}{\cdot}p_{m3}$	0.18	8.1	$p_{m2}{\cdot}B$	0.05	5.5
$p_{m3}{\cdot}p_{m4}$	−0.07	3.9	$p_{m3}{\cdot}B$	−0.03	2.9
$p_{m4}{\cdot}p_{m4}$	0.19	5.8	$p_{m4}{\cdot}B$	0.07	3.7
$q_{r1}{\cdot}q_{r1}$	0.01	0.1	$q_{r1}{\cdot}B$	0.04	0.3
$q_{r1}{\cdot}q_{r2}$	−0.08	0.7	$q_{r2}{\cdot}B$	−0.22	1.4
$q_{r1}{\cdot}q_{r3}$	0.14	1.3	$q_{r3}{\cdot}B$	−0.13	0.5
$q_{r1}{\cdot}q_{r4}$	−0.03	0.5	$q_{r4}{\cdot}B$	−0.26	2.1
$q_{r2}{\cdot}q_{r2}$	0.43	2.7	k_1	0.96	15.5
$q_{r2}{\cdot}q_{r3}$	−0.27	2.1	k_2	0.47	2.1
$q_{r2}{\cdot}q_{r4}$	0.10	1.5	k_3	0.88	8.8
$q_{r3}{\cdot}q_{r3}$	−0.16	0.8	k_4	0.73	3.4
$q_{r3}{\cdot}q_{r4}$	0.20	2.6			
R^2 revenue function	0.95				
R^2 share output 1	0.66				
R^2 share output 2	0.48				
R^2 share output 3	0.62				
R^2 share output 4	0.35				

a *Variables in natural logarithms and scaled around their means. Estimations by LSQ-procedure in TSP, with weights dependent on budget.*

NOTES

[1] Research in this area was part of a research program of the Social and Cultural Planning Office, which started in the mid eighties and intended to improve knowledge about the production structure in various parts of the Dutch public sector.

[2] The original analysis presented here can be found in the thesis of the author (Van Tulder 1994).

[3] Excluding special forces too, with special tasks, like the Koninklijke Marechaussee for guarding the frontiers and protecting members of the royal family.

[4] The weight per type of output is based on the average time needed to process one piece of that output, derived from time keeping research (see Goudriaan, Van Tulder, Blank, Van der Torre, and Kuhry 1989: 150). The steep rise at the end of period is caused to a large extent by a steep rise in the number of processed minor offenses. This has to do with a new way of handling these minor offenses.

[5] In the years after the restructuring some problems in the sphere of management and political control have come clearly into the light, related to some conflicts arising between police chiefs, their subordinates or their superiors (mayors, public prosecutors).

[6] Comparable problems can be found with the production analysis of fire protection. Duncombe and Yinger (1993) use the inverse of the value of fire damages related to the total value of fixed assets as an indicator of C-output in this sphere.

[7] Greenwood, Chaiken and Petersilia (1977) concluded that 97% of crime solutions in Kansas City was independent on criminal investigation activities, but merely based on standard and routine activities. POS (1979) concluded for the Netherlands that the criminal investigation was mainly focused on a small number of more serious crimes.

[8] At least in the short term and without mention to direct prevention (see below).

[9] Of course, the number of recorded crimes is only relatively fixed to the police. Actually not all reported crimes are recorded by the police.

[10] Because the focus is on deriving general characteristics of the production structure, a parametric approach is chosen here (see Chapter 2).

[11] In practice police offices blame sometimes the justice system for releasing suspects very quickly, whereas Public Prosecutors blame the police for processing the wrong cases. So in practice there is certainly a loss.

[12] Punishment had in the years of analysis (1983, 1986) mostly the form of fines or prison sentences. To construct one indicator of severity of punishment we combine the number of unconditional prison days included in the sentence and the unconditional fine (or amount of financial settlement) into one aggregate. With the translation of prison days into fines a common substitution rule used by judges is applied (1 prison day is NLG 50 is approx. USD 25).

[13] There are other types of crime that could be excluded on this ground too, e.g. fencing and some crimes in the sphere of disrupting public order. These types could not be distinguished separately in the analysis (included in a larger group) and are generally small in number. There may also be discussion about the role of police tracing activities with the recording of certain types of crime.

[14] There are indications that on the average 1 in 7 accidents handled by the police involves personal injury.

[15] In 1987 reacted small departments in approx 85% of call for assistance, with large departments this was 80%.

[16] A likelihood ratio test with 27 degrees of freedom gave in a test cfm. Appendix A10.1 a value of 27. The 5% critical value is 40. In tested alternative measures of shadow prices the probability of punishment was explicitly taken into account or prison sentences were given more weight.

[17] In more technical terms: the convexity assumption of the revenue function relating to the prices is sometimes violated.

[18] In technical terms there may be non jointness.

[19] Testing suggests that cases of assistance may be left out. In other respects the relation cannot be simplified. For instance, a Cobb-Douglas approach is too simple and hypotheses of functional inseparability of different endogenous outputs are rejected.

[20] The variables in the analysis describing the autonomous population characteristics are: the proportion of young men, the proportion of divorced men, the proportion of unemployed men, the proportion of population with an income below a certain level, the proportion of non natives and the average income

[21] The other mentioned studies of Dutch police found generally lower estimates of optimal size. So these studies probably would find even larger negative effects of the restructuring.

[22] This has been generally the reaction of policy makers and organization advisory bureaus towards earlier statements of this kind. The region of each department is split into a number of districts.

[23] Large departments are broken down into smaller units, but this was also the case in the pre-restructed departments as well.

References

Anderson, R.W. 1974. Towards a Cost-Benefit Analysis of Police Activity. *Public Finance* 29 (1): 1-17.

Becker, G.S. 1968. Crime and Punishment: An Economic Approach. *Journal of Political Economy* 76 (2): 169-217.

Blank, J.L.T. 1993. *Kosten van Kennis: Een Empirisch Onderzoek naar de Produktiestructuur van het Basisonderwijs in Nederland* (Costs of Knowledge: An Empirical Analysis of the Production Structure of Primary Education in the Netherlands, with English Summary). Rijswijk/Den Haag: Sociaal en Cultureel Planbureau/VUGA (Sociale en Culturele Studie 17).

Carr-Hill, R.A., and N.H. Stern. 1973. An Econometric Model of the Supply and Control of Recorded Offences in England and Wales. *Journal of Public Economics* 2: 289-318.

Carr-Hill, R.A., and N.H. Stern. 1973. *Crime, the Police and Criminal Statistics: An Analysis of Official Statistics for England and Wales Using Econometric Models*. Londen: Academic Press.

Caves, D.W., L.R. Christensen, and J.A. Swanson.1981. Productivity Growth, Scale Economies and Capacity Utilization in US Railroads, 1955-1974. *American Economic Review* 71 (5): 994-1003.

Chapman, J.I., W.Z. Hirsch, and S. Sonenblum. 1975. Crime Prevention, the Police Production Function and Budgeting. *Public Finance* 30: 5.

Christensen, L.R., D.W. Jorgenson, and L.J. Lau. 1971. Conjugate Duality and the Transcendental Logarithmic Functions. *Econometrica* 39 (4): 255-256.

Christensen, L.R., D.W. Jorgenson, and L.J. Lau. 1973. Transcendental Logarithmic Production Frontiers. *Review of Economics and Statistics* 55 (1): 28-45.

Cornes, R. *Duality and Modern Economics*. 1992. Cambridge: Cambridge University Press.

Darrough, M.N., and J.M. Heineke. 1978. The Multi-output Translog Production Cost Function: The Case of Law Enforcement Agencies. In: Heineke, J.M. (ed.), *Economic Models of Criminal Behavior*. Amsterdam: North Holland Publishing Company. (Contributions to Economic Analysis 118).

Darrough, M.N., and J.M. Heineke. 1979. Law Enforcement Agencies as Multi-product Firms: An Econometric Investigation of Production Costs. *Public Finance* 34: 176-195.

Duncombe, W. and J. Yinger. 1993. An Analysis of Returns to Scale in Public Production, with an Application to Fire Protection. *Journal of Public Economics* 52 (1): 49-72.

Ehrlich, I. 1973. Participation in Illegitimate Activities: A Theoretical and Empirical Analysis. *Journal of Political Economy* 81: 521-567.

Finney, M. 1997. Scale Economies and Police Department Consolidation: Evidence from Los Angeles. *Contemporary Economic Policy* 15 (1): 121-127.

Goudriaan, R., F. van Tulder, J. Blank, A. van der Torre, and B. Kuhry. *Doelmatig Dienstverlenen* (Efficient Services). Rijswijk/Alphen aan den Rijn: Sociaal en Cultureel Planbureau/Samsom (Sociale en Culturele Studie 11).

Goudriaan, R., and F. van Tulder. 1993. *Scale and Efficiency of Police Departments.* Paper presented at the 38th International Congress of the Applied Econometrics Association, Athens.

Greenwood, P.W., J.M. Chaiken, and J. Petersilia. 1977. *The Criminal Investigation Process.* Santa Monica: Rand.

Guilkey, D.K., C.A. Knox Lovell, and R.C. Sickles. 1983. A Comparison of the Performance of Three Flexible Functional Forms. *International Economic Review* 24 (3): 591-616.

Gyimah-Brempong, K. 1989. Production of Public Safety: Are Socioeconomic Characteristics of Local Communities Important Factors? *Journal of Applied Econometrics* 4 (1): 57-71.

Lau, L.J. 1978. Applications of Profit Functions. In: Fuss, M., and D. McFadden (eds.), *Production Economics: A Dual Approach to Theory and Applications* (Volume 1). Amsterdam: North Holland Publishing Company. (Contributions to Economic Analysis 110: 133-216).

McFadden, D. 1978. Cost, Revenue and Profit Functions. In: Fuss, M., and D. McFadden (eds.), *Production Economics: A Dual Approach To Theory And Applications* (Volume 1). Amsterdam: North Holland Publishing Company (Contributions to Economic Analysis 110: 3-109).

Popp, D.O., and F.D. Sebold. 1972. Quasi-Returns to Scale in the Provision of Police Services. *Public Finance* 27: 46-61.

POS. 1979. *Projectgroep Organisatie Structuren: Sterkte van de Politie.* Den Haag: Staatsuitgeverij.

Pyle, D.J. 1983. *The Economics of Crime and Law Enforcement.* Londen: MacMillan, 1983.

Pyle, D.J., and D.F. Deadman. 1980. Law Enforcement Agencies as Multi-product Firms: A Comment. *Public Finance* 35 (3): 455-460.

Swimmer, E. 1974. Measurement of Effectiveness of Urban Law Enforcement: A Simultaneous Approach. *Southern Economic Journal* 40 (4): 618-630.

Van Tulder, F.P. 1992. Crime, Detection Rate, and the Police: A Macro Approach. *Journal of Quantitative Criminology* 8 (1): 113-131.

Van Tulder, F.P. 1993. *Law Enforcement, Production and the Decision-making Process: An Analysis of Dutch Police-departments.* Paper presented to the 49th IIPF-Congress on Public Finance and Irregular Activities, Berlin.

Van Tulder, F.P. 1994. *Van misdaad tot straf* (From Crime to Punishment, with English Summary). Rijswijk/ Den Haag: Sociaal en Cultureel Planbureau/VUGA (Sociale en Culturele Studie 21).

Van Tulder, F.P. 1995. *Costs and Benefits of Law Enforcement in the Netherlands.* Paper presented at the Congress of the European Association for Law and Economics, Bern.

Votey, H.L. Jr., and L. Phillips. 1972. Police Effectiveness and the Production Function for Law Enforcement. *Journal of Legal Studies* 1 (2): 423-436.

Walzer, N. 1972. Economies of Scale and Municipal Police Services: The Illinois Experience. *Review of Economics and Statistics* 54 (2): 431-438.

11 EFFICIENCY OF LOCAL POLICE DISTRICTS: A NEW SOUTH WALES EXPERIENCE

Andrew Hughes and Suthathip Yaisawarng[1]

11.1 INTRODUCTION

In recent years, Federal and State governments in Australia have pursued initiatives aimed at improving the efficiency of public sector service provision. These include the development of performance measures that may ultimately facilitate the linking of public sector service providers' budget allocations to agreed output/outcome targets, service quality standards and efficiency indicators. In this context, the New South Wales (NSW) Government is exploring a range of performance measurement techniques including Data Envelopment Analysis (DEA). DEA pilot studies have been undertaken on some of the activities of public sector agencies, including the NSW Police Service.

This chapter provides an analysis of the relative performance of local police districts[2] in New South Wales, illustrating the usefulness of DEA as a management tool. The focus is on managerial or pure productive efficiency, which assesses how well a local police district uses available resources to deliver services as measured in the study. An input oriented model is chosen to identify potential input savings among local police districts and to facilitate the reallocation of resources to high priority areas by the management of the NSW Police Service.

The chapter is organized as follows: Section 11.2 discusses the structure of the NSW Police Service and develops measures of services delivered and resources used for local police districts. Section 11.3 illustrates the DEA technique, presents four alternative model specifications for evaluating the performance of NSW local police districts, and describes procedures for sensitivity tests (see also Chapter 6). Section 11.4 discusses the 1995-1996 sample of local police districts. Section 11.5 provides summary results and illustrates how an inefficient local police district could use DEA results to improve performance. Section 11.6 concludes the chapter.

11.2 STRUCTURE OF NSW LOCAL POLICE DISTRICTS AND MEASURES OF POLICE ACTIVITIES

The NSW Police Service is Australia's oldest and largest police organization and one of the largest in the English speaking world. Table 11.1 shows key employee statistics for the Service in recent years. The number of staff employed by the NSW Police Service increased broadly in line with the State's population growth between 1993-1994 and 1997-1998. In 1997-1998, the Service employed about17,000 staff, including more than 13,000 sworn police officers. In this year, there were about 212 police officers per 100,000 persons.

Table 11.1

KEY EMPLOYEE STATISTICS, NSW POLICE SERVICE, 1993-1994 TO 1997-1998

	1993-1994	1994-1995	1995-1996	1996-1997	1997-1998
Number of employees[a]					
Police officers	12,700	13,100	13,000	13,300	13,400
Administrative, Ministerial & other	3,200	3,300	3,500	3,700	3,800
Total	15,900	16,400	16,600	17,000	17,200
Estimated resident population,					
New South Wales (x million)[b]	6.0	6.1	6.2	6.2	6.3
Number of police per 100,000 persons	211	215	212	213	212

a *New South Wales Police Service* (1998: 64).
b *Australian Bureau of Statistics*, DCRQ.UNERP—NSW.

The NSW Police Service is a non-profit statutory authority funded mainly by the NSW Government. This occurs in most part through the annual State Budget where resources are made available to the Minister for Police to fund the operating and capital activities of the Service. In 1997-1998, total operating expenditure, including depreciation and amortization, was AUD 1,300 million. Employee related costs account for about 80% of total operating expenditure in this year. The book value of the fixed assets owned by the Service was AUD 580 million in 1997-1998.

The NSW Police Service is governed by a number of legislative provisions. The most important statute is the *Police Service Act* 1990. This Act provides for the appointment of the Commissioner of Police by the State Governor, on the recommendation of the Minister for Police. Whilst the Commissioner has a high degree of autonomy in the management and control of the Service, he or she is nevertheless subject to the rule of law and is accountable to Parliament through the Minister for Police.

The operations of the NSW Police Service are organized into two parts: field and specialist. Field operations include local police districts (grouped into eleven regions across the State), prosecution liaison and youth clubs. Local police districts are the primary service delivery units, providing frontline service to the community.[3] For details on the structure and functions of local police districts, see Carrington, Puthucheary, Rose, and Yaisawarng (1997). Specialist operations, including Forensic Services and Traffic Services, provide support services to field operations.

The Steering Committee for the Review of Commonwealth/State Service Provision (1998: 263) lists three main objectives for the provision of police services in Australia: (1) to protect, help and reassure the community; (2) to prevent crime; and (3) to enforce the law . Consistent with these objectives, the NSW Police Service classifies the services of a local police district into two broad categories: law enforcement work and crime prevention activities. The latter category accounts for about 40% of police work.

Following Carrington, Puthucheary, Rose, and Yaisawarng (1997), the analysis in this chapter includes four measures of law enforcement activities: number of incidents, number of charges, number of summons, and number of major car accidents.[4],[5] Two measures of crime prevention activities are used: total kilometers traveled by police cars (see Carrington, Puthucheary, Rose, and Yaisawarng 1997) and an additional measure, number of intelligence reports.[6] The first measure reflects the visibility of police officers in the community, especially for regional districts. The second measure captures activities undertaken by general duty police (i.e., beat police) who patrol urban districts on foot. Beat police officers prepare intelligence reports on local criminal activity. Both measures reflect the presence of police in the community, which are designed to discourage crime.

Local police districts use labor and capital inputs to perform law enforcement and crime prevention services.[7] Labor consists of police and civilian staff. Both perform different but complementary tasks, using different skills. Therefore, uniformed police officers[8] and civilians are separated to reflect their skill differences. They are measured as number of full time equivalent persons and include staff on extended leave (for example, sick leave, long-service leave or on seconded leave to other police units). Three proxies for the capital input exist: number of police cars; number of personal computers; and area of station accommodation. Table 11.2 summarizes the variables used in the study.

Table 11.2

NSW Local Police District Data

Outputs		Units
Law Enforcement Activities		
Y1	Responses to total incidents (excludes major car accidents)	Number in thousands
Y2	Major car accidents	Number in thousands
Y3	Summons served	Number in thousands
Y4	Charges laid	Number in thousands
Crime Prevention Activities		
Y5	Distance traveled by police cars	Kilometers (km) in thousands
Y6	Intelligence reports	Number in thousands
Inputs		*Units*
X1	Police officers	Average number of fulltime equivalent
X2	Civilian employees	Average number of fulltime equivalent
X3	Cars	Average number of fulltime equivalent
X4	Personal computers	Number
X5	Station accommodation area	Square meters (m^2) in thousands

11.3 Methodology

This section aims to provide an intuitive understanding of the DEA efficiency measure and its interpretation, using a simple graphical illustration (see also Section 2.4). The focus is on input saving and managerial efficiency; that is, to what extent can a manager reduce resources without reducing the level of services while maintaining the same scale of operations. The illustration uses a single output or service and two types of inputs. For the case of multiple services and inputs, the same concept can be implemented mathematically using a linear programming approach.[9]

Consider a simplified case where there are four local police districts in the sample. These districts deliver identical levels of services using two inputs: number of police officers (the labor input) and number of police cars (proxy for the capital input). These police districts use different combinations of inputs to deliver services. Figure 11.1 depicts these police districts, where each dot represents the combination of inputs for a local police district.

Figure 11.1

An Input-Oriented DEA

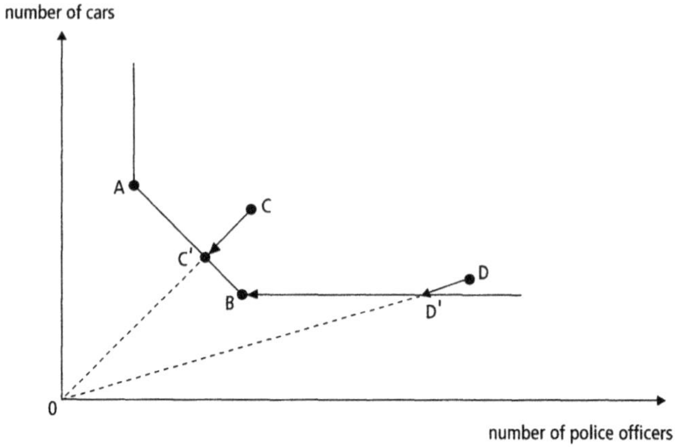

The best practice observations define the frontier, AB and its extensions. Districts A and B use the lowest combination of inputs and are accordingly judged as technically efficient. Districts C and D use more police officers and cars than A and B and are therefore off the best practice observed frontier. These two patrols are inefficient. DEA assigns efficiency scores of one to both A and B. Police districts C and D receive efficiency scores of less than one.

To illustrate how DEA efficiency scores for inefficient units are calculated, first consider District C. The target point for C is C' which lies on the best practice observed frontier and represents the same ratio of police officers to cars as C. An efficiency score for C is computed as OC'/OC which is less than 1, say 0.8. This result indicates that C is 80% efficient, relative to the best districts in the sample. To maintain the same level of service delivery, C could use 20% fewer police officers and cars if the level of resource usage had been equal to that implied by the best practice observed frontier. To do so, C should imitate A and B since C' is a combination of A and B. A and B are efficient peers for C. Since C' is closer to B than A, B has more impact than A. The effect that an *efficient peer* has on the inefficient district is called the *impact (or weight)*. For example, B may have a 60% *impact* and A may have a 40% impact (precise impacts can be obtained if the mathematical approach is used). These impacts are used to compute the target point C'.

Next, consider the other inefficient police district, D. The efficiency score of D is OD'/OD which is less than one, say 0.92. District D could use 92% of its officers and cars currently used to deliver services. This brings D to D'. However, D' is not a target point

for D. As shown in Figure 11.1, number of police officers at D' exceeds that at B while the number of cars is the same. This suggests that, after uniformly reducing both inputs from D to D', it is possible for D to further reduce police officers from D' to B. The difference in number of police officers between B and D' is an *additional reduction (known as non-radial slack)* that is possible if D uses inputs efficiently and achieves the target point B. In this case, B is an efficient peer for D, with a weight of one.

Table 11.3

Specification of Police DEA Models

	Model specifications[a]			
	I	II	III	IV
Outputs				
Law Enforcement Activities				
Y1 Total incidents (excl. major car accidents)	x	x	x	x
Y2 Major car accidents	x	x	x	x
Y3 Summons served	x	x	x	x
Y4 Charges	x	x	x	x
Crime Prevention Activities				
Y5 Kilometers traveled	x	x	x	x
Y6 Intelligence reports	x	x	x	x
Inputs				
X1 Police officers	x	x	x	x
X2 Civilian employees	x	x	x	x
X3 Police cars	x	x	x	x
X4 Personal computers	x	x		
X5 Station area	x		x	

a X indicates that the variable is included in the model.

As discussed in Section 11.2, three alternative measures of the capital input have been considered in the NSW study, none of which individually captures all dimensions of this input. This chapter uses four alternative DEA model specifications, which are presented in Table 11.3. All four specifications include the four measures of law enforcement activities and the two measures of crime prevention activities. The number of inputs varies across specifications in response to number of capital input measures. Model I includes all three measures of the capital input while Model IV includes only

one measure of the capital input, number of police cars. Models II and III comprise the three common inputs of Model IV combined with one additional capital input measure (either number of personal computers or area of station accommodation).

Theoretically, efficiency scores from the model with the larger number of variables should be higher than from the model with smaller number of variables, for a given fixed sample size. Therefore, efficiency scores from Model I should be the highest and those from Model IV should be the lowest. However, the relationship between efficiency scores from Models II and III, which have the same number of model variables, cannot be determined *a priori*.

The majority of published DEA studies either select a single specification *a priori* or experiment with several specifications followed by statistical testing of the mean differences or distributions. In contrast, this chapter develops a suite of sensitivity tests. The objective of the sensitivity testing is to identify the most appropriate model specification; that is, the specification that yields meaningful insights for managerial purposes. A similar approach has been applied in Section 6.5.

Specifically, the sensitivity tests focus on three technical aspects: (1) the overall relationship between each pair of efficiency scores and across all four specifications, (2) variations of individual scores across all four specifications, and (3) assessment of appropriate efficient peers.[10] The Appendix A11 describes the tests in detail.

11.4 Data

The study in this chapter uses a sample of 161 local police districts in 1995-1996.[11] The sample provides four measures of law enforcement activities, two measures of crime prevention activities, and five measures of resources. Inputs are police officers, civilian employees, police cars, personal computers and area of station accommodation. Table 11.4 contains descriptive statistics of the raw data.[12]

Analysis of the samples' descriptive statistics reveals that the number of intelligence reports has the largest variation relative to its mean compared to the other variables.[13] An urban police district known to promote beat police activity and place significant emphasis on the preparation of intelligence reports, is the champion for this output measure. It has the highest number of intelligence reports in the sample, almost four times larger than the second ranked district. The second largest variation relative to its mean is number of summons served. All districts use some civilian employees during

the year with a minimum of 0.37 persons. Capital inputs, especially number of police cars and number of PCs, are relatively uniform across patrols. These measures of the capital input are correlated.[14]

Table 11.4

DESCRIPTIVE STATISTICS FOR NSW LOCAL POLICE DISTRICTS, 1995-1996 (161 DISTRICTS)

	Mean	Std. Dev.	Minimum	Maximum
Law Enforcement Activities				
Total incidents (excl. major car accidents)	3.82	2.45	0.38	13.5
Major car accidents	0.49	0.32	0.03	1.76
Summons served	0.03	0.03	0.00	0.21
Charges	0.64	0.43	0.11	2.24
Crime Prevention Activities				
Kilometers traveled	342	164	39	767
Intelligence reports	0.36	0.56	0.03	6.55
Inputs				
Police officers	57.60	29.60	10.60	150.60
Civilian employees	6.52	5.66	0.37	37.80
Police cars	10.20	4.97	2.00	27.00
Personal computers	28.60	13.50	5.00	105.00
Station area	1.32	1.02	0.11	7.03

The sample includes 97 urban police districts covering the Sydney, Newcastle and Wollongong metropolitan areas and 64 regional districts covering the rest of the State. Generally, the urban police districts had significantly higher outputs for major car accidents, responses to incidents and intelligence reports. The regional police districts as a group had higher outputs for kilometers traveled, reflecting larger patrol territories. Urban districts, on average, have lower numbers of summons served than the regional districts. However, there is a greater variation in summons served across urban districts than across regional ones. On the input side, the urban districts had higher numbers of police and civilian staff but fewer cars than the regional districts. Both had similar numbers of personal computers but the regional districts had significantly more station accommodation despite having fewer staff.

11.5 Results

This section comprises two subsections. Section 11.5.1 presents a summary of the DEA results obtained from four alternative model specifications and discusses the outcomes of the sensitivity analysis. Section 11.5.2 illustrates how DEA results could be used for performance improvement based on the preferred model.

11.5.1 *Summary Results*

Table 11.5 summarizes managerial efficiency scores for all four specifications.[15] Consistent with the theory, the average efficiency score from Model IV is the lowest and that from Model I is the highest. The average efficiency scores range from 0.90 to 0.93 and fall within a 5% range between the models.[16] These scores suggest that it is possible to improve the performance of NSW local police districts. Some districts may be able to improve more than others. For example, there is one urban police district that could potentially reduce its inputs by a significant amount (between 35% and 50%) while providing the same level of services. This district receives the lowest efficiency score in Models III and IV. Its efficiency score of 0.63 in Models I and II just surpasses the next four least efficient urban police districts. It is worth noting that DEA results indicate the potential improvement in relation to best practice performers in the sample. Part of the potential improvement may not be achievable owing to differences in districts' operating environments[17] and/or other police activities not captured by the model variables.

Table 11.5

Summary of Managerial Efficiency

	Model			
	I	II	III	IV
Sample average (161 districts)	0.933	0.921	0.918	0.901
Minimum	0.606	0.606	0.523	0.518
Number of efficient police districts	89	80	77	67

As mentioned in Section 11.3, this chapter uses several tests to check the sensitivity of efficiency results. Broadly speaking, the overall results are not sensitive to different model specifications. The tests provide evidence that the results are positively related and stable across specifications.[18]

Analysis of unit-level variations shows remarkably stable results across the models. The proportion of local police districts that are efficient by default[19] to total number of efficient districts is stable in the 20 to 25% range in each specification. In addition, seven out of the ten most frequent peers in each model are the same. Seventy-eight of the 161 police districts in the sample receive identical efficiency scores across models, some of which are inefficient districts. The remaining 83 police districts receive different scores depending on the model specification. Figure 11.2 shows the difference between maximum and minimum efficiency scores for these 83 districts in descending order across the four model specifications. The length of each bar represents the range of efficiency scores for a district. The longer the length of the bar, the more variation of the particular district's efficiency scores across the four specifications. The vertical position of the bar tells whether the district falls into a high or medium efficiency group. For example, consider the leftmost vertical bar in Figure 11.2. This district receives a maximum efficiency score of one, i.e., it is efficient in one model. Its minimum efficiency score is slightly less than 0.8. Therefore, its efficiency score ranges between 0.8 and 1.0.

Figure 11.2

Ranges of Police Districs' Efficiency Scores Across Models I-IV

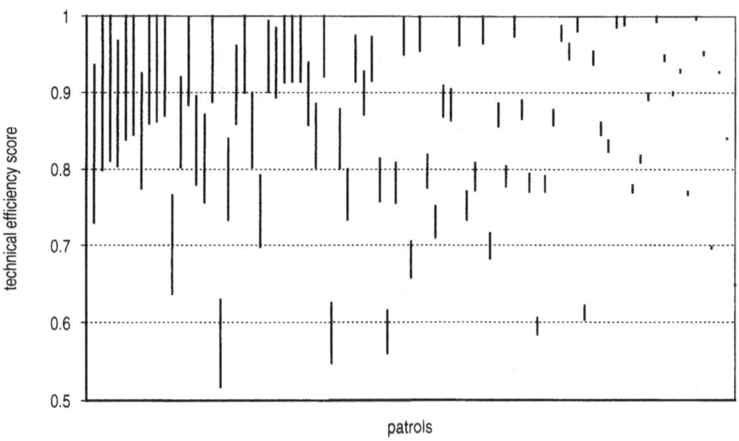

Overall, Figure 11.2 shows that relatively few districts experience a large variation in their efficiency scores. These districts appear to be in the highly efficient group. The majority of the 83 included districts experiencing moderate variations (those with a shorter bar-length). The positions of the bar for these districts scatter in all efficiency groups (see across each row).

Matching the geographic and population profiles of inefficient districts and their efficient peers reveal remarkably similar results across the four specifications (see detailed criteria in the Appendix A11.3). The majority of inefficient patrols have appropriate peers. The number of inefficient districts having at least one inappropriate efficient peer (i.e., problem cases) ranges from seven (Models I and II) to 11 (Model IV). Three cases are common for all four models. They involve two inefficient urban police districts and one inefficient regional police district. Each of these inefficient urban (regional) police districts has one regional (urban) peer with a weight ranging from 12% to 29%. The geographic characteristics of these peers are very different from those of the inefficient districts. A comparison of input and output compositions suggests that inappropriate peers for the remaining problem cases have very different mixes compared to inefficient patrols, especially in respect to total incidents, major car accidents, police officers and police cars. Care must be taken if these peers are used in assisting inefficient districts to improve their efficiency.

Model II is preferable for the following reasons. Since DEA results are relatively stable across these four models, Model I, which includes 11 variables, is the least preferable. Including additional capital proxies do not result in a more accurate measure of efficiency. Instead, they inflate efficiency scores due to an increase in the dimensionality of the respective DEA models and reduce the ability of DEA to pinpoint districts that need improvement. Model IV, which includes the smallest number of model variables among the four specifications, is not the most favorable. Police experts advised that number of cars alone does not adequately capture all dimensions of the capital input. Models II and III include the same number of model variables. However, cars and PCs may be viewed as semi-capital inputs. Both have high depreciation rates (implying short asset lives) compared to buildings as represented by station area. They are subject to greater control by management than area accommodation. More importantly, Model II is more appealing for management than the other three models. Model II provides appropriate peers for most inefficient patrols. Less than 5% of police districts in the sample are classified as problem cases. An illustration of the result in the next section is based on the results from Model II.

11.5.2 *An Illustration of an Inefficient Local Police District*

This section illustrates a potential use of DEA results for developing a performance enhancement strategy. It selects an inefficient local police district and presents DEA results that include an efficiency score, a list of efficient peers and their relative impacts.

Consider an inefficient, small-to-medium urban district (District 5). This police district had slightly higher total incidents, major car accidents, police officers and personal computers in comparison to the 1995-1996 sample average. This district also had more major car accidents and computers than the averages for the urban districts. The other variables are lower than the total and urban averages. The managerial efficiency results for Police District 5 are shown in Table 11.6.

Table 11.6

MANAGERIAL EFFICIENCY RESULTS OF POLICE DISTRICT 5, 1995-1996

	Actual	Possible input reduction (78% of actual inputs)	Additional	Target[a]
Law Enforcement Activities				
Y1 Total incidents (excl. major car accidents)	4.24	4.24		4.24
Y2 Major car accidents	0.71	0.71		0.71
Y3 Summons served	0.01	0.01	0.02	0.02
Y4 Charges	0.47	0.47		0.47
Crime Prevention Activities				
Y5 Kilometers traveled	238	238		238
Y6 Intelligence reports	0.28	0.28	0.17	0.45
Inputs				
X1b Police officers	62.3	48.6		48.6
X2b Civilian employees	5.6	4.4		4.4
X3 Police cars	7.0	5.5		5.5
X4 Personal computers	33.0	25.7		25.7

a The target outputs and inputs are calculated from District 5's efficient peers.

District 5 received a managerial efficiency score of 0.78, meaning that it was 78% efficient relative to all districts included in the sample. This district could have used only 78% of police officers, civilian employees, police cars and personal computers to provide the current level of outputs included in the model if it had operated at observed best practice. In other words, 22% of these measured inputs could have been reduced if District 5 had managed its resources efficiently. Further, it is technologically possible that District 5 could have increased its summons served and number of intelligence reports without requiring additional resources or reducing other services.[20] If District 5 had achieved

the best practice frontier in 1995-1996, it could have used the efficient inputs to produce the target level of outputs.

The target outputs and inputs are calculated from District 5's efficient peers. Peers provide guidance on improving efficiency. In this example, the DEA model suggests six efficient peers for District 5; two of which have impacts of less than 10% each. The remaining four are Districts 129, 114, 76 and 152 with the impacts of 28, 22, 20 and 14%, respectively. These four efficient peers were all urban districts. A comparison of outputs and inputs of these peers with the total and urban averages revealed that Districts 129 and 152 were small-to-medium districts similar to District 5. District 114 was small while District 76 was a relatively large urban district. All peers had a similar input mix to the inefficient district. The first three peers collectively had a 70% impact on District 5. The DEA results provide a starting point for District 5 to explore how these peers operate and to use this information to improve its efficiency.

11.6 CONCLUSIONS

This chapter computes DEA efficiency scores for 161 NSW local police districts in 1995-1996, using four alternative specifications. On average, NSW local police districts were 90 to 93% efficient in the use of resources to deliver services specified in the models. These results are not sensitive to the different sets of variables included in the models, based on several statistical tests developed in this chapter in consultation with police experts. The preferred model comprises four measures of law enforcement activities, two measures of crime prevention activities, and four types of resources. The results suggest that it is possible to reduce the measured inputs while delivering the current level of measured services if all local police districts in the sample could emulate the best districts in 1995-1996. Although most districts are managerially efficient or highly efficient, some districts exhibit room for improvement. Any attempts to improve performance would result in better delivery of police services to the community.

As a management tool, DEA identifies efficient peers that potentially lend insights to areas in which an inefficient police district needs review. This is an important aspect in light of the NSW Police Service's commitment to provide an efficient and effective policing service. The NSW Police Service has recently reorganized its structure "to allow for closer liaison and understanding between stakeholders and decision-makers and to provide more supervisory direction at the front line."[21] (Steering Committee for the Review of Commonwealth/State Service Provision 1998: 288).

Sonnichsen (1997) discusses difficulties in measuring police performance that stem from disagreements on priorities and purposes of policing among various groups including police officers, politicians, judicial officials, public administrators and citizens. Although police activities included in our study are in accordance with well-defined police objectives, we emphasize that the results are a starting point for further investigation. This is in agreement with Whitaker's suggestions, as cited in Sonnichsen (1997: 221), that "measuring police performance is a learning process that undergoes constant change as the nature of police work itself changes over time." Further, it is important that a proper monitoring mechanism be implemented to prevent police commanders from focusing solely on measured services and neglecting other activities that may also promote public safety. Despite the imperfection of performance measurement, it is essential for meaningful public sector reform. It helps to identify areas for review and may subsequently prompt improvements in data collection for outputs and outcomes. Further, in the case of police local districts it may stimulate a reallocation of resources toward solving complex cases such as organized crime and financial crimes.

APPENDIX

This appendix describes sensitivity tests developed in the chapter. The sensitivity tests comprise three aspects.

A11.1 OVERALL RELATIONSHIP

Between Each Pair of Efficiency Scores

The Spearman rank correlation coefficient is used to test the null hypothesis that there is no association between each pair of efficiency ranks. If the results are robust (i.e., they do not vary significantly between the models), then the values of the correlation coefficients are high and statistically significant. The emphasis is placed on the correlation in ranks, rather than in efficiency scores. The ranking is independent of the number of variables included in the DEA model while the efficiency score is not. The ranking of a police district relative to others in the sample tells the standing of the district.

Across All k Specifications

Since efficiency scores from each model specification are not independent across k specifications, the Friedman non-parametric test is used. The null hypothesis is that the probability distributions of efficiency rankings are the same across k specifications. The Friedman test-statistic has a Chi square distribution with (n, k-1) degree of freedom, where n is number of efficiency scores in each specification and k is number of specifications. In our empirical illustration, there are 161 local police districts in the sample (i.e., $n = 161$) and there are four alternative specifications (i.e., $k = 4$). The null hypothesis is rejected if the calculated Chi Square has a p-value less than 0.05; otherwise, it cannot be rejected. A failure to reject the null hypothesis verifies the robustness of the results.

A11.2 VARIATIONS OF INDIVIDUAL SCORES ACROSS K SPECIFICATIONS

Actual efficiency scores for individual police districts are compared across k specifications. If a district receives the same efficiency score regardless of the model specification, its efficiency score is not sensitive to a minor variation in the set of variables included and is robust with respect to the model specification. For those districts that receive different scores across models, the range statistic (computed as the difference between the maximum and the minimum efficiency score for each unit across all models) is used to determine the variations in scores. Districts experiencing relatively large changes in efficiency scores are scrutinized to gain a better understanding of causes of the variations. The procedure involves re-examining the measures of outputs and inputs used in the model, and checking results against other available qualitative information and Police Service expert opinion.

A11.3 ANALYSIS OF APPROPRIATE EFFICIENT PEERS

Determining criteria for assessing appropriate efficient peers should be done on a case by case basis. Mostly it involves evaluation of some special characteristics of the units. These particular features can be determined in consultation with experts in the field. In the case of the NSW Police Service, police experts suggested the use of a geographic and/or demographic profile. This aspect is very important since inefficient districts are able to learn from their relevant efficient peers (i.e., ones providing a similar mix of services) to improve their performance. This also ensures that the model results are realistic.

Specifically, the chapter matches the geographic and/or population characteristics of inefficient districts and their efficient peers. An inefficient urban police district has appropriate peers if one of the following rules is satisfied: (1) all peers are urban districts, or (2) a regional peer carries a weight of less than 10%. If an inefficient urban district has any regional peer with a weight greater than or equal to 10%, geographic characteristics of the area as well as its population size are considered. A regional peer with a population[22] less than or equal to 2,500 persons is an inappropriate peer for an urban police district. Similar rules are used in evaluating peers for inefficient regional districts. Any inefficient district that has at least one inappropriate peer is classified as a 'problem' case. These cases were referred to the NSW Police Service for possible explanation.

Notes

[1] The views expressed in this chapter are the authors' and do not necessarily reflect those of the NSW Treasury, the NSW Police Service or the NSW Government. The authors would like to thank Jim Baldwin of the NSW Police Service for providing us with many helpful comments including the data set. All remaining errors are ours.

[2] A local police district is responsible for providing a policing service in a designated area. In Australia, a local police district is commonly referred to as a police patrol. The terms 'local police district' and 'police patrol' are interchangeable.

[3] A major restructure occurred at the start of the 1997-1998 report year. This saw 11 regions and 80 local area commands replace the four regions, 25 districts and 161 patrols, to redirect resources to areas of greatest need. The restructure has increased both the organizational size and geographic coverage of local police districts in New South Wales.

[4] Previous studies measure efforts to solve existing crime by arrests or clearances (for example, see Darrough and Heineke 1979, Gyimah- Brempong 1987, and Gyapong and Gyimah-Brempong 1988). Others use the reciprocal of the crime rate as a proxy for safety (Gyimah-Brempong 1989, Finney 1997) or crime rate by category as outputs (Levitt 1997). The measures in all of these studies capture only one aspect of the NSW Police Service's law enforcement activities.

[5] Carrington, Puthucheary, Rose, and Yaisawarng (1997) refers to law enforcement activities as reactive policing. The authors provide justifications and caveats associated with these measures.

[6] Carrington, Puthucheary, Rose, and Yaisawarng (1997) refers to crime prevention activities as proactive policing. The authors include only one measure of crime prevention activities because number of intelligence reports by local police districts was not available for the 1994-1995 data set. The NSW Police Service improved its data collection and recording, which made it possible to obtain this output measure in 1995-1996.

[7] Labor and capital are the two major input categories for delivering services. Material supply constitutes a minimal fraction of total resources.

[8] Number of police officers includes general duty police, beat police and detectives.

[9] For detailed discussion of a linear programming formulation, see Färe, Grosskopf, and Lovell (1985) and Steering Committee for the Review of Commonwealth/State Service Provision (1997). See also Section 2.4 of this book.

[10] Peers are the group of organizations on the frontier with which a relatively inefficient organization is compared.

[11] The NSW Police Service provided the data. Excluding one Sydney police district that served as the central jail for offenders waiting to appear before a court, there were 163 local police districts in 1994-1995. In 1995-1996, one local police district acquired another and two districts merged to form a new district. The amalgamations effectively reduce the number of local police districts to 161.

[12] This study computes a boxplot for each possible output-input ratio to identify potential outliers. Any police districts that fall outside the median ± 3 inter-quartile range were referred to the NSW Police Service for further verification of the data accuracy. Any errors in data recording were corrected before the DEA efficiency scores were computed.

[13] For every intelligence report, a district intelligence officer assigns it a security classification, determines the distribution of the report, and makes a judgement of its quality. Whilst an intelligence officer makes a judgement of quality, he or she cannot reject an intelligence report on the grounds that any information, however trivial it may appear, is potentially important for solving a case. The variation in number of reports may be due to differences in emphasis on intelligence report preparation by police districts' commanders.

[14] The pair-wise correlation coefficients between police cars and PCs, police cars and station area, and PCs and station area are 041, 0.66 and 0.31, respectively. These correlations are positively significantly different from zero at the one-percent level.

[15] The Data Envelopment Analysis Program (DEAP) developed by Tim Coelli of the University of New England, Armidale, Australia is used to compute all efficiency scores. These scores are computed relative to a variable returns to scale technology; the technology that exhibits increasing, constant, and decreasing returns to scale at different range of outputs. It reflects a manager's ability to transform resources into measured outputs. The calculated efficiency score is therefore a measure of managerial efficiency. For technical discussion, see Chapter 2 of this book.

[16] Average non-radial slacks including numbers of local police districts having positive non-radial slack are similar across the four model specifications

[17] Possible operating environment variables include proportion of young people and public housing as well as location (urban or regional). Carrington, Puthucheary, Rose, and Yaisawarng (1997) found that the operating environment did not explain differences in efficiency scores across police districts in their sample. Hence, this study does not attempt to incorporate the available proxies of operating environment in the analysis. We believe that difficulty and complexity of incidents, for example, affects the use of resources to deliver services. This information should be incorporated in the analysis when it becomes available.

[18] The Spearman rank correlation coefficients between each pair are high, ranging from 0.81 to 0.96. These correlation coefficients are statistically different from zero at the one-percent level of significance. The Friedman statistical test also fails to reject the null hypothesis that the probability distributions of efficiency ranks are the same across model specifications at the one-percent level of significance. The Chi-square statistic with (161.3) degree of freedom is 4.023, having a p-value of 0.2590.

[19] A local police district is efficient by default if its input and output combination is unique, compared to other districts in the sample. In evaluating the efficiency of this police district, it is compared to itself. The efficient-by-default district does not serve as a peer for other inefficient districts.

[20] Since levels of service are predetermined, the additional summons served may not be feasible in practice. However, if the nearby police districts experience excessive summons, they could arrange for District 5 to deliver such a service without investing in additional resources.

[21] Accidentally, Carrington, Puthucheary, Rose, and Yaisawarng (1997) find that some police districts could potentially use less inputs per unit of service by expanding their level of services. Others could do so by reducing their level of services. The authors suggest redefining police district boundaries on a case by case basis.

[22] That is, population of the main town center in the regional patrol area.

REFERENCES

Carrington, R., N. Puthucheary, D. Rose, and S. Yaisawarng. 1997. Performance Measurement in Government Service Provision: The Case of Police Services in New South Wales. *Journal of Productivity Analysis* 8: 415- 430.

Darrough, M.N., and J.M. Heineke. 1979. Law Enforcement Agencies as Multiproduct Firms: An Econometric Investigation of Production Costs. *Public Finance* 34: 176-195.

Färe, R., S. Grosskopf, and C.A.K. Lovell. 1985. *The Measurement of Efficiency of Production*. Boston: Kluwer-Nijhoff Publishing.

Finney, M. 1997. Scale Economies and Police Department Consolidation: Evidence from Los Angeles. *Contemporary Economic Policy* 15 (1): 121-127.

Gyapong, A.O., and K. Gyimah-Brempong. 1988. Factor Substitution, Price Elasticity of Factor Demand and Returns to Scale in Police Production: Evidence from Michigan. *The Economic Journal* 54: 863-878.

Gyimah-Brempong, K. 1987. Economies of Scale in Municipal Police Departments: The Case of Florida. *The Review of Economics and Statistics* 69: 352-356.

Gyimah-Brempong, K. 1989. Production of Public Safety: Are Socioeconomic Characteristics of Local Communities Important Factors? *Journal of Applied Econometrics* 4: 57-71.

Levitt, S.D. 1997. Using Electoral Cycles in Police Hiring to Estimate the Effect of Police on Crime. *American Economic Review* 87 (3): 270-290.

New South Wales Police Service. 1998. *Annual Report 1997-1998*.

Sonnichsen, R.C. 1997. Measuring Police Performance. In: Mayne, J., and E. Zapico-Goni (eds.), *Monitoring Performance in the Public Sector: Future Directions from International Experience*. New Brunswick: Transaction Publishers.

Steering Committee for the Review of Commonwealth/State Service Provision. 1997. *Data Envelopment Analysis: A Technique for Measuring the Efficiency of Government Service Delivery*. Melbourne: Productivity Commission - Secretariat for the Steering Committee.

Steering Committee for the Review of Commonwealth/State Service Provision. 1998. *Report on Government Services*. Melbourne: Productivity Commission - Secretariat for the Steering Committee.

Part Five: Regional Services

12 What Is Known about Municipal Efficiency? The Belgian Case and Beyond

Bruno De Borger and Kristiaan Kerstens [1]

12.1 Introduction

Although other alternatives exist for the evaluation of the performance of municipal governments (e.g., financial performance), most economists are convinced that technical and cost efficiency are relevant criteria for performance measurement at the local level. The purpose of this chapter is to briefly discuss the difficulties involved in capturing municipal efficiency, and to summarize the most important results of recent research on the efficiency of municipalities in Belgium. More specifically, we discuss the difficulties to benchmark local governments, we summarize the determinants of municipal efficiency in Belgium, and we contrast our findings to results available in the literature. Finally, we try to assess the policy implications resulting from this research.

Structure of the chapter is as follows. By way of introduction, we start out in Section 12.2 with a brief discussion of the role of municipal governments. We review the allocative and distributive role of municipalities as it is typically described in the economic literature, and we summarize the actual functions assigned to local governments in Belgium and in a number of other European countries. We also emphasize the importance of efficiency measurement as a partial but crucial assessment of municipal performance. In Section 12.3 we discuss the conceptual, methodological and practical problems in benchmarking local governments. The next section introduces the data and sample utilized. Available empirical results on the efficiency of Belgian municipalities are reported in Section 12.5. Both global performance evaluations and efficiency analyses for specific municipal services are reviewed. In addition, we provide some new evidence related to scale efficiencies. For purposes of comparison, a number of findings for Belgian municipalities are contrasted with results on municipal efficiency in other countries, as reported in the literature. Finally, Section 12.6 briefly reiterates what is known about municipal efficiency and, equally importantly, it points out what remains to be done.

12.2 The Role of Local Governments and the Need for Performance Measurement

Any evaluation of the performance of municipalities should keep in mind the role and objectives of local governments. We therefore begin by briefly reviewing the allocative and distributive roles typically assigned to municipalities. We then provide some information on the specific functions assigned to sub-central governments. Finally, we emphasize the need for performance evaluation in general and for efficiency measurement in particular.

12.2.1 *The Allocative and Distributive Roles of Local Governments*

From an allocative viewpoint the early analyses by Olson (1969) and Oates (1972) have emphasized that, at least in theory, efficient decentralization involves a perfect correspondence between the spatial impact of any given public service and the jurisdiction responsible for decision-making with respect to, and financing of, the service considered. Of course, since the theoretical ideal of optimal jurisdictions and optimal decentralization is not feasible, assigning certain functions to the municipal level should be based on a comparison of the costs and benefits of decentralized decision-making. The benefits of decentralization include, among others, the close association between preferences, decision power, and financing responsibility, some possible informational advantages, and the efficiency-enhancing potential associated with mobility of the population at the local level. The disadvantages of decentralization are related to the spillover effects of some local public services, to possible inefficient tax competition between municipalities, to potential tax exporting behavior by local governments, and to the possible existence of economies of scale in the provision of local public services.[2]

The literature has convincingly shown that the relative magnitude of costs and benefits of decentralization and the related optimal division of responsibilities between different levels of government should depend on: (i) the degree to which preferences for public services differ across jurisdictions; (ii) the spatial impact of local public goods; (iii) the mobility of production factors (e.g., labor and capital) and of the tax base across jurisdictions; (iv) the potential existence of economies of scale in the production of local public services; and (v) the informational advantages associated with decentralized decision-making.

Whereas the literature generally agrees on the factors that determine the precise allocative role of local governments in providing local public services, there is less agreement on the potential for municipalities to engage in redistribution policies. However, most economists agree that redistribution is difficult due to the mobility of the population across

jurisdictions. This generates a problem of adverse selection. For example, if low-income people hope to benefit from redistribution they are attracted to municipalities pursuing an active redistribution policy towards the poor. High-income earners on the other hand, expecting to pay the bill for this policy, may prefer to migrate to municipalities with little redistributional activities. It is often argued, therefore, that municipalities have no formal role via distributive local taxation but that they should focus on the implementation of overall national distributional policies via local tax progressivity, local tax credits etc. Moreover, it is clear that from the viewpoint of the national government distributional issues may justify grants to correct for income inequalities between municipalities, unequal provision of public services or differences in the capacity to generate tax revenues.

12.2.2 *The Functions Assigned to Local Governments*

It is clear that there may be substantial differences in the way different countries implement the above broad principles in practice. In this subsection we therefore consider some information on the assignment of functions to and the financing of sub-central governments in a number of European countries, with a specific focus on the relative position of Belgium.

Comparative studies concerning tasks accomplished by sub-central authorities reveal large similarities between different countries, although some remarkable differences remain. This can be illustrated by considering Table 12.1, adapted from King (1992). Note that this table deals with all sub-central government levels and not just municipal governments. It includes a broad variety of different institutions situated at different tiers, whose precise functions and interrelationships largely depend on historical contingencies. For instance, in Belgium one can distinguish regional, provincial and municipal authorities, public centers for social welfare, intermunicipal agencies, etc.
Despite the level of aggregation underlying Table 12.1, it does provide some relevant information. *First*, it is notable that sub-central authorities not only provide local public goods (parks, roads, police, justice, town planning, etc.) but also some merit goods (various categories of education, sports and leisure pursuits, etc.).[3] Some activities can even be grouped under a regional development function, since they focus on stimulating economic activity in the area. *Second*, remarkable differences can be observed for justice, ports, airports and electricity, among others. *Third*, Belgium appears quite decentralized on the basis of this table. However, this is mainly due to the importance of the regions in the Belgian institutional framework. It certainly does not imply a large assignment of power to municipal governments because many of the functions included in Table 12.1 are assigned to the regional and not the municipal governments. This is confirmed by looking at public sector employment data.

Table 12.1

FUNCTIONS OF SUB-CENTRAL AUTHORITIES IN WESTERN EUROPE

	Austria	*Belgium*	*Denmark*	*France*	*Germany*	*Ireland*	*Italy*	*Luxembourg*	*Netherlands*	*Norway*	*Portugal*	*Spain*	*Sweden*	*Switzerland*	*UK*
Slaughterhouses	•	•	•	•	•	•	•	•	•	•		•	•	•	•
Theaters, concert halls	•	•	•	•			•	•	•	•	•	•	•	•	•
Urban road transport	•		•	•	•		•	•	•	•	•	•	•	•	•
Ports		•			•		•		•	•	•		•	•	
Airports					•		•		•	•			•	•	•
District heating	•	•	•	•	•		•		•				•	•	
Agriculture, forestry, fishing, hunting	•	•		•	•		•		•	•	•	•	•	•	•
Electricity		•	•		•		•		•	•	•	•	•	•	
Commerce	•	•		•	•		•		•		•	•	•	•	
Tourism	•	•		•	•		•		•	•	•	•	•	•	•
Security, Police	•	•		•	•		•	•	•			•		•	•
Justice					•		•		•			•			•
Pre-school education	•	•	•	•	•		•	•	•	•		•	•	•	•
Primary & secondary education	•	•	•	•	•		•	•	•	•		•	•	•	•
Vocational & technical training	•	•		•	•		•		•	•		•	•	•	•
Higher education		•			•		•		•			•	•	•	•
Adult education	•	•	•	•	•		•		•	•		•	•	•	•
Hospitals & convalescent homes	•	•	•	•	•		•	•	•	•		•	•	•	
Personal health	•	•	•	•	•		•	•	•	•		•	•	•	
Family welfare services	•	•	•	•	•		•		•	•		•	•	•	•
Welfare homes	•	•	•	•	•		•	•	•	•		•	•	•	•
Housing	•	•		•	•	•	•	•	•	•	•	•	•	•	•

Common to all countries: refuse collection and disposal; museums, art galleries and libraries, parks and open spaces; sports and leisure pursuits; roads; water supply; town planning; fire protection.

Source: Adapted from King (1992: Table 3.1)

While about 25% of all civil servants are employed within the overall sub-central public sector, only about half of these actually work for municipalities (Heyndels 1996).

Sub-central authorities across Western Europe also differ in their share in government spending and in their capacity to generate tax revenues from different sources (see also Owens 1992). While a huge variety of local financing resources exist, apparently many countries rely predominantly on just a few tax sources. In Belgium, income and property taxes are the most important local taxes. Moreover, conditional or general grants from higher tiers of government provide another major revenue source. As reported in Heyndels (1996), slightly less than half of Belgian municipal financial resources are received under the form of grants, mostly unconditional ones. This number is not exceptional among European countries. Over time grants do seem to be a declining source of revenue.

In Table 12.2 we report the attribution of tax revenues to municipalities. This provides an indication of the degree of tax autonomy of municipalities in individual countries. Notice the large variability in the shares attributed to municipal governments among countries. Belgian local government tax shares are at the lower end of the distribution, reflecting rather limited tax autonomy of municipalities. This is not surprising, given the focus of the federalization process in Belgium on the regional government level. Moreover, it confirms that in a European context, decentralization towards Belgian local governments is quite limited.

12.2.3 *Local Government Objectives, Control, and Efficiency*

Given the role and tasks assigned to the local public sector in providing public services and generating tax revenues, there is an obvious problem of control. Local politicians (managers) can be viewed as agents responsible for implementing efficient policies facing two principals: on the one hand the national government, and on the other hand local residents. The public choice literature (see e.g. Mueller 1989) suggested a large number of sources of inefficiency that may result from this principal-agent relation. We enumerate, without aiming at completeness, a series of potential problem areas. First, there are the differences in objectives between principal and agent, whereby municipal politicians and officials pursue goals (e.g., budget maximization, emoluments of all kinds, etc.) independent of the preferences of their constituency. Second, the local political agenda may dictate compromises. Third, there is a lack of monitoring instruments and the cost of setting up adequate control mechanisms may be non-negligible. Fourth, there may exist informational asymmetries inhibiting a perfect control by either central government or residents.

Table 12.2

SHARE OF TOTAL TAX REVENUES ATTRIBUTED TO MUNICIPALITIES 1975-1995 (PERCENTAGE OF TOTAL TAX REVENUE)

	1975	1985	1995
Federal countries			
Australia	4.2	3.7	3.4
Austria	12.4	10.7	10.8
Belgium	4.7	5.2	5.7
Canada	9.9	9.3	9.5
Germany	9.0	8.9	7.3
Switzerland	19.5	16.7	15.8
United States	14.7	12.6	13.0
Unitary countries			
Czech Republic	-	-	10.7
Denmark	29.8	28.3	31.6
Finland	23.6	22.4	21.9
France	7.6	8.7	9.9
Greece	3.4	1.3	1.2
Hungary	-	-	3.6
Iceland	18.7	18.6	20.4
Ireland	7.3	2.3	2.4
Italy	0.9	2.3	4.8
Japan	25.6	26.0	24.2
Korea	10.1	-	19.5
Luxembourg	6.6	6.6	6.4
Netherlands	1.2	2.4	2.6
New Zealand	7.7	6.5	5.2
Norway	22.4	17.7	19.7
Poland	-	-	7.5
Portugal	-	3.5	5.6
Spain	4.3	11.2	12.8
Sweden	29.2	30.4	32.4
Turkey	-	10.2	12.8
United Kingdom	11.0	10.1	3.9

Source: OECD (1997, Table 137: 210)

Public sector organizations should ideally achieve a variety of goals simultaneously. Among the social objectives of the public sector one finds technical efficiency, allocative efficiency, redistribution, and macroeconomic objectives (unemployment absorption, price stability, etc.). Of course, not all organizations in the public sector necessarily contribute to all of these goals. Municipalities, for instance, should mainly focus on efficiency goals, and, as argued above, have little role to play in redistribution. Marchand, Pestieau and Tulkens (1984) show that the interaction of different objectives leads to complex decision rules that are hard to evaluate in practice. They suggest a performance approach to public sector analysis that gives a pre-emptive priority to technical efficiency, since it is the only goal compatible with all others.

Therefore, the first step in identifying appropriate institutional structures and their impact on efficiency is to develop proper measures of technical inefficiencies and carefully analyzing their determinants. The next section is entirely devoted to the issue of measuring technical efficiency at the municipal level. A second, consequent step is to implement incentive structures that include corrective measures based on efficiency scores in order to reduce observed technical inefficiencies.

12.3 Efficiency Measurement at the Municipal Level: Difficult but Necessary

The performance of municipalities can be evaluated using a wide variety of performance indicators, including financial measures (debt ratios, expenditures per resident, tax revenue per resident, etc.), efficiency measures and indices of effectiveness (i.e., of the degree to which the objectives have been realized). We focus on the measurement of efficiency because, unlike the other concepts, it is a requirement that any society has good reason to impose on local governments. Moreover, since the definitions of various efficiency measures (e.g., technical efficiency, scale efficiency, cost efficiency) have been spelled out elsewhere in this book (see Chapter 2), we assume the reader is familiar with the meaning of these concepts.

12.3.1 *Efficiency Measurement: Informational Requirements*

In principle, measurement of all efficiency concepts is easy if reliable data are available on both quantities and prices of all inputs and outputs of the production process. In many cases, however, the unavailability of some necessary information places restrictions on the type of efficiency that can be measured and on the quality of

the resulting performance indicators. For example, technical efficiency can in principle be estimated using input and output data only, and it does not require price information. Allocative efficiency can only be estimated if price information is available in addition to information on both inputs and outputs. If for some reason input data are unavailable but only total costs and output information can be obtained, then overall cost efficiency can be measured, but it is impossible to separate it into technical and allocative efficiency. Consequently, the type of information available determines the flexibility of performance measurement (see Färe and Primont 1988 for details).

Of course, a prerequisite to estimate any of the efficiency notions is to translate the production process at the local government level in terms of the economist's standard notion of transforming inputs into outputs. This is not a trivial problem. In general, one distinguishes three stages in the production process (see, e.g., Bradford, Malt and Oates 1969, De Groot and Goudriaan 1991, and Levitt and Joyce 1987). A first stage is the transformation of primary inputs (labor, equipment) into activities, which can be considered intermediate outputs (e.g., teaching in municipal schools, police surveillance, administrative work, etc.). Associated with this transformation of primary inputs into quantities of the intermediate outputs (hours of teaching, number of police patrols, etc.) one can define process indicators that reflect the efficiency of the production process. Second, the transformation of intermediate outputs into direct outputs (D-outputs) for 'consumption' (e.g., number of graduating students, number of crimes solved, etc.) can be measured by performance indicators reflecting technical efficiency in the transformation of activities in outputs. Third, the transformation of direct outputs into their ultimate welfare effects on consumers (e.g., increased knowledge in society due to education, increase in safety as perceived by inhabitants, etc.) is captured by effect indicators. They reflect the degree to which the direct outputs of municipal activities translate into welfare improvements.

Clearly, in principle efficiency measures can be defined at each stage of the production process, but in practice data are typically unavailable to distinguish between primary inputs, intermediate outputs, direct outputs, and final welfare effects. In many cases, the analysis is limited to the relation between primary inputs or activities and direct outputs. The third step in the production process is rarely studied. One exception is the Lovell, Walters and Wood (1994) study developing a stratified model of the educational production process.[4]

The next question is what constitutes reasonable indicators of inputs and direct outputs of the municipal production process. On the input side the problem is to find reasonable measures of the production factors labor, capital equipment, and other inputs (e.g.,

various materials). Neither of these inputs is homogeneous, however. Although one could argue in favor of using labor input in terms of full-time equivalent person-hours, it is clear that the personnel of municipalities form a heterogeneous group in terms of educational background, age, and overall productivity. It is unclear how to correct labor hours for quality differences. For the capital input, measurement of the usage of municipal capital services is hampered by the lack of systematic registration of municipal capital accounts in many countries. Moreover, heterogeneity in quality and vintages of capital inputs should ideally be accounted for. On the output side severe measurement problems arise as well. Although for very specific services reasonable direct output measures may be available (e.g., the number of loans in public libraries) in other cases no clear-cut output can be defined (e.g., how to measure the output of purely administrative departments).

Information on input prices suffers from similar measurement problems. First, in many European countries there is limited flexibility for municipalities to determine wage structures. Seniority and hierarchical level are by far the most important determinants of wages. Therefore, even if they are available (in many cases they are not) average municipal wages mainly capture institutional factors, differences in average seniority, and in labor quality rather than pure wage differentials. Prices for capital goods typically do not vary across municipalities at all, since municipalities operate on the same financial markets and obtain the same conditions.

Finally, even if reasonable data are available the question remains to what extent variations in the relation between inputs and outputs really reflect variability in performance. Indeed, in many cases municipal output is demand-driven. In case the inputs are insufficiently flexible, municipalities may have to adjust output quality in response to unexpected demand variations. For example, an unexpected increase in the number of students at municipal schools may result in larger average class sizes rather than in an adjustment of inputs. Although a simple comparison of inputs and outputs will indicate an increase in efficiency, this may reflect a quality reduction rather than a true efficiency gain. This example underscores the necessity to obtain a sufficiently rich and complete description of technology.

12.3.2 *Global Versus Specific Efficiency Measurement*

At this point it is instructive to note the difference between efficiency measurement on a global scale and evaluating the efficiency in the provision of a specific municipal service (such as green services, road construction and maintenance, fire protection,

public libraries, education, etc). In the latter case, it may be somewhat easier to obtain good proxies for the relevant inputs and outputs. For example, although one can disagree about the best set of output indicators for a municipal library, it seems logical that the number of loans should be one important element. However, if one is interested in evaluating the performance of a municipality in the global provision of services, then the problem of defining a comprehensive set of reasonable output indicators fully describing municipal activities is much more difficult. What are appropriate outputs for a municipality as a whole? How to define proxies for the various aspects of local production in the fields of education, administrative services, social and recreational services, transportation, etc?

One reaction to these differences in data problems is to argue in favor of limiting performance measurement to specific local public goods. Not surprisingly, many studies in the literature have focused on efficiency of municipalities in providing very specific goods. However, although such studies may yield valuable information for local governments, from a policy viewpoint evaluating global performance may be much more important. For example, to improve the design of appropriate grant mechanisms between central and local governments an evaluation of the performance of municipalities is required. If current municipal production is not evaluated, grant mechanisms may well be compensating, and in fact rewarding, inefficiencies instead of promoting economy of resource use. Another example for the need of global efficiency measures relates to the appropriate size of municipal production. This cannot be ascertained on the basis of specific efficiency studies.

In view of the above discussion and despite the difficulties involved in defining inputs and outputs, it is extremely important to provide a global evaluation of local government performance. The best one can do is to use proxies that are strongly positively correlated with the true, but unobservable outputs, and to include output categories that as much as possible capture all major categories of service provision. Of course, it goes without saying that this global analysis can benefit from being complemented by detailed performance studies focusing on specific local public goods.

12.3.3 *Efficiency Estimation: Alternative Methodologies*

The efficiency measures discussed in the empirical part of the chapter have been estimated using non-parametric and econometric approaches. For readers unfamiliar with these various approaches, we very briefly provide the thinking behind these methods. For more details the reader can consult, among many others, the excellent paper by

Lovell (1993) and Chapter 2 in this book. He also discusses the merits and disadvantages of various methods.

Econometric approaches specify parametric functional relations between outputs and inputs (technical efficiency) or between costs, outputs and input prices (cost efficiency). The production (cost) frontier is determined by estimating the parameters of the specified functional form using appropriate econometric techniques. The deviations of the observations from the estimated frontier form the basis for the calculation of efficiency scores for individual observations. The specified parametric frontiers can either ignore or allow for measurement error: the former is labeled deterministic; the latter approach yields stochastic frontiers. Deterministic parametric frontiers interpret any deviation from the frontier as inefficiency. Stochastic parametric frontiers try to separate the effects of measurement error and inefficiency.

The non-parametric approaches known as Free Disposal Hull (FDH) and Data Envelopment Analysis (DEA) are not based on the specification of a functional form for the frontier. Instead the frontier is determined by 'enveloping' the data as tightly as possible. This results in piecewise linear frontiers. Imposing the least assumptions, FDH yields the closest envelopment of the data. Graphical representation of the frontier yields a staircase shape in input-output space. Contrary to FDH, the DEA-approach assumes convexity of the production possibilities set. Graphically, the DEA-frontier 'convexifies' the staircase FDH-frontier. In both cases, inefficiency is again measured in terms of the deviation of an individual observation from the frontier.

12.4 Data Description

In this section we provide some information on the type of data used to study the global efficiency of Belgian municipalities. Of course different studies in the literature have used somewhat different data sets and tested a wide variety of specifications. Apart from the seminal paper by Tulkens (1990), the most relevant studies of global efficiency of Belgian municipalities are Vanden Eeckaut, Tulkens, and Jamar (1993), De Borger, Kerstens, Moesen, Vanneste (1994) and De Borger and Kerstens (1996a, b). The former two studies were limited to the 235 municipalities of the Walloon region, whereas the others analyzed the performance of all 589 Belgian municipalities. Moreover, some studies focused on technical efficiency using explicit input and output indicators (e.g., De Borger, Kerstens, Moesen, Vanneste 1994), whereas others analyzed cost efficiency (e.g., Vanden Eeckaut, Tulkens, and Jamar 1993 and De Borger and Kerstens 1996a). Finally, a variety of efficiency measures have been used. For example, the analy-

sis in Tulkens (1990) and De Borger, Kerstens, Moesen, Vanneste (1994) is restricted to FDH only, while De Borger and Kerstens (1996a) studied cost efficiency using a variety of frontier methodologies.

To avoid going into a detailed description of the precise data used in the various studies, we focus in this section on a standard data set used in De Borger and Kerstens (1996a). It serves to illustrate both the strength and the shortcomings of efficiency measurement. This same data set has also been used in the empirical work on scale efficiency reported below.

The sample consists of observations on total current municipal expenditures and on various output indicators for each of the 589 local governments in 1985. This combination of expenditures and outputs implies that our focus is on cost efficiency. The output indicators used include the number of beneficiaries of minimal subsistence grants, the number of students enlisted in local primary schools, the surface of public recreational facilities, the total population, and the fraction of the population older than 65. Descriptive statistics for all outputs and costs are provided in Table 12.3.

Table 12.3

Descriptive Statistics on Outputs and Input

	Mean	*Stand. Dev.*	*Minimum*	*Maximum*
Number of beneficiaries of subsistence grants	74.2	269.2	0.99	4,430
Number of students	3,070	7,500	1	1,050
Surface public recreational facilities (x 1,000)	1,020	1,020	6	10,200
Total population	16,700	28,800	100	483,000
% Population above 65	0.18	0.03	0.07	0.29
Total expenditures (million BEF)	450	2,020	1.14	38,500

Notice the wide range and large standard deviations for almost all variables. This simply reflects the important size differences among municipalities. All data come from a more elaborate database constructed at the research institute CADEPS (Free University of Brussels) on the basis of information from the Nationaal Instituut voor de Statistiek (NIS) and from the Gemeentekrediet van België (GKB).

As previously emphasized, many of these 'output' variables are to be considered as crude proxies for the services delivered by municipalities. For example, population is assumed to proxy for the various administrative tasks (e.g., maintaining the register of births, marriages, and deaths; issuing certificates, passports, etc.) performed by municipal governments, but it is clearly not a direct output of local production. Similarly, the fraction

of the population older than 65 is a crude proxy for the supply of social services to the elderly (e.g., retirement homes, medical services in public hospitals, general assistance, etc.). Minimal subsistence grants are related to the services provided to low income families, where it should be noted that Belgian local governments cover 50% of the costs.

The quality of these output indicators is less than desirable, as they are rather loosely related to the services delivered by municipal governments. This illustrates the general problem inherent in the definition of inputs and outputs for the public sector alluded to in the previous section. Also note that it was not possible to correct for unobservable variations in quality, since no reliable quality indicators were available. Fortunately, the focus on cost efficiency slightly reduces the severity of the latter problem. Indeed, using total costs at least guarantees that all inputs are accounted for. Although variations in quality are hard to capture in explicit input indicators, they should at least partly show up in the price being paid (information that is unknown) and consequently in the budget.

With few exceptions studies on the efficiency of Belgian municipalities deal with cost efficiency.[5] There are good reasons for this tendency. Estimating cost efficiency implies that the efficiency measures capture a conglomerate of pure technical and allocative inefficiency. However, one could argue that the risk of misspecifying the inputs is more important than the eventual bias of mixing up technical and allocative inefficiency. Using total expenditure at least assures no input is ignored. Furthermore, in the Belgian context, it is not unreasonable to assume there is no input price variability: there is little wage flexibility as salary scales of municipal personnel are completely fixed, and all municipalities in fact obtain most funds from a single specialized financial institution.[6] Therefore, cost efficiency is the main focus of the global efficiency results in this article.

12.5 Results of Efficiency Studies

The purpose of this section is twofold. First, we summarize the results and the policy implications of studies analyzing the efficiency of Belgian municipalities. Second, we compare our results to other studies available in the literature to see how robust these findings are. We thereby initially focus on results derived from the analysis of global efficiency of local governments (Section 12.5.1). We then briefly report some new results on scale efficiency (Section 12.5.2), comparing it with available evidence. Finally, we summarize a small but representative sample of studies that look at the performance of municipalities in producing specific services (Section 12.5.3).

12.5.1 Global Efficiency of Municipalities

Let us first consider studies dealing with global efficiency of municipal governments. Our initial focus is on Belgian studies, part of which have been introduced in Section 12.4. Furthermore, also a brief comparison with the few existing papers on overall efficiency of municipalities in other countries is provided. Table 12.4 summarizes a number of relevant references and it provides some additional information on the data and methodology used in each of these analyses as well as on their most important findings.

Focusing on the data set presented in Section 12.4, the cost efficiency measures were estimated using FDH, DEA, and econometric approaches. Several frontier methods were estimated to check for the robustness of results. Mean efficiency was estimated at 0.94 in the FDH case, 0.73 for the DEA application, a remarkably low 0.57 when the deterministic parametric frontier method was used and, finally, about 0.78-0.81 in the case stochastic parametric frontiers were estimated. A detailed analysis revealed that parametric and non-parametric approaches differed markedly in terms of both efficiency levels and their implied rankings. Among non-parametric methods, it was confirmed that FDH and convex DEA models differ in important ways, in line with Tulkens (1990) and Vanden Eeckaut, Tulkens, and Jamar (1993). Given our ignorance about public sector production in general and about municipal production at the aggregate in particular, these findings have, among non-parametric methods, favored the FDH.

The resulting efficiency scores were further used in an explanatory analysis. In general, it was found that the fiscal revenue capacity, the financing mechanisms of local governments, and the political characteristics of local governments are important determinants of inefficiencies. To be slightly more specific, it is necessary to consider the results of the explanatory analysis in more detail.

First, it was consistently found (both for the explanation of cost efficiency and technical efficiency) that higher per capita incomes and wealth of citizens decreased estimated efficiencies of municipal governments. A potential explanation is that high incomes reduce the incentives of both politicians and taxpayers to monitor expenditures. Politicians have fewer incentives to control expenditures because higher incomes increase the fiscal capacity of municipalities (see e.g. De Groot and Van der Sluis 1987, Spann 1977, Silkman and Young 1982, and Wyckoff 1990). Citizens may be less motivated to carefully monitor expenditures due to the high opportunity costs of time and effort.

A second systematic finding is that the way local public services are financed is important for performance. In particular, local tax rates and the size of intergovernmental grants

seem to matter a great deal. The literature has pointed out that high tax prices alert voters to control public expenditures, especially if cost comparisons between municipalities are easy (see e.g. Spann 1977). In line with this observation we consistently find that the local income tax contributes positively to efficiency. The size of the per capita block grant yields a negative impact on performance. This is not surprising in the Belgian institutional environment. Quite a lot of these funds take the form of unconditional block grants, and there exists relatively little ex-post control on actual spending. The cost of inefficient behavior is increasingly shared by a broader constituency (national taxpayers) as the proportion of outside funding increases, so that grants reduce efficiency (see, in another context, De Groot and Van der Sluis 1987, Silkman and Young 1982, and Wyckoff 1990). Apart from a 'flypaper' effect in the traditional sense (see e.g. Hamilton 1983) we indeed consistently find a negative relation between efficiency and per capita grant size.

Third, although they were introduced in various different ways, the studies on municipal efficiency in Belgium consistently find that political indicator variables are relevant in explaining efficiency. The literature offers several avenues of explanation. Indeed, both the property rights and principal-agent literature suggest reasons why politicians and public managers lack proper incentives to effectively audit and control expenditures. For example, the process of political decision making itself may impede the effective control of the public sector (Mueller 1989, Bartel and Schneider 1991). In particular, efficiency may be affected by the size and composition of political coalitions, as arbitrage in the bargaining process may require explicit or implicit side payments (e.g., logrolling) depending on the number and nature of the coalition partners. The results for Belgium suggest that the number of coalition parties may have a negative effect on efficiency. Moreover, it is found that the presence of the socialist party has a positive effect. The effect of liberals in the coalition is not clear, as the sign of the coefficient was found not to be robust across specifications.

A fourth systematic finding is related to the fact that the political participation of citizens themselves seems to enhance the performance of a municipality. Though difficult to quantify directly, evidence in the literature suggests that political participation is closely related to education (Mueller 1989). The educational level of the adult population was found to consistently improve cost and technical efficiency. Although this variable may to some extent also capture the impact of population characteristics on costs (Schwab and Oates 1991) it provides partial support for the hypothesis formulated above.

In view of this discussion the determinants of the global efficiency of Belgian municipalities are easily summarized. The most relevant factors seem to be the fiscal revenue capacity (proxied by income), the financing mechanisms of local governments (local

taxes and grants), the political characteristics of local governments and the political participation of the local population. A relevant question is whether these Belgian results are in line with research in other countries. Some information to investigate this question has been included in Table 12.4. It summarizes a number of relevant references: it provides some additional information on the data and methodology used in each of these analyses, and it briefly describes their most important findings.

Some remarkable results are the following. Focusing on a sample of Greek municipalities, Athanassopoulos and Triantis (1998) find mean cost efficiencies somewhat below ours. In their explanatory analysis, a high share of fees and charges in municipal income improves performance, while the share of grants does the opposite. This is line with our own results. The positive effect of the investment share may be related to the unmeasured quality of services provided. Population density affects cost efficiency negatively, which may be interpreted as evidence that some of the municipalities in their sample suffer from scale diseconomies. This study is of importance, since Greece is, together with Belgium, among the most centralized countries in Europe.

In their US study Hayes, Razzolini and Ross (1998) analyzed the determinants of slack using variables representing socioeconomic characteristics, costs, and measures of municipal competition for residents (tax competition). The results suggested that inefficient behavior was associated with richer communities, lower education levels, and a lack of competition for residents among municipalities. The former two findings are entirely consistent with the Belgian results. The latter finding raises an interesting issue that could not be studied in the Belgian context, viz. the relation between tax competition and technical efficiency. Although Hayes, Razzolini and Ross (1998) suggest that competition reduces efficiency, the evidence is mixed. In an earlier study, Hoyt (1990) did not find a significant relation between tax competition and efficiency. While not explicitly measuring technical efficiency, he tested whether the flight of capital and residents affected local public service expenditures and did not find evidence for a limiting impact of municipal competition. Since the evidence is even mixed for the USA, it seems implausible that the much lower population mobility in Europe exerts a strong disciplining effect on efficiency.

Little evidence was found in the international literature on the effects of purely political factors. In the Greek case, Athanassopoulos and Triantis (1998) report that parties affiliated to the central government perform poorly at the local level. Hayes and Chang (1990) study whether or not the city manager form of government is more efficient than the mayor-council form of government in formulating and implementing public policies. The theoretical model shows that depending on compensation and the labor mar-

Table 12.4

Global Performance of Municipalities: Description of Frontier Models and Key Results

Study	Database Sample Size Country	Frontier Method	Inputs (I) Outputs (O)	Key Results
Athanassopoulos and Triantis 1998	Cross-section 1986 N = 172 municipalities Greece	DEA-CRS, DEA-VRS, Stochastic parametric cost frontier	Total current expenditures I Actual households O Average house area O Built-up area O Heavy industrial use area O Tourist area O	- mean CE varies between 0.50 and 0.88, depending on the frontier methodology - a high % of fees and charges in income improves CE positively, as does a high investment share - grants and population density affect CE negatively - parties similar to central government perform poorly
Conceição Sampaio de Sousa and Ramos 1999	Cross-section 1991 or 1992 N= 701 & 402 municipalities in states Minas Gerais & Bahia Brazil	DEA-CRS, DEA-VRS, FDH	Total current expenditures I Actual residents O Domiciles with safe water O Domiciles with garbage collection O Illiterate population O # students primary & secondary schools O	- mean CE varies between 0.56 and 0.91 (0.55 and 0.75), depending on the frontier methodology for Minas Gerais (Bahia) - most municipalities enjoy increasing returns to scale - TE smallesr among small municipalities - more technical inefficiency than scale inefficiency
De Borger, Kerstens, Moesen and Vanneste 1994	Cross-section 1985 N = 589 municipalities Belgium	FDH	Total personnel I [alternative: Blue collar workers I; White collar workers I; Municipal building surface I] # min. subsistence grants O # students primary schools O Surface recreational facilities O Total population O % of population > 65 O	- mean TE varies between 0.86 and 0.99, depending upon specification - scale and education affects TE positively - per capita block grant and income affect TE negatively.

De Borger and Kerstens 1996a	Cross-section 1985 N = 589 municipalities Belgium	FDH, DEA-VRS, COLS, Stochastic parametric cost frontier	Total current expenditures I # min. subsistence grants O # students primary schools O Surface recreational facilities O Total population O % of population > 65 O	- mean CE varies between 0.57 and 0.94, depending on the frontier methodology - local tax rates and education affect CE positively - per capita block grant and income affect CE negatively - methodological: differences among nonparametric deterministic methods are as large as between parametric and nonparametric methods
Hayes and Chang 1990	Cross-section Year: n.a. N = 191 municipalities USA	Stochastic parametric cost frontier	Police protection O Fire protection O Refuse collection O Labor price I Capital price I % owner occupied housing E % minorities E % population over 25 E fire rating E	- current costs only include the 3 services - CE: mean = 0.81; mode = 0.84 - mayor-council and city manager forms of government perform equally well
Hayes, Razzolini and Ross 1998	Cross-section Year: n.a. N = n.a. municipalities USA	Deterministic indirect output distance frontier	Ratio of population to crimes O Median housing value O Input information: n.a.	- high median income and population growth decreases TE - more high school graduates affect TE positively - suburban municipalities are more efficient (Tiebout effect)
Vanden Eeckaut, Tulkens and Jamar 1993	Cross-section 1986 N = 235 Walloon municipalities Belgium	FDH, DEA-VRS	Total current expenditures I # min. subsistence grants O # students primary schools O Road length O Crime rate O Total population O % of population > 65 O	- strongly decreasing returns to scale - municipalities run by nationwide parties are more efficient, while composite majorities are less efficient - methodological: convex nonparametric deterministic methods yield less credible results than FDH

ket for city managers, either of both government forms can be more efficient. Empirically, they find no statistically significant difference in efficiency. Finally, Liner (1994) provides evidence on the role of institutional restrictions on efficiency. For a sample of cities from 43 US states, he studies how changes in institutions, in particular restrictive versus nonrestrictive annexation laws, influence municipal expenditures and employment. Municipalities operating under municipal-determination annexation laws are not found to experience significantly different growth rates in costs and employment per resident from that of cities operating under annexation laws that imposed a greater number of barriers to annexation.

12.5.2 *Scale Efficiency of Belgian Municipalities*

We now briefly turn to a rather under-explored issue in municipal performance gauging, i.e., the question of scale efficiency. The data set is the one described earlier in Section 12.4 of the chapter. Results using FDH and DEA type of technologies are reported in Tables 12.5 and 12.6.[7] Average scale efficiency based on FDH is estimated at about 0.77, which is much lower than average technical efficiency.[8] In the DEA-model the opposite is observed: average scale efficiency is larger than average technical efficiency. These differences between FDH and DEA should be interpreted in view of different maintained assumptions. A second finding is that the FDH analysis suggests that an overwhelming majority of Belgian municipalities seem to be operating under decreasing returns to scale (84%). Only 6% of municipalities enjoy increasing returns to scale, while about 10% are situated in a region of constant returns. For DEA the results are qualitatively similar, but slightly more extreme (93%, 5% and 2% operates under decreasing, increasing and constant returns to scale, respectively). Attaching most weight to the FDH findings, this implies that the poor performance of Belgian municipalities can only be partly remedied in the short run by improving management. More fundamental, structural solutions require reconsidering the current scale of production.

Table 12.5

TECHNICAL (TE) AND SCALE EFFICIENCY (SCE) RESULTS

	FDH		DEA	
	TE	SCE	TE	SCE
Average	0.94	0.77	0.78	0.85
Standard deviation	0.12	0.16	0.17	0.15
Minimum	0.44	0.20	0.32	0.18
Number of efficient observations	391	57	64	14

Table 12.6

Returns to Scale Results per Type of Technology

	FDH		DEA	
	Number	%	Number	%
Increasing returns to scale	36	6.11	28	4.75
Constant returns to scale	57	9.68	14	2.38
Decreasing returns to scale	496	84.21	547	92.87

These scale findings are in line with parametric Cobb-Douglas and DEA results reported in Vanden Eeckaut, Tulkens and Jamar (1993) for the Walloon region. More recently, Vanden Eeckaut (1997) reported similar FDH-based scale results for the same region. However, the results are not necessarily consistent with the interesting conclusion reported in Moesen (1990). In his study of the 1977 municipal territorial reform (i.e., a merging operation of on average four to five municipalities to one), he concludes that this operation was successful in terms of reducing municipal costs per capita. Although this may have been the case, our evidence, based on a different data set and a different time period, suggests that this reform operation may have created entities that are too large, on average. Casual evidence in favor of this view, though it definitely merits further study, is that Belgian municipalities are, in terms of population, on average among the bigger ones in Europe (see Moesen 1990). The only other study reporting scale findings is Conceição Sampaio de Sousa and Ramos (1999) who find for Brazil exactly the opposite problem: most municipalities are far too small. Clearly, there is a definite need for additional studies to clarify this matter.

12.5.3 *Efficiency of Municipalities in the Provision of Specific Local Services*

In addition to studies of the global performance of municipalities, a series of contributions has explored the efficiency of specific municipal services. They are briefly summarized in Table 12.7, together with a number of relevant international studies. The selection of papers is quite restrictive; in certain areas many more articles are available. Though it is clear that not all local goods and services have been covered in the literature, there are studies in existence on waste collection, public libraries, green services, police, civil registries, etc.[9]

Table 12.7

Performance in Specific Municipal Services: Description of Frontier Models and Key Results

Study	*Database* *Sample Size* *Country*	*Frontier Method*	*Inputs (I)* *Outputs (O)*	*Key Results*
Waste Collection				
Burgat and Jeanrenaud 1990, 1994	Cross-section Survey 1987 N = 98 Switzerland	COLS FDH (primal)	Labor in hours I Capacity (m3) I Tons/year O	- distance between collection and disposal site influences TE negatively (COLS) - TE (inputs): 67 inefficient municipalities with range 0.33-0.98 (ibid. for output orientation)
	Cross-section Survey 1987 N = 88 Switzerland	COLS (dual)	See above	- cost efficiency for 53% of DMUs within range 0.4 and 0.6 - impact environmental variables: outsourcing to private companies increases cost efficiency (e.g. by reduced wages). Using user charges (instead of general revenue) has the same effect for smaller municipalities
Courcelle, Kestemont, Tyteca and Installé 1998	Panel 1990-1994 N = 9 programs Different European countries	FDH	Extra cost of waste management program I Diversion ratio (% material recovered) O Residue ratio (% residue in sorting process) O	- 5 to 12 efficient observations depending upon specification
Distexhe 1993	Cross-section 1990 N = 176 municipalities Belgium	DEA-VRS, FDH and costfunction	Labor in hours I Capacity I Tons/year O Frequency O	- mean TE (inputs) = 0.72 for 2I/2O DEA model - results TE depend on specification of production process - private companies and competitive tendering increase TE

Police				
Davis and Hayes 1993	Cross-section 1982-1986 N = 141 IL municipalities USA	Stochastic parametric cost frontier	1/per capita crime rate O Labor price I Capital price I Community median income E % minorities E % high school graduates E	- mean CE: 0.76 - high tax rates increase CE - population size affects CE nonlinearly - high homeownership increases CE - urban status and mayor versus city manager have no impact
Grosskopf, Hayes and Hirschberg 1955	Time Series 1977-1987 Monthly data Dallas Police Dept. USA	Deterministic parametric input distance frontier	Corrected arrest rates for car theft O Corrected arrest rates for murder O Reported car thefts I Reported murders I Police officers I Sergeants I Civilians I	- estimated shadow price ratios relative to observed prices ratios indicate AE - budget tightening and escalating crime rates improved allocative inefficiencies early in the period
Hayes and Wood 1995	Cross-section 1982-1986 N = 319 IL municipalities USA	Stochastic parametric money metric utility frontier	Police expenditures O Uniformed labor price I Civilian labor price I Capital price I	- CE: mean = 0.96 - less inefficiency in rural areas - high % of owner occupied housing yields less inefficiency - other potential determinants like education and tax rates are insignificant
Road Maintenance				
Deller and Nelson 1991	Cross-section 1984 N = 446 USA (IL, MI, WI)	DEA-VRS	Personnel I Road graders I 1-axle trucks I Surfacing material I Length roads (3 types) O	- mean TE (inputs) = 0.975 (radial) & 0.488 (non-radial) - mean SE (inputs) = 0.691 - mean AE (inputs) = 0.349 - positive relation between size and efficiency
Deller and Halstead 1994	Cross-section 1987 and 1990 N = 104 USA	Stochastic parametric cost frontier	Personnel I Road graders I 1-axle trucks I Length roads (3 types) O	- increasing returns to scale - mean cost E (inputs) = 0.594 - 50% of observations is in range 0.40-0.75 - decision makers with formal training perform better

Road Maintenance				
Hjalmarsson and Odeck 1996	Cross-section 1983-1985 N = 72 departments Norway	DEA-CRS DEA-VRS	Labor Cost I Energy Cost I Cost Rubber Materials I Maintenance Cost I Vehicle-km O [alternative: Effective production time O]	- mean TE (inputs) = 0.76 for Vehicle-km and 0.88 for Effective production time - mean SE (inputs) = 0.85 for Vehicle-km and 0.94 for Effective production time - no differences between brands and age of vehicles - regional differences in terms of climate (snowclearing in the inland)
Civil Registry				
Bouckaert 1993	Cross-section 1989-90 N = 11 regional cities Belgium	FDH	Personnel I Delivered documents O [alternative: Population O]	- 3 cities are technically inefficient - mean TE (inputs) = 0.839 for Delivered documents and 0.913 for Population[a] - quality indicator little effect
Fire Services				
Bouckaert 1992	Cross-section 1987 N = 13 regional cities Belgium	FDH	Tot. Expenses I Fire related interventions O [alternative: Other emergency interventions O]	- 5 to 6 cities are cost inefficient - mean CE (inputs) = 0.919 for fire-related interventions and 0.963 for all emergency interventions[a]
Public Libraries				
Depré, Bouckaert and De Smet 1991	Cross-section 1990 N = 11 regional cities Belgium	FDH	Labor in hours I Loans and inter-library loans O	- mean TE (inputs) = 0.94[a] - 3 cities are technically inefficient and also deliver a poor quality (quality indicator in terms of location, opening hours, etc.)
Vitaliano 1997	Cross-section 1992 N = 235 public libraries (72 municipal ones) USA	Stochastic parametric cost frontier	Labor Cost I Total circulation O Total hours opened O New collection O	- mean TE (inputs) = 0.79 - government run libraries are more efficient - % of local funding and gifts has a negative effect

Study	Data	Method	Inputs/Outputs	Results
Green Services				
Depré, Bouckaert and Wijns 1991	Cross-section 1990 N = 11 regional cities Belgium	FDH	Labor Cost I Surface in ha (weighted by frequency) O	- 4 to 5 cities are technically inefficient depending on whether temporary (subsidized) personnel is included or not - mean TE (inputs) = 0.89/0.85 with/without temporary personnel[a] - quality indicator little effect

a Own calculations.

With few exceptions, the Belgian contributions are best considered case studies because of their small sample sizes. Examples of this approach include Bouckaert (1992, 1993), Depré, Bouckaert and De Smet (1991), Depré, Bouckaert and Wijns (1991), and Courcelle, Kestemont, Tyteca, and Installé (1998). Most of these studies are based on a small sample of medium-sized Flemish cities and study their efficiency in the provision of fire services, civil registration services, in the operation of municipal libraries, and in the supply of green services. They have the merit of considering the production process in quite some detail and thereby provide interesting information for the municipalities. In fact, some of these studies, e.g. Bouckaert (1993), were validated by the benchmarked units and led to effective behavioral changes. But given the small sample sizes, these studies deliver limited general information on underlying causal mechanisms.

In addition, they are useful from another perspective, viz., they confirm the intrinsic difficulties of evaluating municipal performance. Probably the best example is Bouckaert (1994), who, in a follow up study to Depré, Bouckaert and Wijns (1991), analyses in great detail the functioning of green services in 11 regional cities in Belgium. Enumerating in great detail their activities, he observes an enormous heterogeneity in the amount provided (e.g. whether the unit buys or grows its own plants, whether it is responsible for green inside municipal buildings or not, etc.) and in the organization of these services (e.g. whether all or parts are outsourced). This microscopic view reveals the large number of decisions that underlie aggregate behavior that may not be observable in terms of inputs (labor, capital, etc.) or costs.

The only Belgian study that is based on a relatively large sample of municipalities is Distexhe (1993). She investigated waste collection in the Walloon region (the French speaking region of Belgium) using FDH, DEA, and econometric approaches. Interestingly, she finds that both private provision of waste collection services and a for-

mal process of competitive tendering increase efficiency. Her results are broadly consistent with e.g. Burgat and Jeanrenaud (1990, 1994) and Tickner and McDavid (1986), who also indicate that contracting out is more efficient.

Some systematic findings in this more focused literature can be linked to the global performance evaluation above. For example, in view of the Belgian results it is interesting to note that Davis and Hayes (1993) found strong evidence of a positive relation between tax rates and monitoring effort in the US police system. This suggests that the impact of tax rates on efficiency might indeed be related to monitoring. The same study confirmed the earlier results that mayors versus city managers forms of government have no impact on efficiency.

12.6 What Have We Learnt and What Remains to Be Done?

This chapter has emphasized the necessity and the difficulty of global performance evaluation at the level of municipal governments. While in our view the global efficiency measures should not be used for ranking individual municipalities or to formulate policies towards specific local services, the attempts to unveil causal mechanisms is certainly worth going for. However, despite huge problems in defining proper inputs and outputs of the municipal production process some consistent results do follow from the available empirical studies. At the global level the fiscal revenue capacity, the financing mechanisms of local governments, the political characteristics of local governments and the political participation of the local population all are important determinants of efficiency. Specifically, we consistently found that citizens' average income and the size of per capita grants negatively affect municipal performance. Local tax rates and the educational achievements of the population stimulate efficiency. Moreover, it was shown that political factors are important as well.

As to the Belgian literature on specific municipal services which have been subjected to performance gauging, it seems clear that their analysis so far has contributed more to managerial remedies than to a general body of knowledge about causal influences. However, both the analysis of the global and specific performance of municipal services is seriously hampered by the lack of readily available data. Both Bouckaert (1996) and Moesen (1994) emphasized the poor situation in Belgium with respect to performance auditing. Bouckaert and Van Reeth (1998) underline that the Flemish region at least accepted the principle of moving towards a performance-centered budgeting and auditing process, but that other countries are moving faster in this respect. Only when a performance perspective will be ingrained into the public sector, the data will become

available to comprehensively assess municipalities as well as other parts of the public sector.

We conclude by pointing to three important avenues for further research. First, so far no published study used panel data to trace down the evolution of global municipal performance over time. This could provide a firmer test ground for the suggested causal links. Second, a fundamental question that hitherto received too little attention is about the proper production model for municipalities, and public sector organizations in general. It has specifically been argued that the cost indirect output correspondence is well suited to model public sector behavior. It assumes that organizations maximize outputs, but, by including a budget constraint, it recognizes that most organizations have some degree of freedom in choosing their input mix. Recent applications of this indirect production model in the public sector include Blank, Eggink and Merkies (1998), Glass, McKillop and O'Rourke (1998), Grosskopf, Hayes, Taylor and Weber (1997) and Chapters 4 and 6 of this volume. Clearly, more work needs to be done to assess its merits and drawbacks.

Finally, most studies have focused rather narrowly on performance gauging. A fruitful area for future research is to integrate performance measurement into more general planning procedures at the local and regional level. A first effort in this direction is available. Indeed, a recent study (Athanassopoulos 1995) integrated the application of frontier methods into a planning model including equity concerns and resource constraints. It links DEA models to multi-level planning problems and applies the model to a sample of Greek local authorities. It evaluates municipal efficiencies, meanwhile accounting for equity and effectiveness objectives and common resource constraints (in this case intergovernmental transfers) at the aggregate level.[10] More research in this field may over time result in a much-needed practical policy instrument.

Notes

[1] We thank J. Blank for his detailed comments. We bear sole responsibility for the final text.

[2] Note that some of the disadvantages of decentralization can be reduced or even eliminated by the introduction of centralized tax collection, by imposing certain restrictions on local tax rates, and by the introduction of a grant system from higher to lower level governments.

[3] For Belgium, the variability in the expenditure shares of nine functional categories of local services, covering a wide variety of the above functions, has recently been studied in Bastiaens, De Borger and Vanneste (1999).

[4] A highly indirect way to evaluate final welfare effects of municipal services is to assess their performance by comparing aggregate property values among municipalities (Grossman, Mavros and Wassmer, 1999). This assumes the desirability of the municipal mix of services is fully capitalized into the value of the stock of housing and other properties. In a somewhat similar vein, Prieto and Zofio (2000) recently combine need variables (population and housing) with detailed direct outputs to assess the provision performance of Spanish municipalities. Assuming that the state applies the same financing rules to each demand by municipalities, differences in provision indirectly reveal eventual differences in efficiency in production.

[5] De Borger, Kerstens, Moesen, Vanneste (1994) studied technical efficiency on the basis of a three input-five output model. The inputs included white-collar staff, blue-collar workers and capital services (crudely proxied by the surface of buildings owned by the municipality). The outputs were as defined above, except that the two population variables were replaced by the surface of municipal roads and a proxy for the services delivered to residents of other municipalities. so as to capture spillovers of local services to other municipalities.

[6] If the assumption of identical input prices were not valid, then Färe and Primont (1988) show that our estimates of cost efficiency would provide a lower bound to the true technical efficiency.

[7] To avoid conflating scale and technical efficiencies, scale efficiency is defined as the ratio between an efficiency measure calculated on a constant returns to scale technology and one computed on a variable returns to scale technology. The constant returns to scale model provides a long run competitive equilibrium benchmark. The volume of the resulting production possibility set is always larger (never smaller) than in the variable returns to scale case. Qualitative scale information can be obtained by further comparing these technologies to a third, non-increasing returns to scale technology. Traditionally, this is done using non-parametric, convex DEA technologies, but Kerstens and Vanden Eeckaut (1998) developed an analogous method based on FDH-based technologies. We prefer the latter method given its weaker assumptions, although, for comparative purposes, we also report the DEA findings.

[8] Computing scale efficiency using costs instead of proper inputs yields normally different results. Färe and Grosskopf (1985) indicate conditions under which these dual and primal approaches to scale efficiency coincide.

[9] Additional studies and new results for police efficiency can also be found in Chapters 10 and 11. One topic we will pay no attention to education, since no studies are available for the Belgian case. Existing studies on school performance are summarized in Chapters 7 to 9.

[10] In the same vein, but less systematically, Ali, Lerme and Nakosteen (1993) used DEA to assess US municipalities in terms of both their needs and fiscal efforts in order to suggest changes in the current patterns of inter-governmental transfers.

References

Ali, I., C. Lerme, and R. Nakosteen. 1993. Assessment of Intergovernmental Revenue Transfers. *Socio- Economic Planning Sciences* 27 (2): 109-118.

Athanassopoulos, A. 1995. Goal Programming & Data Envelopment Analysis (GoDEA) for Target-Based Multi-Level Planning: Allocating Central Grants to the Greek Local Authorities. *European Journal of Operational Research* 87 (3): 535-550.

Athanassopoulos, A., and K. Triantis. 1998. Assessing Aggregate Cost Efficiency and the Related Policy Implications for Greek Local Municipalities. *INFOR* 36 (3): 66-83.

Bartel, R., and F. Schneider. 1991. The "Mess" of the Public Industrial Production in Austria: A Typical Case of Public Sector Inefficiency? *Public Choice* 68 (1): 17-40.

Bastiaens, E., B. De Borger, and J. Vanneste. 1999. Crowding Out Effects in the Local Budget and the Allocation of Local Public Consumption. *Tijdschrift voor Economie en Management* (forthcoming).

Blank, J., E. Eggink, and A.H.Q.M. Merkies. 1998. *The Choice of an Appropriate Economic Model for the Structure of Production: The Case of Dutch General Hospitals*. Den Haag: Sociaal en Cultureel Planbureau (Werkdocument 49).

Bouckaert, G. 1992. Productivity Analysis in the Public Sector: The Case of Fire Service. *International Review of Administrative Sciences* 58 (2): 175-200.

Bouckaert, G. 1993. Efficiency Measurement from a Management Perspective: A Case of the Civil Registry Office in Flanders. *International Review of Administrative Sciences* 59 (1): 11-27.

Bouckaert, G. 1994. *Vormen van gemeentelijke dienstverlening: Project regionale steden*. Leuven: Katholieke Universiteit Leuven (Faculteit der Sociale Wetenschappen, VCOB).

Bouckaert, G. 1996. Measurement of Public Sector Performance: Some European Perspectives. In: Halamachi, A., and G. Bouckaert (eds.), *Organizational Performance and Measurement in the Public Sector: Towards Service, Effort and Accomplishment Reporting*. Westport: Quorum Books, 223-237.

Bouckaert, G., and W. Van Reeth. 1998. Budget Modeling for Efficiency and Effectiveness: The Case of the Flemish Government. In: De Graan, G.B.K., and F.G. Volmer (eds.), *Performance Budgeting: A Perspective on Modeling and Strategic Planning in the Public Sector in Holland*. Delft: Eburon. 43-53.

Bradford, D., R. Malt, and W. Oates. 1969. The Rising Cost of Local Public Services: Some Evidence and Reflections. *National Tax Journal* 22 (1): 185-202.

Burgat, P., and C. Jeanrenaud. 1990. *Mesure de l'Efficacité Productive et de l'Efficacité Cout: Cas des Dechets Menagers en Suise*. Neuchâtel: Université de Neuchâtel (IRER DP nr 9002).

Burgat, P., and C. Jeanrenaud. 1994. Technical Efficiency and Institutional Variables. *Swiss Journal of Economics and Statistics* 130 (4): 709-717.

Conceição Sampaio de Sousa, M., and F.S. Ramos. 1999. Measuring Public Spending Efficiency in Brazilian Municipalities: A Nonparametric Approach. In: Westermann, G. (ed.), *Data Envelopment Analysis in the Service Sector*. Wiesbaden: Gabler Edition Wissenschaft. 237-267.

Courcelle, C., M.P. Kestemont, D. Tyteca, and M. Installé. 1998. Assessing the Economic and Environmental Performance of Municipal Solid Waste Collection and Sorting Programmes. *Waste Management & Research* 16 (3): 253-263.

Davis, M., and K. Hayes. 1993. The Demand for Good Government. *Review of Economics and Statistics* 75 (1): 148-152.

De Borger, B., and K. Kerstens. 1996a. Cost Efficiency of Belgian Local Governments: A Comparative Analysis of FDH, DEA, and Econometric Approaches. *Regional Science and Urban Economics* 26 (2): 145-170.

De Borger, B., and K. Kerstens. 1996b. Radial and Nonradial Measures of Technical Efficiency: An Empirical Illustration for Belgian Local Governments using an FDH Reference Technology. *Journal of Productivity Analysis* 7 (1): 41-62.

De Borger, B., K. Kerstens, W. Moesen, and J. Vanneste. 1994. Explaining Differences in Productive Efficiency: An Application to Belgian Municipalities. *Public Choice* 80 (3-4): 339-358.

De Groot, H., and R. Goudriaan. 1991. *De Produktiviteit van de overheid: Over prestaties, personeel en uitgaven in de publieke sector*. Schoonhoven: Academic Service.

De Groot, H., and J. Van der Sluis. 1987. Bureaucracy Response to a Budget Cut: An Economic Model. *Kyklos* 40 (1): 103-109.

Deller, S., and J. Halstead. 1994. Efficiency in the Production of Rural Road Services: The Case of New England Towns. *Land Economics* 70 (2): 247-259.

Deller, S., and C. Nelson. 1991. Measuring the Economic Efficiency of Producing Rural Road Services. *American Journal of Agricultural Economics* 72 (1): 194-201.

Depré, R., G. Bouckaert, and F. De Smet. 1991. *Uitdieping naar efficiëntie en effectiviteit van het menselijk potentieel: Bibliotheek*. Leuven: Katholieke Universiteit Leuven (Faculteit der Sociale Wetenschappen, VCOB).

Depré, R., G. Bouckaert, and J. Wijns. 1991. *Uitdieping naar efficiëntie en effectiviteit van het menselijk potentieel: Groendienst*. Leuven: Katholieke Universiteit Leuven (Faculteit der Sociale Wetenschappen, VCOB).

Distexhe, V. 1993. L'Efficacité productive des services d'enlèvement des immondices en Wallonie, *Cahiers Économiques de Bruxelles* 34 (137): 119-138.

Färe, R., and S. Grosskopf. 1985. A Nonparametric Cost Approach to Scale Efficiency. *Scandinavian Journal of Economics* 87 (4): 594-604.

Färe, R., and D. Primont. 1988. Efficiency Measures for Multiplant Firms with Limited Data. In: Eichhorn, W. (ed.), *Measurement in Economics*. Heidelberg: Physica-Verlag. 177-186.

Glass, J., D. McKillop, and G. O'Rourke. 1998. A Cost Indirect Evaluation of Productivity Change in UK Universities. *Journal of Productivity Analysis* 10 (2): 153-175.

Grosskopf, S., K. Hayes, and J. Hirschberg. 1995. Fiscal Stress and the Production of Public Safety: A Distance Function Approach, *Journal of Public Economics* 57 (2): 277-296.

Grosskopf, S., K. Hayes, L. Taylor, and W. Weber. 1997. Budget-Constrained Frontier Measures of Fiscal Equality and Efficiency in Schooling. *Review of Economics and Statistics* 79 (1): 116-124.

Grossman, P.J., P. Mavros, and R.W. Wassmer. 1999. Public Sector Technical Inefficiency in Large U.S. Cities. *Journal of Urban Economics* 46 (2): 278-299.

Hamilton, B.W. 1983. The Flypaper Effect and Other Anomalies. *Journal of Public Economics* 22 (2-3): 347- 361.

Hayes, K., and S. Chang. 1990. The Relative Efficiency of City Manager and Mayor-Council Forms of Government. Southern *Economic Journal* 57 (1): 167-177.

Hayes, K., and L. Wood. 1995. Utility Maximizing Bureaucrats: The Bureaucrat's Point of View. *Public Choice* 82 (1-2): 69-83.

Hayes, K., L. Razzolini, and L. Ross. 1998. Bureaucratic Choice and Nonoptimal Provision of Public Goods: Theory and Evidence. *Public Choice* 94 (1-2): 1-20.

Hjalmarsson, L., and J. Odeck. 1996. Efficiency of Trucks in Road Construction and Maintenance: An Evaluation with Data Envelopment Analysis. *Computers and Operations Research* 23 (4): 393-404.

Heyndels, B. 1996. De bevoegdheden en begrotingen van de provincies, de gemeenten en de OCMW's in België. In: Vanneste, J., and D. van Reeth (eds.), *Openbare Financiën* 2. Brussel: STOHO. 172-190.

Hoyt, W. 1990. Local Government Inefficiency and the Tiebout Hypothesis: Does Competition Among Municipalities Limit Local Government Inefficiency? *Southern Economic Journal* 57 (2): 481-496.

Kerstens, K., and P. Vanden Eeckaut. 1998. Distinguishing Technical and Scale Efficiency on Non-Convex and Convex Technologies: Theoretical Analysis and Empirical Illustrations. Louvain-la-Neuve: UCL (CORE Discussion Paper 9855).

King, D. 1992. Current Issues in the Theory of Fiscal Federalism. In: King, D. (ed.), *Local Government Economics in Theory and Practice*. London: Routledge. 23-42.

Levitt, M., and M. Joyce. 1987. *The Growth and Efficiency of Public Spending*. Cambridge: Cambridge University Press.

Liner, G.H. 1994. Institutional Constraints, Annexation and Municipal Efficiency in the 1960s. *Public Choice* 79 (3-4): 305-323.

Lovell, C.A.K. 1993. Production Frontiers and Productive Efficiency. In: Fried, H., C.A.K. Lovell, and S. Schmidt (eds.), *The Measurement of Productive Efficiency: Techniques and Applications*. Oxford: Oxford University Press, 3-67.

Lovell, C.A.K., L. Walters, and L. Wood. 1994. Stratified Models of Education Production Using Modified DEA and Regression Analysis. In: Charnes, A., W.W. Cooper, A.Y. Lewin, and L. M. Seiford (eds.), *Data Envelopment Analysis: Theory, Methodology and Applications*. Boston: Kluwer, 329-352.

Marchand, M., P. Pestieau, and H. Tulkens. 1984. The Performance of Public Enterprises: Normative, Positive and Empirical Issues. In: Marchand, M., P. Pestieau, and H. Tulkens (eds.), *The Performance of Public Enterprises: Concepts and Measurement*. Amsterdam: Elsevier, 3-42.

Moesen, W.A. 1990. Budgetary Neutrality, The Internalization of Spill-Overs and the Municipal Mergers in Belgium. Leuven: Katholieke Universiteit Leuven (CES Public Economics Research Paper Nr. 8).

Moesen, W.A. 1994. The Need for Performance Auditing in the Public Sector and the Best Practice Frontier. *European Journal of Law and Economics* 1 (4): 263-274.

Mueller, D. 1989. *Public Choice II*. Cambridge: Cambridge University Press

Oates, W. 1972. *Fiscal Federalism*. New York: Harcourt, Brace and Jovanovich.

OECD. 1997. Revenue Statistics. Paris: OECD.

Olson, M. 1969. The Principle of Fiscal Equivalence: The Division of Responsibilities among Different Levels of Government. *American Economic Review* 59 (2): 479-487.

Owens, J. 1992. Financing Local Government: An International Comparison with Particular Reference to Local Taxation. In: King, D. (ed.), *Local Government Economics in Theory and Practice*. London: Routledge. 56-86.

Prieto, A., and J. Zofio. 2000. Evaluating Efficiency in Public provision of Infrastructure and Equipment: The Case of Spanish Municipalities. *Journal of Productivity Analysis* (forthcoming).

Schwab, R., and W. Oates. 1991. Community Composition and the Provision of Local Public Goods. *Journal of Public Economics* 44 (2): 217-237.

Silkman, R., and D.R. Young. 1982. X-Efficiency and State Formula Grants. *National Tax Journal* 35 (3): 383- 397.

Spann, R. 1977. Public Versus Private Provision of Governmental Services. In: Borcherding T. (ed.), *Budgets and Bureaucrats: The Sources of Government Growth*. Durham: Duke University Press, 71-89.

Tickner, G., and J.C. McDavid. 1986. Effects of Scale and Market Structure on the Costs of Residential Solid Waste Collection in Canadian Cities. *Public Finance Quarterly* 14 (4): 371-393.

Tulkens, H. 1990. Non-Parametric Efficiency Analyses in Four Service Activities: Retail Banking, Municipalities, Courts and Urban Transit. Louvain-la-Neuve: UCL (CORE Discussion Paper 9050).

Vanden Eeckaut, P. 1997. *The Free Disposal Hull and Measurement of Efficiency: Theory, Applications and Software*. Louvain-la-Neuve: CIACO (PhD).

Vanden Eeckaut, P., H. Tulkens, and M. Jamar. 1993. Cost Efficiency in Belgian Municipalities. In: Fried, H., C.A.K. Lovell, and S. Schmidt (eds.), *The Measurement of Productive Efficiency: Techniques and Applications*. Oxford: Oxford University Press, 300-334.

Vitaliano, D. 1997. X-Inefficiency in the Public Sector: The Case of Libraries. *Public Finance Review* 25 (6): 629-643.

Wyckoff, P.G. 1990. Bureaucracy, Inefficiency, and Time. *Public Choice* 67 (2): 169-179.

13 Efficiencies in United States Metropolitan Areas

Harold O. Fried and J. Douglas Klein [1]

13.1 Introduction

The Capital Region of New York State is one of 332 metropolitan areas (MAs) in the United States. People, resources and firms are free to move among these and other areas of the country. This implies that just as firms compete, so do locations, as people and businesses vote with their feet. The result is that some locations benefit and thrive, whereas others lose and decline.

There are numerous determinants of regional economic growth. Some are outside the control of government (exogenous), and others can be influenced by government action (endogenous). Very few variables are purely exogenous, perhaps the best example is the weather. However many variables are more or less exogenous, depending upon the time frame. With enough time, government can influence many of the major determinants of growth and location choice, such as taxes, utility costs, recreational amenities, education, transportation services, and so on.

Given competition between regions, prosperity, growth and even survival depend upon government actively striving to provide the most attractive environment for people to live and for firms to do business (Dilorenzo 1983). The stakes are high: failure to act will result in losses to other areas. The objective of this paper is to demonstrate an approach a region can use to identify policies to improve the effectiveness of the provision of government services. This approach is applied to the Capital Region of New York State.[2]

Often with the assistance of external consultants or professional associations, government agencies monitor performance using a long list of partial productivity ratios.[3] Conferences and journals are devoted to the search for the best partial productivity performers around the world. The underlying premise is that these stellar performers

are doing something that is transferable and has the potential to improve performance elsewhere.

We take a different tack. Rather than begin with the individual government agency, we begin with the region. Using multiple rather than partial measures of productivity, we first identify regions that dominate the performance of the Capital Region. This identifies a set of potential role model regions that the Capital Region can draw upon to enhance its performance. Since the performance of these regions is the result of the performance of the individual parts, the second stage considers specific practices used by agencies operating in the dominating regions. The purpose of the initial screen at the regional level is to ensure that the overall environment in which these individual agencies operate is similar to that of the Capital Region. By controlling for the overall operating environment, we raise the likelihood that agency level practices are indeed transferable to the Capital Region.

Our approach involves the specification of a regional model of the production of government services, the identification of regions which are like but which dominate the Capital Region, and the search for transferable practices at the agency level. Throughout this process, we interact with local government practitioners. It is only with the assistance of front-line managers that we are likely to get the specification of inputs and outputs right, and that we are able to choose the most appropriate regions for detailed case studies. Moreover, we assign the search for agency level transferable practices largely to practitioners. We can identify individual agencies operating in dominating regions and suggest some practices that appear to explain performance, but this is merely a catalyst. It is the individual agency managers who are most qualified to interact with their counterparts. Based upon years of operational experience, only they have the insight to ask the specific questions necessary to implement these practices. The final product is a combination of our academic skills as researchers and the complementary skills of practitioners.

To be more specific, we use Data Envelopment Analysis (DEA) to evaluate the Capital Region of New York State (Albany-Schenectady-Troy MA) relative to a sample of 290 MAs throughout the country. We find the Capital Region to lie inside the frontier. By evaluating the Capital Region's efficiency in a succession of models as described below, we identify a long list of dominating regions. These regions are like the Capital Region in the sense that they produce similar levels of government services, but unlike the Capital Region in that they produce these services with lower expenditure. In consultation with New York State and Capital Region local government officials, we selected five regions for intensive case studies in order to identify specific operational practices

that might underlie their superior performance, and which might successfully transfer to the Capital Region. The likelihood that these practices are relevant and transferable to the Capital Region is high because of the way we identify the role model regions.

The role model regions are typically similar in the production of government services to the Capital Region. This is necessary, but not sufficient, for their selection to be role model regions. They are also required to be more efficient in that production. These conditions eliminate from consideration efficient regions that provide more government services and spend more. These regions are bigger. These conditions also eliminate regions that provide fewer government services and spend less. These regions are smaller. And finally these conditions eliminate regions that provide a similar level of government service and spend more. These regions perform worse than the Capital Region. Moreover, government officials helped to select the role model regions. They provided important institutional input that ensured that the selected regions are similar in their overall governmental structure.

Section 13.2 of the paper describes the role of regions in the analysis. Section 13.3 reviews some related research and discusses our methodology. Section 13.4 describes the data and specifies the regional models for the production of government services. Section 13.5 is an application of our approach to the Capital Region of New York State. Section 13.6 concludes the chapter by addressing some typical questions raised by practitioners.

13.2 The Role of regions

As we use the term, a region is a metropolitan area (MA). The U.S. Census Bureau recognizes 332 MAs in the United States, and defines an MA as "a large population nucleus, together with adjacent communities that have a high degree of economic and social integration with that nucleus" (U.S. Bureau of the Census 1998: B1). Although the actual definition is complicated, an MA is essentially a municipality of at least 50,000 persons and its surrounding area.[4] An MA is only a statistical construct. It does not have political, taxing, or service functions, and generally spans governmental service jurisdictions such as police, schools, roads, water, sewer, public assistance and so on.

The organization of government in the United States varies across states. Consider New York State, for example. The entire state is divided into municipalities and towns. In addition, there are counties, villages, and special districts. Special school districts

provide education. Special fire districts provide fire protection. Municipalities, towns and the state maintain roads. Municipalities, towns, counties, villages and the state provide police protection and parks. Municipalities and towns provide water and sewer. Counties administer public assistance. Counties, towns, municipalities, special districts and the state are each able to levy taxes. In other parts of the country, the organization of government is very different. For example, counties provide education in the South, while towns and municipalities provide education in New England.

This paper evaluates the performance of metropolitan regions. Since these are statistical constructs, measures of the resources and the services are calculated by aggregating the resources and the services of the government entities that operate within the metropolitan region. For example, number of students is an educational output. This output for a metropolitan region is the total number of students in the region, regardless of the specific government organization for education in that region. Some regions perform better than others. Since regions are constructed from underlying governmental entities, regions must perform well because the underlying special districts, municipalities, counties, towns, and villages perform well in providing government services. Similarly, regions perform poorly because the underlying components perform poorly. For poorly performing regions, this paper identifies dominating regions that are similar in terms of resources, but superior in providing government services.

The results can be used to improve the performance of government service entities that operate in poorly performing regions. For example, a school district in an under performing region can examine practices by school districts that operate in dominating regions. However, the form in which government services are provided varies across regions. For this reason, a school district that operates in an under-performing region might select benchmark partners from regions with a similar form of governmental organization. This is likely to be possible if the procedure generates a large number of dominating regions to choose from. The "peeling the onion" procedure used in this paper is designed to serve this purpose.

The regional analytic framework is flexible. If the form of governmental organization is deemed to be important, then an inefficient region can screen its list of more efficient role models to include only regions with a similar form of governmental organization. However, regions with diverse forms of governmental organization may also have something to learn from one another. A concept like year-round-schooling, for example, would work as well (or as poorly) in a county-run school system as in a special-school-district system. In this case, it is useful to have access to the full set of potential role model regions. Moreover, it may be the form of governmental organiza-

tion itself that is the focus of the analysis. Some regions provide services more centrally than others. Perhaps the regions that are configured centrally are more efficient because there are economies of scope in the production of government services. An inefficient region can study the jurisdictional organization of its more efficient role models to learn the implications of governmental re-structuring.

13.3 METHODOLOGY

Measuring the Efficiency of Local Governments

Benchmarking, performance measurement, efficiency and productivity have become watchwords in the administration of government at all levels. Concerns over rising taxes and the competition to attract and retain both people and employers require governments to provide public services effectively and efficiently.

Most benchmarking exercises take a partial productivity approach by comparing the ratio of an output to an input. Labor productivity is a partial productivity measure. For example, consider tons of refuse collected per employee hour, or miles of road cleared of snow per dollar expended (National Center for Public Productivity 1999). The state of Oregon has pioneered the use of benchmarks in evaluating government performance, and many of their targets are partial productivity measures (Oregon Progress Board 1997). Partial measures ignore other factors that contribute to productivity and fail to provide an overall measure of performance.

Ratings guides to metropolitan areas similarly rely on partial productivity measures, but add two additional features. First, they often include factors beyond the control of local government officials, such as climate and outdoor recreation opportunities, and second they assign weights in order to combine rankings in a variety of areas into a single final number.[5]

A host of studies measure the efficiency of individual government functions using a variety of efficiency measurement tools that account for multiple inputs and multiple outputs simultaneously. Efficient performance is typically implemented as best-practice. Efficient units provide more service with the same or fewer resources than other units in the sample. Many government functions have been studied using this approach: police protection (Carrington, Puthucheary, Rose, and Yaisawarng 1997, Davis and Hayes, 1993); fire protection (Ahlbrandt, 1973); public education (McCarty and Yaisawarng, 1993, Grosskopf, Hayes, and Taylor 1997 Rongen, 1995); courts (Kittelsen and Forsund, 1992, Pedraja-Chaparro and Salinas-Jimenez, 1996, Tulkens,

1993); solid waste collection (Edwards and Stevens, 1978, Carroll, 1995); mass transit (Tulkens, 1993, Nolan, 1996); public utilities such as water and publicly provided electricity (Bhattacharyya, Harris, Narayanan, and Raffiee 1995, Färe, Grosskopf, and Logan 1985, Cote, 1989); highway maintenance (Deller, Nelson, and Walzer 1992, Deller and Halstead, 1994, Rouse, Putterill, and Ryan 1997); libraries (Vitaliano, 1998); hospitals and health care (Ferrier and Valdmanis, 1996, Linna, 1998; Mirmirani, and Li 1995, Valdmanis, 1992); and others.

Our objective is to take an even wider view, to measure the overall performance of regions. Similar efforts have been made to rank the efficiency or productivity of nations (Färe, Grosskopf, Norris, and Zhang 1994; Hirschberg, Maasoumi, and Slo 1991), and a body of work assesses the efficiency of Belgian municipalities (Tulkens, 1993, Vanden Eeckaut, Tulkens, and Jamar 1993, De Borger, Kerstens, and Moesen, 1994, De Borger, and Kerstens 1996).[6] See also Chapter 12 by De Borger and Kerstens. In our study, we aggregate across all governmental units included in each census-defined Metropolitan Statistical Area.

One feature of benchmarking that should not be overlooked is the distinction between efficiency and effectiveness (Deller, Nelson, and Walzer 1992). Efficiency relates to the technical production of the service, while effectiveness is more value-laden, and addresses the degree to which the needs of the public are met. Ammons (1995) makes a similar distinction between outputs (e.g. arrests; garbage collected) and outcomes (e.g. public safety, clean streets). This point is also stressed by Van Tulder (see Section 10.3) and De Borger and Kerstens (see Section 12.3). Outcome, or quality data is at best difficult to obtain, and often not available at all. One solution to the quality problem is to follow up any benchmarking exercise with case studies, or contacts with the practitioners in the field who are in a position to interpret and understand the results. The Money rankings employ this (Fried, Marable, and Nance-Nash1996). Bouckaert (1993) suggests that when the evaluative process is more open and inclusive, the results will be more readily and widely accepted.

The efficiency analysis in this paper serves only as an input to selecting the case studies that follow. The analysis produced a ranking of potential role-model partners for the New York State Capital Region. Specific regions were selected for case study by consulting with government officials. The case studies were the source for practices that could be transferred to the Capital Region to improve its performance.

The Production Function for a Region

A region uses multiple sources of revenue (property, sales, income taxes, fees, state and federal aid) to provide numerous services (education, transportation, safety and health, infrastructure, etc.). We envision this as a production function in which governments take expenditures or inputs and convert them to services or outputs. As a first cut, define the production unit as a metropolitan area (MA). In order to treat an MA as the unit of analysis, we aggregate both expenditures and quantities of service to the MA level.

Identifying All Possible Role Models

There are 332 MAs in the United States, all of which under varying circumstances are engaged in converting expenditures into governmental services. Some regions perform these functions better than others do. Best practice regions anchor the efficient frontier. Inefficient regions lie inside the frontier. For the inefficient regions, we obtain a quantitative measure of inefficiency. In addition, the analysis identifies efficient regions responsible for rendering another region inefficient. These efficient regions constitute ideal benchmark or role model partners to examine in formulating a program to enhance performance in the inefficient region in order to enable it to move closer to the best practice frontier.

The starting point is data on inputs and outcomes for each metropolitan region in the country. Each observation represents the input-outcome combination for a region. We use a linear programming technique known as data envelopment analysis (DEA) to obtain a best practice frontier composed of linear segments that connect the outermost data points. For an introductory discussion of Data Envelopment Analysis, see Lovell (1993) and Section 2.4 of this volume. Regions on this frontier are efficient and perform at the level of best practice; regions inside this frontier are inefficient and perform below best practice. In fact, DEA provides a measure of the extent to which an inefficient region lies below the frontier. This measure is the percentage by which an inefficient region's expenditures could be reduced while holding outcome levels constant, were the region to perform up to best practice.

Measuring a region's efficiency is only a first step. Far more exciting is the subsequent analysis that, for inefficient regions, identifies the specific best practice role model regions on the frontier. It is these efficient regions which form the frontier and render other regions inefficient. These regions are a lot like the inefficient region which they dominate in that they provide a similar level of services,[7] but they are unlike it in that they provide this similar level of services at lower cost. These benchmark efficient regions are therefore a valuable source for governmental innovations that could enable

the inefficient region to provide governmental services at lower cost. This is the technical implementation of the process loosely and intuitively described in the introduction.

The role models generated by this procedure for an inefficient region are on the frontier. Inefficient regions are only compared to efficient regions, but for our purposes this may be overly restrictive. There may be *inefficient* regions that dominate other more inefficient regions and which would also be interesting and useful role models. To identify these relatively efficient regions, we use an iterative procedure to identify the full set (efficient and inefficient) of role models for the Capital Region. First, we solve the described above, then we remove the best role models for the Capital Region from the data set and solve the problem again. This identifies second order role models. New role model regions are successively uncovered as the frontier shrinks inward toward the Capital Region. We repeat this process until the Capital Region itself lies on the frontier.[8]

13.4 Data Specification of Inputs and Outcomes

The outputs of a region are the provision of government services. For our analysis, we aggregate these to the regional level. The resources or inputs of a region are the expenditures on these government services. These are also aggregated to the regional level.

The Specification of Outcomes

The objective is to identify regions that are able to provide government services at levels similar to the Capital Region while spending less. The selection of output measures of government services is very important. For example, it would be a mistake to measure education output with a variable which directly proxies spending (e.g. public education budget). If inputs were simply expenditures, this would result in output justifying expenditure by construction. For this reason, we need to measure the *outcomes* of government services rather than using expenditure as output. In the case of education, for example, our outcome measure is number of students educated.

Table 13.1

Local Government Function by Share of Expenditure (percentages)

	National Average	Capital Region
Five included functions	57.3	69.8
Education, K-12	40.8	47.6
Police services	6.5	3.9
Welfare administration	4.4	11.3
Highways	3.1	5.0
Sewer	2.5	2.1
Excluded functions	42.7	30.2
Health care and hospitals	8.9	2.5
Fire protection	3.1	2.8
Education, higher, veterans	2.7	3.1
Housing services	2.6	1.6
Parks and recreation	2.3	1.2
Solid waste	2.1	1.6
Correctional system	2.0	2.5
Libraries	0.8	0.5
Airports	0.8	0.3
Protective inspections	0.5	0.2
Natural resources	0.4	0.2
Mass transit	0.2	0.1
Water transportation	0.2	0.1
Parking	0.1	0.1
Government services, other	16.1	13.6

Source: Computed from U.S. Bureau of the Census (1992)

Ideally, we would like physical quality and quantity measures of all government services at the regional (MA) level for all 332 metropolitan regions in the country. In practice, we succeeded in obtaining data on five functional areas of government service: public education (kindergarten-12), welfare administration, police, public sewer, and roads for 290 MAs.[9] As shown in Table 13.1, these five services account for 69.8 percent of expenditures for the Capital Region, and 57.3 percent nationally.[10]

Our outcome measures are listed below, along with the sources of the data:

- Public education: Number of children in public education (kindergarten-12). Source: 1990 U.S. Census of Population (U.S. Bureau of the Census, 1992).

- Welfare administration services, measured by number of families receiving welfare benefits Source: 1990 U.S. Census of Population (U.S. Bureau of the Census, 1992).[11]
- Police services, measured by the inverse of the number of crimes. The inverse is used in order that a lower number constitutes better police performance.[12] Source: U.S. Federal Bureau of Investigation, Uniform Crime Reporting Program (U.S. Federal Bureau of Investigation, annual).
- Public sewer services, measured by the number of housing units connected to public sewer. Source: 1990 U.S. Census of Population (U.S. Bureau of the Census, 1992).
- Roads and highways, measured by the center lane miles of road and highway. Source: U.S. Federal Highway Administration, 1993 Highway Statistics. (U.S. Federal Highway Administration, 1993).

The Specification of Inputs

We use two alternative input specifications. The first input specification consists of a single measure: total regional government expenditures on the outcomes included in a particular model plus a pro-rated share of general government expenditures. General government expenditures are pro-rated according to the share of total expenditures on the government services included in our outputs to total expenditures. The second input specification divides the first into two parts: (1) revenue raised through property taxes, pro-rated by the share of included functional expenditures in total regional expenditures, and (2) the residual expenditure on the included functions.[13] All of the local government expenditure data are computed from the 1992 Census of Governments and aggregated to the MSA level using the Census Bureau 1992 MSA definitions. (U.S. Bureau of the Census, 1992).[14]

Alternative Model Specifications

The evaluation of the performance of the Capital Region in terms of the provision of government services and the set of role model regions generated is sensitive to the specification and measurement of inputs and outputs. For this reason, we report results for five alternative model specifications. We consider models with one input measure and two input measures. We consider models with all five outcomes and also models that omit welfare and police, since we have the least confidence in the data for these two functions.[15] Finally, we consider models with inputs and outcomes specified in absolute terms and models specified in per capita terms. Table 13.2 gives our five alternative specifications. Table 13.3 provides descriptive statistics for inputs and outcomes.

Table 13.2

FIVE MODEL SPECIFICATIONS

	Absolute values of variables	*Variables in per capita terms*	*One input (Total expend.)*	*Two inputs (Prop. Tax and residual)*	*Five outcomes (see text)*	*Three outcomes*
1	•		•		•	
2	•			•	•	
3	•			•		•
4		•		•	•	
5		•		•		•

Table 13.3

DESRIPTIVE STATISTICS 290 METROPOLITAN AREAS

	Mean	*St. Dev.*	*Min.*	*Max.*
Inputs-Five Output Model (USD x 1,000)				
Total Expenditures on Included Services	824	1,610	66	16,600
Pro-Rated Property Tax	269	538	5	4,140
Residual Expenditure	555	1,130	40	12,500
Inputs - Three Output Model (USD x 1,000)				
Total Expenditures On Included Services	668	1,170	56	9,620
Pro-Rated Property Tax	220	420	5	3,270
Residual Expenditure	447	793	37	7,230
Outcomes				
Public School Enroll. K-12 (USD x 1,000)	100	160	10	1,430
Welfare Households (USD x 1,000)[a]	18	36	1	391
Inverse of Crime Rate (USD x 1,000)[a]	92	89	1	604
Highway Miles	1,450	2,130	93	16,900
Public Sewer Households (USD x 1,000)	222	398	17	3,330

a Excluded in 3-outcome models.

Source: U.S. Bureau of the Census

13.5 AN APPLICATION TO THE CAPITAL REGION OF NEW YORK STATE

Evaluating the Performance of the Capital Region

The first step is to evaluate the performance of the Capital Region relative to the sample of 290 MAs for the five model specifications. Table 13.4 summarizes these results. The low efficiency scores of the Capital Region suggest that the subsequent onion peeling exercise is likely to expand the set of potential role model regions substantially.

Table 13.4

RESULTS OF DEA, MODELS 1-5[a]

	Model 1	Model 2	Model 3	Model 4	Model 5
Average Efficiency Score	0.76	0.80	0.77	0.78	0.76
Number of Efficient MSAs	31	47	36	50	22
Efficiency Score of Capital Region	0.42	0.45	0.49	0.45	0.49

a *See Table 13.2 for specification of the five models.*

We note that in spite of the difference among the specifications of the models, there is remarkable consistency in the overall average efficiency scores, and the efficiency scores for the Capital Region. We also note that the Capital Region's efficiency scores range between 0.42 and 0.49, indicating that other regions are able to produce the Capital Region's level of output with less than half the expenditure. While a large number of MAs receive efficiency scores of 1.0, indicating that they lie on the frontier, only four to six were role models for the Capital Region, depending upon the model. We turn next to identifying other possible role models.

Peeling the Onion

For each of the five models, the MAs which form the efficient reference set for the Capital Region are dropped from the data set and Capital Region is re-evaluated. This generates a new set of efficiency scores for the Capital Region and a new set of potential role models. These newly "efficient" MSAs are removed from the data set and the Capital Region is evaluated again. The process is repeated for each model until the Capital Region is rendered efficient, as discussed in Section 13.3.

Generating the Set of Potential Role Models

Five model specifications combined with our iterative procedure for identifying role models generate a very long list of role models for the Capital Region. In fact, only 21 of the 289 other MAs failed to show up as a role model for the Capital District at some stage in at least one of the five models. The number of MAs that became efficient before the Capital Region is given in Table 13.5.

Table 13.5

POTENTIAL ROLE MODELS

	Model 1	Model 2	Model 3	Model 4	Model 5
Number of layers before Capital Region becomes efficient	30	30	34	83	99
Number of potential role-models emerging from the onion peeling process	142	145	129	256	252

In order to create a manageable and reasonable list from which to choose MAs for our case studies, we applied the following criteria, which produced a list of 48 role model regions:

- The Capital Region had to receive an efficiency score of 0.75 *or less* relative to the reference set containing the potential role model region, on average across all five model specifications. This insures that the Capital Region has something to learn from the provision of government services in the role model region. The loose interpretation of an efficiency score of 0.75 or less is that the role model region is able to provide a level of government services similar to the Capital Region at an expenditure level of 75 percent or less of that of the Capital Region.
- The region had to appear as a role model for the Capital Region in at least four out of the five model specifications.[16]
- The average weight that a region received as a role model for the Capital Region had to be at least 0.05.[17]
- The average scaling factor attached to a role model region had to be less than three. The scaling factor is an indication of the size of the role model (measured in terms of the input(s)) compared to the Capital Region. Roughly speaking, this criterion eliminated regions smaller than one-third the size of the Capital Region.

Selecting the Five Role Model Regions for Case Study

Although all 48 role models for the Capital Region are potentially interesting and useful, some are more useful than others. For example, the top three role model regions are in Texas, followed by two entries from Tennessee and one from Florida. However, local governments in Texas, Tennessee and Florida operate in very different social, historical and climactic environments than the Capital Region, tempering their relevance. The purpose of the secondary screen is to take these important institutional features into account. We discussed the list of 48 role model regions with government officials from the Capital Region. With their help, we selected five particularly appropriate regions for further study.[18]

Learning from the Role Model Regions

The five role model regions selected are similar to the Capital Region in important respects, and different in the important respect that they provide government services more effectively. This result is based upon a regional model, aggregated to the regional level from the local government components. To discover the reasons for the superior performance at the regional level and to identify transferable policies at the functional level, we returned to the local components and examined operating practices.

We provided Capital Region managers with information on agencies such as police, health care, schools in the role model regions, which appeared to be particularly innovative. The purpose was to stimulate an interchange at the managerial level and to foster the transfer of ideas and practices. Our hope is that the information inspired agency managers in the Capital Region to contact their counterparts in the role model regions to learn more. It is the front-line practitioners who are most qualified to delve into the intricacies of practices to determine if they can be implemented elsewhere.[19]

13.6 Conclusions and Policy Issues

Our regional models of government service production are based on aggregated local data on outcomes and expenditures for education, police, public welfare administration, roads and public sewer hookups. These regional models are used to evaluate the performance of the Capital Region of New York State relative to the other regions in the country. A long list of potential role model regions is generated by "peeling the onion", using five different input-outcome specifications. This long list is reduced to a much smaller set of five regions through consultation with local government officials. We then conducted case studies for these five regions and identified potentially transferable practices at the local agency level. Implementation involves an

active interchange between local agency managers in the Capital Region and their counterparts in the role model regions. By establishing such a dialogue between front line managers, the right questions can be asked and answered. Our goal is to facilitate this interchange and let it take on a life of its own.

We conclude this conclusion with a series of questions and answers from the perspective of the consumer of this information:

How do you know these practices are transferable to the Capital Region? We do not know with certainty that they are transferable. However, we do know that these practices come from areas that are doing things right according to our regional models and share some important features with the Capital Region. These commonalities make these practices particularly relevant and useful.

The Capital Region consists of the City of Albany, the Village of Voorheesville and everything in between. How can these practices be relevant to such a diverse set of local governments? We recognize that the Capital Region is heterogeneous and so are the other five case study regions. Particular practices are not expected to be transferable to all local jurisdictions. It is for this reason that we were careful to select practices from localities spanning a wide range in terms of size. Local officials should apply their own screen to our lists and pursue opportunities where they perceive a good match.

Are any of these innovative practices being applied in the Capital Region now? And even if they are not, don't you think managers in this region already know about them? Some of these practices may very well be implemented in the Capital Region now, and some local officials may very well be aware of these practices. However, we doubt that all local opportunities to benefit from these practices have been exhausted, and we doubt that all local officials are aware of them. Running local government is often an exercise in daily crisis management, with little opportunity to examine the big picture for new approaches and to rethink current practices. This takes effort and time, and both are scarce. Our objective is to make some new ideas more accessible and to facilitate their consideration.

Do the conclusions of this paper suggest that local government officials are doing a poor job? The answer to that question is NO. In fact, the approach is predicated on the belief that local government officials are dedicated to providing the best possible service to taxpayers. No matter how good a job they are doing, it can always be done better. The best managers recognize this and view better practices in effect elsewhere as an exciting opportunity, not as a missed opportunity that has to be defended.

Will the implementation of some of these practices guarantee a reduction in local property taxes? Nothing is guaranteed. However, to the extent that practices are implemented which reduce the costs of providing a given level of public service, there is an opportunity to reduce property taxes. There is also an opportunity to spend more elsewhere and to improve the quality of services. Local government officials have to make the choice.

Appendix

City Administration Function - Indianapolis

- A major effort towards privatization of city functions is spearheaded by an advisory group from the private sector called Service, Efficiency, and Lower Taxes for Indianapolis Commission (SELTC). It is composed of nine businessmen, each heads a committee of volunteers with responsibility for a city function.

  60 government services have been privatized over four years. Examples include microfilming (61% cost saving), printing (47% cost saving), sewer bill collections ($1.8 million in saving). Total saving from competition are estimated to be between $10 million and $20 million.

  Savings have been realized even when the city continues to perform an operation. City units have reduced costs in order win competitive bids against private firms.

- Activity based costing is used to accurately identify the costs associated with a particular operation such as filling a pothole.

  This permits cost comparisons across city road maintenance crews.

  It is also the basis for competition with the private sector. In road maintenance there were 92 truck drivers and 32 supervisors. Based on activity based costing, the union truck drivers recognized that they would lose a bid against private firms because of the overhead. They asked the mayor to lay off the supervisors. The mayor agreed, the city won the contract.

- The Indianapolis Neighborhoods Initiative shifts responsibility for many functions from the city to the neighborhood level. This includes the following:

  The Indianapolis Neighborhood Housing Partnership (INHP) is a nonprofit organization that mobilizes private and public support for purchasing a home.

  The Indianapolis Neighborhood Resource Center (INRC) provides financial and advisory support for community organizations.

  Community Development Corporations mobilize resources to construct affordable housing.

  Neighborhood Crime Watch Block Clubs organize the community to assist the police in crime prevention.

  The Indianapolis Public Housing Authority administers publicly provided housing.

- Neighborhoods organize groups to bid on contracts to maintain their own neighborhood parks.

- The city has embarked on a half billion dollar program over three years to repair the city's infrastructure without having to increase taxes.

- A Regulatory Study Commission is examining development codes, zoning, licensing provisions, occupancy requirements, and health and safety requirements.

- Since 1992, the city budget has decreased and tax rates have remained constant. Savings from privatization have been used to fund infrastructure repairs, police and fire.

- The City has a tax increment financing program. This permits depressed areas to use tax revenues generated from local new businesses to improve infrastructure.

- Businesses locating in the Indianapolis Urban Enterprise Zone receive tax concessions.

- Indianapolis has embarked on a Rebuilding Families Initiative. This includes encouraging responsible fatherhood, discouraging teen pregnancy and improving support for teenage mothers.

For further information:

- City of Indianapolis Mayor's Office
  Mayor Stephen Goldsmith
  317-327-3601
- Urban Enterprise Association
  317-541-2740
  Krista Rush
  Social Policy Advisor
  317-327-3601
- "Goldsmith plays cards close to his chest,"
  http://www.starnews.com/news/election/96/nov/1101SN_goldsmith.html
- "Cool Facts About Indianapolis," http: //www.mibor.net/sites/bintz/linkp2.html
- "Creating a Competitive City,"
  http: //www.ci.indianapolis.in.us/features/competition/
  http: //www.ci.indianapolis.in.us/features/competition/creating_02.html
  http: //www.ci.indianapolis.in.us/features/competition/creating_03.html
  http: //www.ci.indianapolis.in.us/features/competition/creating_04.htm
- "State of the City," January 22, 1997, http: //www.ci.indianapolis.in.us/mayo/sotc1997.html
- "Tax Increment Financing," http: //www.ci.indianapolis.in.us/edvt/svc-tif.htm
- "Indianapolis Urban Enterprise Zone," http: //www.ci.indianapolis.in.us/edvt/svc-entz.htm
- "The Rebuilding Families Initiative," http: //www.ci.indianapolis.in.us/features/fam/
- "The Neighborhoods Initiative," http: //www.ci.indianapolis.in.us/features/neig/

Notes

[1] Research funded in part by the New York State Commission on the Capital Region. We acknowledge the research assistance of Meg Boericke, Theresa Carr, Karen Chamberlain, John Colvert, Martin Strosberg, and Kira Walle. Dennis Rapp kept us focused on the hard practical issues that matter to government officials. Thank you everyone. Any errors are ours.

[2] The Capital Region of New York State consists of six counties, and three distinct cities of more than 50,000 people (Albany, the capital of New York State, Schenectady, and Troy). A description of the Capital Region may be found in the first report of the New York State Commission on the Capital Region, 1996/005/00 New York State Commission on the Capital Region (1996).

[3] Organizations involved in benchmarking and the measurement of public performance include: the American Productivity and Quality Center (APQC); the Government Finance Officers Association (GFOA); the International City/County Management Association (ICMA); the National Association of Development Organizations (NDTO); the National Performance Review (NPR); the National Center for Public Productivity (NCPP); the US. Council of Mayors; and the Urban Institute. This list does not include federal, state, and local productivity organizations.

[4] For the details, see *U.S. Bureau of the Census* (1998: Appendix B, Geographic Concepts and Codes).

[5] *Money* magazine's ratings (Fried, Marable, and Nance-Nash, 1996) use weights generated from a survey of readers; *Forbes* (Oliver, 1996) relies on a consultant to combine four costs of doing business in a city. Other examples include Savageau and Loftus (1997), and Heubusch, and Thomas (1997).

[6] These papers use (and compare) a variety of efficiency measurement tools, including Data Envelopment Analysis, Free Disposal Hull (FDH, a technique related to DEA), and econometric estimation of a best-practice frontier. They have in common the specification of a list of municipal government outputs, while considering government expenditure as the input to the production process. In each of the cases, the results were designed to be more illustrative of the techniques used, than of immediate policy relevance.

[7] In fact, they are the nearest efficient regions to A in terms of levels of output.

[8] Ordinarily, efficiency analysis ends with the construction of the first frontier, and each region's efficiency score is measured against the best observed regions, which make up the frontier. The iterative procedure which we use enables us to identify *all* regions which could serve as role models for a given target region, not just the initially efficient regions. We could have used the free-disposal hull (FDH) methodology of Vanden Eeckaut and Tulkens (1993), but this would have generated a more restricted list of role-model peers.

[9] The 20 MSAs omitted had one or more data items unavailable.

[10] In measuring these percentages, we exclude capital expenditures, since these are "lumpy" and may give misleading expenditure impressions across MAs. We also exclude expenditures related to public utilities. Some municipalities operate their own electric or gas utility, while others rely on the private sector. There is a wide variation in expenditures and revenues associated with the operation of utilities, so these data are not included. Technically, our evaluation of regions is partial since it does not include 100 percent of all government services. However, it should not be confused with a partial productivity ratio approach that is based upon the ratio of one output to one input. We do judge a MA's efficiency while simultaneously considering all of our included outputs.

[11] In computing the input expenditure going toward welfare administration we explicitly exclude welfare benefits, which vary widely across regions, and include only welfare administrative costs.

[12] This is a crude measure of the effectiveness of public safety expenditures since it fails to control for regional differences in the socioeconomic environment. However, this can be incorporated in the second stage in our discussions with civic leaders. A large number of dominating regions are identified for an under performing region. Some of these dominating regions are more useful than others. In particular, it might be important to study intensively a dominating region with a similar socio-economic environment.

[13] Throughout the U.S. there is great interest in ways to cut property taxes, and therefore it would be desirable to be able to link expenditures with the specific source of revenues. Unfortunately, the Census does not permit us to make this link. Since tax dollars are fungible, this shortcoming is perhaps not so important.

[14] In order to obtain expenditure data at the MSA level, we aggregated expenditures across all of the local government units in the MSA, as reported by the Bureau of the Census. MSAs include counties, cities, other municipalities, towns, townships, school districts, and special districts (e.g. water, fire, sewer). For example, the Capital Region MSA consists of 295 separate governmental units.

[15] Using three outcomes instead of five reduced the average percentage of expenditures covered from 60 percent to about 50 percent.

[16] All of the 48 regions appeared as role models in all five models, with the exception of Ann Arbor, which appeared in four out of the five.

[17] The weights represent the importance of each region as a role model, and can range from 0 to 1. The Capital Region generally had four or five peers at a time, so weights on any one averaged 0.2 to 0.25.

[18] The five MSAs selected for further study were as follows. Numbers in parentheses indicate the Capital District's average efficiency score in the five models in the onion layer where the given MSA was an efficient role model. Grand Rapids-Muskegon-Holland, MI (.698), Indianapolis, IN (.666), Raleigh-Durham-Chapel Hill, NC (.678), Scranton-Wilkes-Barre-Hazleton, PA (.653), and Springfield, MA (.689).

[19] The Appendix contains an example from our case studies.

References

Ahlbrandt, R.S. Jr. 1973. Efficiency in the Provision of Fire Services. *Public Choice* 16: 1-15.

Ammons, D. N. 1995. Overcoming the Inadequacies of Performance Measurement in Local Government: The Case of Libraries and Leisure Services. *Public Administration Review* 55 (1): 37-47.

Bhattacharyya, A., T.R. Harris, R. Narayanan, and K. Raffiee. 1995. Technical Efficiency of Rural Water Utilities. *Journal of Agricultural and Resource Economics* 20 (2): 373-391.

Bouckaert, G.. 1993. Measurement and Meaningful Management. *Public Productivity & Management Review* 17 (1): 31-42.

Carrington, R., N. Puthucheary, D. Rose, and S. Yaisawarng. 1997. Performance Measurement in Government Service Provision: The Case of Police Services in New South Wales. *Journal of Productivity Analysis* 8 (4): 415-430.

Carroll, W.. 1995. The Organization and Efficiency of Residential Recycling Services. *Eastern Economic Journal* 21 (2): 215-225.

Cote, D.O. 1989. Firm Efficiency and Ownership Structure: The Case of U.S. Electric Utilities Using Panel Data. *Annals of Public and Co-operative Economy* 60 (4): 431-450.

Davis, M.L. and K.J. Hayes. 1993. The Demand for Good Government. *Review of Economics and Statistics* 75 (1): 148-152.

De Borger, B., K. Kerstens, W.A. Moesen, and J. Vaneste. 1994. Explaining Differences in Productive Efficiency: An Application to Belgian Municipalities. *Public Choice* 80 (3-4): 339-358.

De Borger, B., and K. Kerstens. 1996. Cost Efficiency of Belgian Local Governments: A Comparative Analysis of FDH, DEA, and Econometric Approaches. *Regional Science and Urban Economics* 26 (2): 145-170.

Deller, S.C., and J.M. Halstead. 1994. Efficiency in the Production of Rural Road Services: the Case of New-England Towns. *Land Economics* 70 (2): 247-259.

Deller, S.C., C.H. Nelson, and N. Walzer. 1992. Measuring Managerial Efficiency in Rural Government. *Public Productivity & Management Review* 15 (3): 355-366.

Dilorenzo, T.J. 1983. Economic Competition and Political Competition: An Empirical Note. *Public Choice* 40: 203-209.

Edwards, F.R., and B.J. Stevens. 1978. The Provision of Municipal Sanitation Services by Private Firms: An Empirical Analysis of the Efficiency of Alternative Market Structures and Regulatory Arrangements. *Journal of Industrial Economics* 27 (2): 133-147.

Färe, R., S. Grosskopf, and J. Logan. 1985. The Relative Performance of Publicly Owned and Privately Owned Electric Utilities. *Journal of Public Economics* 26 (1): 89-106.

Färe, R., S. Grosskopf, M. Norris, and Z. Zhang. 1994. Productivity Growth, Technical Progress, and Efficiency Change in Industrialized Countries. *American Economic Review* 84 (1): 66-83.

Ferrier, G.D., and V. Valdmanis. 1996. Rural Hospital Performance and its Correlates. *Journal of Productivity Analysis* 7 (1): 63-80.

Fried, C., L.M. Marable, and S. Nance-Nash. 1996. Best Places to Live in America: Our 10th Annual Survey. *Money* (July): 66-95.

Grosskopf, S., K.J. Hayes, L.L. Taylor. 1997. Budget-Constrained Frontier Measures of Fiscal Equality and Efficiency in Schooling. *Review of Economics and Statistics* 79 (1): 116-124.

Heubusch, K. and G.S. Thomas. 1997. *The New Rating Guide to Life in America's Small Cities*, 2nd. Ed. Amherst: Prometheus Books.

Hirschberg, J.G., E. Maasoumi, and D.J. Slottje. 1991. Cluster Analysis for Measuring Welfare and Quality of Life Across Countries . *Journal of Econometrics* 50 (1-2): 131-150.

Kittelsen, S.A.C., and F.R. Forsund. 1992. Efficiency Analysis of Norwegian District Courts. *Journal of Productivity Analysis* 3 (3): 277-306.

Linna, M.. 1998. Measuring Hospital Cost Efficiency With Panel Data Models. *Health Economics* 7 (5): 415- 427.

Lovell, C.A.K.. 1993. Production Frontiers and Productive Efficiency. In: Fried H.O., C.A.K. Lovell, and S.S. Schmidt (eds.), *The Measurement of Productive Efficiency*. Oxford: Oxford University Press, 3-67.

McCarty, T., and S. Yaisawarng. 1993. Technical Efficiency in New Jersey School Districts. In: Fried H.O., C.A.K. Lovell, and S.S. Schmidt (eds.), *The Measurement of Productive Efficiency*. Oxford: Oxford University Press. 271-287.

Mirmirani, S., and H.C. Li. 1995. Health Care Efficiency Measurement: An Application of Data Envelopment Analysis. *Rivista Internazionale Di Scienze Economiche e Commerciali* 42 (3): 217-229.

National Center for Public Productivity. Feb 1999. A Brief Guide for Performance Measurement in Local Government. Accessed March 1999. Available at http://newark.rutgers.edu/~ncpp/cdgp/Manual.htm

New York State Commission on the Capital Region. 1996. *Growing Together within the Capital Region - The Report of the State Commission on the Capital Region*. Albany: New York State Commission on the Capital Region.

Nolan, J. F. 1996. Determinants of Productivity Efficiency in Urban Transit. *Logistics and Transportation Review* 32 (3): 319-342.

Oliver, S.. 1996. It's the Costs, Stupid. *Forbes* (October 21): 252-270.

Oregon Progress Board. 1997. Oregon Benchmarks. Accessed February 16, 1998. http://www.econ.state.or.us/opb/97tables/benchmarks_list.htm

Pedraja C., F., and J. Salinas-Jimenez. 1996. An Assessment of the Efficiency of Spanish Courts Using DEA. *Applied Economics* 28 (11): 1391-1403.

Rongen, G.. 1995. Efficiency in the Provision of Local Public Goods in Norway. *European Journal of Political Economy* 11 (2): 253-264.

Rouse, P., M. Putterill, and D. Ryan. 1997. Towards a General Managerial Framework for Performance Measurement: A Comprehensive Highway Maintenance Application. *Journal of Productivity Analysis* 8 (2): 127-149.

Savageau, D. and G. Loftus. 1997. *Places Rated Almanac: Your Guide to Finding the Best Places to Live in North America.* Macmillan General Reference.

Tulkens, H. 1993. On FDH Efficiency Analysis: Some Methodological Issues and Applications to Retail Banking, Courts, and Urban Transit. *Journal of Productivity Analysis* 4 (1-2): 183-210.

U.S. Bureau of the Census. 1992. *Census of Governments, 1992.* Washington, D.C.: U.S. Bureau of the Census.

U.S. Bureau of the Census. 1998. *State and Metropolitan Area Data Book. A Statistical Abstract Supplement*, 5th Edition 1997-98. Washington, D.C.: Bureau of the Census.

U.S. Federal Bureau of Investigation. Annual. *United States Federal Bureau of Investigation's Uniform Crime Reporting Program.* Available from the Inter-University Consortium for Political and Social Research at: http://www.icpsr.umich.edu/NACJD/archive.html.

U.S. Federal Highway Administration. 1993. *Highway Statistics.* Washington, D.C: U.S. Department of Transportation, Federal Highway Administration.

Valdmanis, V. 1992. Sensitivity Analysis for DEA Models: An Empirical Example Using Public vs. NFP Hospitals. *Journal of Public Economics* 48 (2): 185-205.

Vanden Eeckaut, P., H. Tulkens, and M.A. Jamar. 1993. Cost Efficiency in Belgian Municipalities. In: Fried, H.O., C.A.K. Lovell, and S.S. Schmidt (eds.), *The Measurement of Productive Efficiency.* Oxford: Oxford University Press, 300-374.

Vitaliano, D. F. 1998. Assessing Public Library Efficiency Using Data Envelopment Analysis. *Annals of Public and Cooperative Economics* 69 (1): 107-122.

Part Six:
Concluding Remarks

14 Efficiency Research in the Public Sector: Some Final Considerations

Jos L.T. Blank [1]

14.1 Introduction

Public services are a vital part of the economy as a whole. Government is responsible for infrastructure services as water and electricity supply, public transportation and human services such as health care and education. In fact, government produces throughputs, such as gas, water, electricity, roads, and healthy and skilled labor, for other sectors of the economy. Low productivity in public services implies higher taxation, which on its turn reduces the allocative efficiency in the economy. To a large extent, the quality of public services also determines the international competitiveness of the private sector. In most countries, the public sector is a large and growing fraction of the national economy. For example, between 1970 and 1995, taxes as a percentage of GDP increased from 30 to 37% in all OECD countries.

Public sector services face different competitive conditions, different ownership structure, different objectives and different constraints than private firms. In particular, public service providers are not subject to the discipline of the competitive market. Public firms can operate inefficiently and survive. Governments are experimenting with privatization and deregulation to improve efficiency and to reduce the cost of the public sector. However, it seems that these policy experiments are more a result of ideological standpoints than of thorough theoretical and empirical evidence.

Academic researchers have studied the problem of public sector inefficiencies. They have developed techniques to measure efficiency, to evaluate and to improve performance. There are academic studies of specific public services, though there are few examples of actually transferring the recommendations of the academic research to the policy arena. Academics and policy makers are working on the same

problems, but rarely work together. There is a communication gap between academics and policy makers. Which this book aims to narrow by presenting the most important methodologies and applications in the academic literature in a form that is accessible to policy makers. In particular, the objectives of this book are:

1. to summarize recent research;
2. to create an accessible overview of methods;
3. to develop better insight into the underlying factors potentially affecting productivity and efficiency;
4. to put the findings of efficiency analyses into perspective;
5. to demonstrate the potentials of translating these findings into policy.

Put simply, efficiency measurement involves a comparison between observed and optimal values of services and resources. It can be defined as the ratio of observed to maximum feasible service provision obtained from given resources, or vice versa.

One of the major problems of efficiency analysis has probably been the inaccessibility of the research to non-specialists. The complexity of the underlying theory and the advanced statistical methods have frightened off policymakers. The book tries to 'popularize' efficiency research and to underline its importance. This requires improving the accessibility of the current research and increasing the familiarity with the subject in policy and management circles.

Chapter 1 presents the framework for a typical efficiency analysis of a public sector operation. We summarize four components of this framework here.

The first important issue is the measurement of resources, service delivery and their corresponding prices. Measuring service delivery is a difficult and controversial problem. It is not always clear what a public sector service provides. Traditional quantity measures may ignore effectiveness and quality dimensions. Aspects referring to effectiveness, like the improvement of health or quality play an important role in this matter. Because public sector firms may not operate in a private market, service delivery prices often do not exist.

The second important issue is the economic environment of the public firm. In contrast to a private firm, managers of a public firm may be more constrained by government regulations, and the objective is more nebulous than the maximization of profit. The economic model of the public firm has to reflect these circumstances.

A third important issue is the sensitivity of the results to the assumptions of the model and the methodology used. Ideally, a policy maker should consult the results of different studies that use alternative methods and different data sets. A single study could be misleading. Next to comparing results of various methods, results of one analysis can also be used to complement the other analysis.

A fourth important issue is to translate the results of an efficiency study into a program for efficiency improvement. This is the area in which the work of academics has the most to offer the policy maker. Instruments like consolidation, financing systems, build-in incentive schemes have been widely discussed.

The remainder of this chapter illustrates these four issues with case studies. Section 14.2 considers problems measuring resource usage and service delivery. Section 14.3 focuses on the objectives of public firms and the constraints they face. Section 14.4 is about sensitivity analysis. Section 14.5 explains how to use efficiency analysis to generate policy recommendations to improve performance and provides some examples of applying and implementing efficiency analyses. And Section 14.6 concludes the book with some general comments.

14.2 Measuremant Issues

14.2.1 *Measurement of Service Delivery*

The distinction between efficiency and effectiveness plays an important role in defining service delivery. Effectiveness refers to the extent to which service deliveries of service providers fulfill the objectives formulated by government (public health, public safety etc.). These objectives (or desired outcomes) are only indirectly achieved by producing products and services. For example, road safety is an important objective of Central Government. The number of speeding tickets by the police may be a service delivery that may contribute to road safety. Since desired outcomes are influenced by regulation and other environmental aspects it us useful to separate service deliveries from outcomes. Services deliveries must therefore be defined in terms of quantities of products and services produced of a given quality.

Thus, measuring service provision depends on choices based upon government desires or societal preferences. This implies that not all services or quality of services need to be accounted for. Before any performance measurement can be executed, consensus should therefore exist on the service deliveries by the providers. The

reports of the Victorian Department of Treasury (Victoria DTF 1997) and the Steering Committee (SCRCSSP 1999) discuss this issue extensively.

Many public firms produce a heterogeneous set of services. Hospitals are a good example. From a conceptual standpoint this is not a problem. However, problems caused by lack of data may arise when analysing performance. The number of observations limit the number of distinct services and other relevant variables that can be included in the analysis. If there is insufficient data, it may be necessary to aggregate services by some weighting scheme based on *a priori* assumptions or external information. It may also be possible to group variables that are closely related based upon statistical techniques. For instance, the number of patients that visits a cardiologist and the number of patients that need heart surgery may be sufficiently connected, so there is no need to treat both categories separately in the analysis.

Measuring the quality of service delivery is probably the most difficult task in performance analysis. The definition gives rise to serious debate, although there may also be some concensus on this issue. Quality may refer to the effectiveness of a program (improvement in health or in skills), to the accessibility of service (waiting lists); to the treatment of customers (satisfaction), to the convenience of the service (distance to the nearest service provider), and to the appropriateness of the services provided (service delivery relates compared to customer needs). This is important because there may be a trade-off between efficiency and quality. If quality is not properly taken into account, it is the low quality providers that might set the best-practice standard. An inadequate measurement of quality therefore may cause inaccurate efficiency estimates. Unfortunately, in most countries and for most services, adequate data on quality of service provision does not exist. Sometimes, for example in the Netherlands, data are collected, but are not available for reasons of confidentiality.

Service delivery prices are frequently missing or distorted to serve other objectives, such as equity or affordability. Sometimes service delivery prices can be derived in a indirect manner. In Chapter 10 Van Tulder has used the severity of punishment to derive virtual service delivery prices for police services. Chapter 4 describes how hospital service prices can be derived from the Dutch operative budgeting system. However, distorted or missing service delivery prices may not be a problem if the mix of services of the service provider is constrained by regulations and politics. Inefficiency resulting from misallocation in service provision can therefore be ignored.

14.2.2 *Measurement of Resource Usage*

The definition of resource usage does not cause many debates amongst researchers. In general, three resources are specified: labor, material supplies and capital. In many cases these resources are further disaggregated. In hospitals, for example, it is common practice to disaggregate labor into nursing staff, medical staff and other personnel. The greater the disaggregation of resources, the greater is the possibility of identifying a non-optimal composition of resource usage. However, data availability limits the extent to which resources can be disaggregated since this increases the number of variables in the model.

However, in spite of concensus on the definition of resources, there are plenty of conceptual problems. These conceptual issues are often neglected or underestimated. Various measurement methods may lead to serious differences in efficiencies calculated. These conceptual differences are directly related to the break down of cost into quantity and resource price and the available data. In the following paragraphs, we comment on these issues for labor, material supplies and capital separately.

Production theory assumes that resources are homogeneous. This is not true in practice. For example, hospital nurses and physicians are different resources; it might not be appropriate to combine the two with equal weights and call it labor. Moreover, even within the group of nurses, productive capabilities might differ substantially. Seniority and education affect labor productivity and labor cost. If resource usage is measured by the number of full-time personnel, a provider with highly-qualified and well-trained personnel might be deemed efficient, despite the fact that labor cost might be high due to high salaries which might not be offset by higher labor productivity. If resource usage is measured by labor cost however, this same provider might be deemed inefficient.

In a cost analysis framework we could face the same problem. A common practice amongst many researchers is the use of so-called unit values: labor cost per full time equivalent employee. In our view, this is a dubious practice. As the number of employees is a decision variable for the providers, the resulting unit values depend on this decision and the resource price exogeneity requirement described by Lovell (see Section 2.2) will be violated. From an economic point of view, the unit value is also questionable, since in practice two neighbouring providers operating under the same resource market conditions may face different resource prices which seems rather odd. If data on seniority, skill or educational level are available con-

structing a labor index, based on *a priori* weighting scheme (index concept), should be preferred. Deflating labor cost by a region and time varying labor price index is another option (cost concept).

Since material supplies are even more heterogenous than labor, the unit value concept can usually not be applied at all. Therefore, most researchers apply the cost concept by deflating material cost by a consumer price index. As regional variations in the consumer price index in many cases fail to exist the usage of material is measured by its cost. Only inter-temporal variation of the consumer price index then remains. High cost of living regions are then penalized in the efficiency analysis as a result.

The measurement of capital services and its price has puzzled economists since Smith (1776) and Ricardo (1951-1995: vol. I). There are many conceptual problems such as the actual use of capital in a year (depreciation rate) and the vintage composition of capital (quality). It is beyond the scope of this book to provide a thorough discussion of the subject. Readers are referred to the standard work by Jorgenson (1996). In practice, three approaches are used.

The first approach measures capital by physical quantities, like square metres of floor space. This approach ignores the quality dimension, which may affect the use of other resources. Modern buildings, for example, are better insulated and require less fuel for heating.

The second approach values the capital usage by the book value, which may be biased as a result of administrative procedures.[2] A direct approach is reading the book value of capital assets from the provider's balance accounts. An indirect variant of this approach is to use insurance premiums on furniture and premises as a proxy.

The third approach, known as the Perpetual Inventory Method (PIM), calculates the capital stock as the sum of past investments, corrected for depreciation and investments prices. This method is probably the most accurate, but can not often be applied due to insufficient data on time series of past investments.

Although there may be no serious disagreements on the resources involved in the analyses, researchers do operationalize resource usage and resource prices in many different ways. This problem seems sometimes to have escaped from researcher's attention in efficiency analysis. Chapters 4 and 6 show that different measurement

methods may affect efficiency scores substantially. Policymakers and management should be aware of this problem and be conscious of how it has been dealt with in research.

14.2.3 *Measurement of Environmental Characteristics*

Aside from services and resources, prices and budgets, there is another category of variables that affect efficiency. These are the uncontrollable features of the operating environment that influence the ability of management to convert resources into services and have an impact on the measurement of performance. Since these environmental features may vary across service providers and influence service delivery performance, they must be accounted for. If differences in the operating environment are neglected, producers that operate under unfavorable circumstances are evaluated poorly when they might be performing well and vice versa. Managers who perceive that they are evaluated unfairly, may choose not to cooperate. Some examples are the socio-economic status, the age distribution and population density. Sexton, Sleeper, and Taggart (1994) give a nice illustration of how environmental features affect efficiency of pupil transportation in North Carolina and how the efficiency scores should be corrected to present a clear picture of managerial inefficiency. Pupil density, road density, elevation of the county seat, unpaved road miles are examples of such environmental features.

In most cases the external environment can only be measured indirectly. For example, in health care, the age distribution is only a proxy for the severity of illness, or the patient's ability to recover. In education, socio-economic status, measured by ethnic background or income, is only a proxy for incentives and stimulation pupils receive from their surroundings. In some cases, there might not even be data on suitable proxies for the external environment.

14.2.4 *Data Availability*

It is clear that good data on service provision, resource usage, and the external environment are crucial for a proper efficiency analysis. In general, there are huge voids in information provision. Efficiency analysis may have a strong impulse on improving information services and data availability. Unfortunately, the lack of good data may sometimes induce a reluctance against performance analysis amongst service providers. In their minds, the relative poor data will result in inac-

curate efficiency measures. Therefore, no performance gauging should be conducted at all. In particular, when information on quality and environmental aspects are lacking, then low quality and favorable environmental conditions determine best practice. A cooperative attitude from service providers may in the end backfire on them in the form, for example, budget cuts. This attitude causes a vicious circle: the lack of good data leads to non-cooperative behavior, which in turn prevents starting a new process of improving information.

It might be necessary to enact legislation to procure the relevant information. The resistance and the skepticism of service providers could be overcome by guaranteeing that the results will not be used by the government as an excuse to reduce budget allocations. Outcomes will only be used for reallocating resources within the sector and not between sectors. School consolidation in the Netherlands in the nineties illustrated that this kind of policy works effectively. Savings from exploiting economies of scale were returned to the school to improve quality of education (See also Chapter 8).

The strong wishes from efficiency analysists to improve data and their availability may coincide with a tendency in New Public Management and Adminstration. From this discipline there are strong recommendations for adapting and improving accounting schemes in public services.

14.3 Economic Behavior and Economic Constraints

In Chapter 2 Lovell provides a menu of analytical frameworks within which the efficiency of public sector service provision can be evaluated. The selection of an analytical framework leads to the specification of a theoretical "best practice" service delivery standard. There are three issues to consider.

The first issue concerns the choice of a *technical approach or economic approach*. In a technical approach only the relation between quantities of service delivery and resource usage is analyzed. In an economic approach cost or revenue is the analysis' starting point. A "wrong" mixture of service deliveries or resources used might also contribute to inefficiency.

The second issue to consider is the *behavioral objective* assigned to service providers. One pair of possibilities is that providers are expected to maximize either service provision or service revenue. Another pair of possibilities is that providers are

expected to minimize either resource consumption or resource expenditure . In the former pair of cases, it is appropriate to adopt a service-oriented approach to the evaluation of provider performance, while in the latter pair of cases it is appropriate to adopt a resource-oriented approach to the evaluation of provider performance.

The third issue to consider is whether or not service providers must attempt to satisfy an *operating budget constraint* or must attempt to meet a *service revenue target*. If they are not constrained this way, then a direct approach is appropriate. While if they are financially so constrained, then an indirect approach is appropriate.

Collecting these three sets of alternatives, a total of eight approaches to service delivery performance evaluation are potentially available. The choice among them depends on data availability, the behavioral objective, and the relevance of targets. It is important to select an analytical framework that is consistent with the behavioral objective assigned to service providers, although inadequacies of the available data frequently constrain the selection to a "second best" framework. Fortunately, the cost of settling for a second best framework is likely to be a less rigorous performance evaluation than is desired and not necessarily an inaccurate performance evaluation.

In some circumstances, the behavioral objective is unclear. In this case, it is useful to estimate efficiency using various frameworks. A comparison of results and some empirical testing normally give some clues as to which framework fits the data best. In Chapter 4, Blank, Eggink and Merkies illustrated this approach for Dutch hospitals.[3]

14.4 Methodology: Efficiency Measurement, Robustness and Reliability

14.4.1 *Methods of Calculating Efficiency*

In the past two decades, a variety of analytical techniques have been developed to measure efficiency. In Section 2.4 Lovell distinguishes two main streams for measuring efficiency. Data Envelopment Analysis (DEA) and Stochastic Frontier Analysis (SFA) are techniques to construct a best practice service delivery standard against which the performance of providers is evaluated. DEA is a deterministic programming technique that has been applied frequently. DEA can be simply explained to managers and user-friendly software is available.[4] SFA is an econometric technique. It requires more statistical knowledge than DEA.

The choice of empirical technique depends on the perceived accuracy of the data. If the data are reasonably accurate then both techniques can be applied. However, if the data are rather noisy (measurement errors etc.), then SFA should be preferred. SFA accounts for the noise of the data, while computing efficiency scores, whereas DEA attaches any noise to measured efficiency. So DEA easily confuses the effects of inaccurate data with variations in managerial performance.

Although many researchers express their doubts on the accuracy of the underlying data, they do apply DEA to compute efficiency. This may lead to misleading information on the performance of management of service providers. Since data are always noisy, one might be tempted to conclude that SFA should always be preferred to DEA. However, SFA also has its drawbacks. SFA is a more cumbersome technique, software is less accessible, and the results can be ambiguous. Gyimah-Brempong, for example, shows in Chapter 9 that efficiency scores are sensitive to the underlying assumptions on the distribution of the noise. Furthermore, SFA is less adept at accommodating multiple resources and services simultaneously. We should bear in mind the fact that we are not measuring temperature or wind speed, but phenomena resulting from complex individual interactions with a wide variety of technical contexts.

14.2.2 *Techincal and Allocativer Inefficiencies*

Generally speaking, two types of inefficiency can be distinguished: technical and allocative inefficiency. Technical efficiency refers to the extent by which services can be expanded without increasing the resources. Alternatively, it measures the amount by which the resources can be contracted without reducing the services. Allocative efficiency refers to the extent to which a service bundle can be expanded by selecting another mix of resources (at given costs) or by which a resource bundle can be contracted by varying the mix of service deliveries (at given revenue).

While the choice for a resource contraction or service expanding oriented efficiency measure depends on the behavioral objective of service providers (see Section 14.3), the calculation of allocative efficiency depends crucially on the availability of price data. When adequate price data are non-existent, researchers have to limit themselves to the measurement of technical efficiency solely. Though the resulting efficiency analysis yields poorer information, it is still valuable. For a complete overview, the reader is referred to Table 2.1 and the corresponding discussion by Lovell.

14.4.3 *Economics of Scale, Economics of Diversification and Technical Change*

Besides the efficiency of individual providers, the economic characteristics of the best practice technology may be of great interest as well. Even in the hypothetical situation that all providers operate efficiently, the sector as a whole may be able to increase efficiency in at least three ways.

The first way is by exploiting (dis)economies of scale. (Dis)economies of scale means that a service provider is capable of expanding services by proportionally more, or less, as it expands its resources. By merging and splitting up providers the aggregate service provision can be attained by fewer resources, or service provision can be enhanced with the same amount of resources.

The second way of improving overall efficiency is by exploiting (dis)economies of diversification. Economies of diversification are present if the cost of providing a range of services collectively is less (or more) than the collective cost of providing each service or a subset of services individually. Specialization in services or widening the scope of services may enhance total service provision.

Technical progress is the way service delivery technology improves through time. Technical progress is a shifting of the best practice through time. Its diffusion through the sector depends on decisions made by the management. The cost of investments in new technology are compared to the expected benefits. The central government is, to a certain extent, capable of influencing the speed and diffusion of new technology by subsidizing implementations of innovations or by dictating new standards. Market conditions and economic constraints may also be a (dis)incentive for the introduction of new technology.

14.4.4 *Robustness and Sensitivity*

Section 14.2 discussed the alternatives for measuring service delivery and resource use. Especially, service delivery measurement may be an issue of serious debate. In Section 14.4.1 we pointed out that the choice of computational technique can influence efficiency scores. The chosen framework regarding economic objectives and constraints (Section 14.3) also has its impact on computed efficiency results. So, there are at least three choices for researchers. Each choice can lead to a different set of efficiency scores. Therefore, we highly recommend using various definitions, techniques and frameworks, and investigating the robustness of the

resulting outcomes. Chapter 3 by Webster, Kennedy and Johnson, Chapter 4 by Blank, Eggink and Merkies and Chapter 11 by Hughes and Yaisawarng conclude some good illustrations of this sensitivity analysis.

Sensitivity analysis does not merely indicate the reliability of the efficiency scores. It may also be used to measure the costs of regulatory constraints on service delivery and other government policies. In Chapter 7, Grosskopf, Hayes, Taylor and Weber show how the inefficiency due to resource use regulations in the Texas school system can be calculated by comparing the outcomes from two different frameworks.

14.5 Performance Improvement: Policy Recommendations and Implementation

In this section we discuss some policy issues related to efficiency improvement, though the list of topics covered may be incomplete. It will, nevertheless, give an impression of the possibilities for efficiency improvement by diverse policy measures.

In discussing the possibilities of performance improvement we distinguish two alternatives: external and internal performance improvements. Internal performance improvements are based on changes in variables over which management is assumed to have control. Examples of such management decision variables include the hierarchical structure, the size of departments, the composition of personnel, and the extent of contracting-out. Benchmarking is an important tool in identifying these management variables. External performance improvements are based on changes in variables that are outside the control of management, but under the control of government. Capacity planning, financing systems, property rights, market conditions, and other regulations are examples of external performance variables.

The distinction between internal and external improvement can be blurred in some contexts, but it is still a convenient categorization. The financing system, for example, might lack incentives for efficient behavior. Benchmarking is a method that can trace down these kind of inefficiencies.

14.5.1 *Benchmarking*

Government could initialize a system of benchmarking, which can be defined as a system of identifying and implementing best practice. According to Helgason (1997) "The basic idea behind benchmarking is simple:

1. Find an organisation that is best at what your own organisation does;
2. Study how it achieves such results;
3. Make plans for improving your own performance;
4. Implementing the plans; and
5. Monitor and evaluate the results."

Benchmarking is primarily a management information tool. Besides its management decision support function, benchmarking may sometimes also reveal the effects of government interference. We discuss this issue in the following paragraphs.

Fried and Klein show in Chapter 13 how a process of benchmarking can be conducted. They produce a long list of potential benchmark partners by consulting with local government officials. The final set of benchmark partners can be used to identify transferable practices to improve the performance at the local agency level. In Chapter 11 Hughes and Yaisawarng also illustrate the method of identifying peers for New South Wales police districts.

Striking examples of successful benchmarking are the studies of Bessent, Bessent, Elam, and Long (1984), and Norlander and Roos (1998). Bessent, Bessent, Elam, and Long (1984) report on their three years experience in the Educational Productivity Council with resulting improvements in effectiveness and efficiency for 25 Texan school districts. Norlander and Roos describe their experiences with the National Corporation of Swedish pharmacies (NCS) from 1989 until 1996. Over these years the NCS was given information on pharmacies that performed better than others and also on their characteristics.

Helgason (1997) distinguishes two benchmarking strategies: the top-down and the bottom-up approach. The top-down approach is imposed externally, for example by the Ministry of Finance. In the bottom-up approach individual providers develop their own benchmarking projects. The top-down approach has some advantages compared to the bottom-up approach. It secures participation from many organisations, it ensures a better overview of the effects, ensure standarization of methods and links the results to the budget process and other crucial decision making processes. Potential drawbacks of the external approach are resistance from service

providers, lack of detailed knowledge about the service provision, and non-usage of the results by the service providers. Helgason (1997) recommends using both methods. Norlander and Roos (1998) also stress the importance of a strong interaction between the Central Agency and individual service providers. They underline that the connection between theoretical models and operational instruments, and the ability to perceive and identify the need for change, are essential for successful implementation. They also stress that good dissemination of information is as important as the implementation itself.

14.5.2 *Capacity Planning*

The long term effects of capital investments in terms of cost (risk) and the need to meet demand for public services (equity, accessibility) often force governments to play an important role in capacity planning. This is the case for many public services and in many countries. There are various methods for controlling and regulating capacities. It is beyond the scope of this book to discuss all of these methods extensively. Explicit legislation on the maximum or minimum number of beds (operating theatres, cells etc.) or earmarked capital cost budgets are examples, but much more variants can be distinguished.

The effect of capacity planning in terms of cost refers to the sector efficiency. The existence of (dis)economies of scale and diversification motivates the various regulations on capacity planning. The effects of scale have been an important research issue by academics and policy. Less attention has been paid to the cost effects of diversification and the allocation of types of capitals. Dervaux, Kerstens and Leleu show in Chapter 5 how the sector efficiency can be improved by reallocating surgery rooms amongst hospitals. In Chapter 8 Merkies illustrates how efficiency results were successfully implemented in a Dutch policy towards consolidation of primary schools.

14.5.3 *Financing Systems*

Governments use various financing systems and procedures depending on their objectives regarding efficiency, equity, and access.[5] Financing systems can be characterized according to the extent providers control the available resources. Some systems dictate the exact amounts of resource usage, other systems provide the service suppliers with a great deal of freedom. A full expenses reimbursement

system, that covers all costs made by the service provider, is at one extreme. Some financing systems are expressed in nominal terms, other systems are expressed in physical quantities.

There is a wide range of hybrid systems. Dutch primary education is a good example of such a mixture of systems. Schools use buildings and classrooms provided by the local government. The number of teachers is determined by legislation depending on the number of pupils and their social backgrounds. Material supplies are financed by a fixed *ex-ante* money-budget.

Each financing system has its own merits, own incentives and disincentives. There is an extensive theoretical literature available on these matters. The reader is referred to the literature on principal-agent theory, public choice theory and game theory. For a survey see, for example, Ricketts (1994) and Foss (1995).The effects on efficiency of various financing systems can be established empirically by comparing the efficiency outcomes of different systems. Apparently, we need two sets of observations: one for each system. Since two systems rarely exist simultaneously, this implies we need to conduct inter-regional or inter-temporal comparisons. However, we should be aware that other differences between regions or across time can bias the outcomes. However, in Chapter 7 Grosskopf, Hayes, Taylor and Weber illustrate an alternative way of evaluating financial reforms. By relaxing some of the model constraints, they are able to evaluate a financing system without access data on the alternative system.

Most budgeting systems are cost-based, which means that expected service delivery is being converted to a money-budget. The central question is at which tariffs or budget parameters this conversion should be made. Efficiency analysis may be a useful instrument to determine the appropriate trade-offs. From the cost function we can easily derive marginal costs for each service delivery. By using the marginal cost derived from the best-practice cost function we are able to settle for a cost covering trade-off charge for an efficient provider. Since it covers the efficient cost, this system has a strong build-in incentive to promote efficient behavior. An excellent example of implementation of this system can be found in Sexton, Sleeper, and Taggart (1994). They advised the North Carolina's State Department of Public Instruction to review and to revise reimbursements for cost incurred in transporting students for educational purposes. Between 1991-1993, the new reimbursement system lead to cost savings of 15% annually. One striking feature of this study is the efforts the authors have put in correcting the inefficiencies for environmental char-

acteristics. They cleaned up the calculated inefficiencies such that merely managerial capabilities are accounted for.

14.5.4 *Property Rights: Privatizing*

Privatizing has been an important subject in government policy debates in the last two decades. Efficiency arguments are frequently used to support privatizations of public services. More in general, it is assumed that property rights or ownership affect public provision's efficiency. The central economic issue is whether public corporations or "private" foundations or private enterprises tend to use resources more efficiently. In Chapter 9, Gyimah-Brempong concludes that private prisons in Florida do not perform better than state prisons. In Chapter 4, Blank, Eggink and Merkies could not find any hard evidence that privately founded hospitals operate more or less efficiently than local or central government hospitals.

The argument that privatization promotes efficiency is based upon the built-in incentives and discipline provided by competition. The previous examples suggest that this is not always the case. Even in a privatized situation, economic behavior is being influenced by regulations concerning financing, planning and quality. These constraints could affect efficiency more than the ownership. Together with the property rights, market conditions are also subject of a debate. The effect of market conditions is often more substantial (see Tulkens and Pestieau 1993) and is discussed in the next section. It is therefore necessary to conduct a performance evaluation with every policy measure concerning privatizing or market liberatizations. The studies of the UK Privatization Program by Boussofiane and Parker (1997), of BAA by Parker (1999), and of electric utilities by Pollitt (1995) are good examples of ex-post evaluations. However, we must ascertain that not many ex-post evaluations have been conducted or have been used in the policy decision making process so far.

In analyzing the effects of ownership researchers should guard against artificial influences such as differences in quality of service provision or users' selection that may correlate with the ownership of the providers. For example, private schools might select pupils, whereas public schools have to give access to all pupils. Private clinics might conduct only standard surgery, whereas public hospitals treat with patients with serious conditions. Byrnes (1991), for example, discusses the effects of the presence of selectivity bias in the case of ownership and efficiency of water utilities.

14.5.5 *Market Conditions*

Competition and the conditions for market-entry also affect provider's behavior. Competition and market-entry, as an indicator for future competition, influence efficiency. From standard economic theory, we know that moreased competition leads to more efficient behavior. In the struggle to survive, providers lower service prices and consequently decrease cost by improving efficiency. In Chapter 4, however, Blank, Eggink and Merkies show that in Dutch hospital industry, increased competition leads to more inefficiency. As service prices are fixed by the central government, hospitals do not compete on the basis of service prices, but on the basis of quality of services provided. These quality differences are measured partly as inefficiencies.

Governments have a number of instruments to affect the degree of market competition. In Section 14.5.2 we already discussed the role of government in planning capacities. Government may rule capacities either by fixing exact capacity quantities or in a more liberal way by defining upper and lower capacity boundaries. Anti-trust legislation and an anti-trust authority can be used to ensure a competitive environment.

The effect of market conditions, as illustrated by the hospital example, is not clear *a priori*. The combination of market conditions, property-rights and other regulations finally determine the overall effect on efficiency.

14.5.6 *Other Regulations*

Aside from the above mentioned regulations concerning capacity planning, financing, and market conditions, there is a whole range of other government regulations affecting a provider's productive performance. Examples include regulations on wages, service prices, the composition of service delivery, the composition of resources, the skill level of labor and the quality of service provision. Since each rule limits the space in which providers endeavour to reach their goals, it can be expected affecting provider's performance in a negative sense. However, these policies might promote equity, quality, and access. Moreover, regulations can improve the sector's efficiency at the expense of individual provider's efficiency. For example, the prohibition of an efficiency improving merger between two providers by anti-trust legislation may be considered as contra-productive. However, anti-trust legislation might prevent arising monopolies which could increase inefficiency on the long term.

To my knowledge, little research has been done in this area of evaluating government interventions. Analyzing efficiency differences between two situations may not cause any serious problems. The efficiency analysis, however, should be integrated with an analysis of effectiveness of some rule. In Chapter 8, Merkies discusses a model (besides the cost model) which describes the demand for primary education. From this model he is able to account for the consequences of school consolidation in terms of religious preference and access. Therefore, he was able to make a sound evaluation of school consolidation in the Netherlands.

14.6 CONCLUDING REMARKS

Public services are a vital part of the economy as a whole. Government is directly or indirectly responsible for infrastructure services, such as water and electricity supply, public transportation and human services such as health care and education. Law enforcement, for instance, is of great importance not only for reasons of justice, but also for economic development. In fact, government is in the business of producing throughputs, such as gas, water, electricity, roads, and healthy and skilled labor, for other sectors of the economy. Higher taxes are the result of low productivity in public services and these taxes can further reduce allocative efficiency in other sectors.

As stated extensively in Chapter 1, the lack of the market mechanism and the imperfect information on service delivery efficiency leads to a tendency of overspending. By improving efficiency the overspending can be used to deliver more services or returned to the tax payers. In Australia, for example, the Industry Commission estimates that reforms associated with the National Policy Competition will increase GDP by 5,5% in the long run. Although it is hazardous to draw conclusions from the limited information presented in this book, it seems that by exploiting all efficiency improvements GDP growth could be increased by at least 5% in most countries. Most studies report inefficiencies in the order of magnitude of 20% to 30%. Part of these inefficiencies may always remain as a result of social constraints, such as access of services (the late night bus for example) and must be accepted as a fact of life. Nevertheless, major savings can be attained by changing the planning of capacities, by altering financing systems, by changing property rights, by other deregulations and improving information for the management of service providers.

In spite of the fact that policy makers want us to believe otherwise, the effects of policy measures on efficiency are not always clear. Privatizing and increasing market competition are favourite subjects in circles of policymakers. This book contains two examples which find no evidence at all to support the view that privatization promotes efficiency. Policy measures should therefore be evaluated for each public provision and each economic context separately. Consequently, ideological arguments can be dropped from debates on policy measures and be replaced by more rational motives.

Improvement of management information to service providers can contribute to increased efficiency. This improved management information can result from comparing service providers and explaining efficiency differences. This process of benchmarking faces some hurdles, such as the hampering of good data and poor cooperation by service providers. Government should strive to overcome these problems. For the acquisition of good data, government could require information delivery as a condition for public financing. Government can take away some of the reluctance by guaranteeing that the information for efficiency analysis will not be used for general budget cuts, but merely for expanding services or improving quality.

It is important to conduct efficiency studies correctly. Efficiency measurement is characterized by complexity of theories and advanced statistical methods and it can be misused by a novice. However, it is of great importance in recognizing the value of concepts such as mulitple-resources multiple-services, best-practice, service or resource orientation and economic constraints. Less sophisticated and perhaps more transparent approaches such as partial productivity measures, are not an acceptable alternative.

Furthermore, we recommend that research on efficiency and productivity be centralized or at least coordinated. In the Netherlands, research on efficiency is conducted by government agencies, a health research institute, and various private consulting firms. As a result, methods are not uniform and resources are wasted: there is inefficiency! This probably applies to many other countries as well. It seems that Australia with its Steering Committee is an enlightened exception and serves as a best-practice benchmark.

NOTES

[1] The author thanks Kris Kerstens and Hal Fried for their helpful comments on an earlier draft of this chapter. Special thanks also to Ian, Joy, and Bas, who were a great inspiration to me.

[2] In the Netherlands it is common practice to depreciate capital value based on historical cost pricing.

[3] Blank, Eggink, and Merkies (1998) describe the process of choice of the "right" framework in detail.

[4] A nice overview of DEA can be found in Seiford (1999).

[5] Access should be interpreted in a broad sense. It refers to opening hours, waiting time, travel time and so on.

REFERENCES

Bessent, A., W. Bessent, J. Elam, and D. Long. 1984. Educational Productivity Council Employs Management Science Methods to Improve Educational Quality. *Interfaces* 14 (6): 1-8.

Blank, J.L.T., E. Eggink, and A.H.Q.M. Merkies. 1998. *The Choice of an Appropriate Economic Model for the Structure of Production: The Case of Dutch General Hospitals*. Den Haag: Sociaal en Cultureel Planbureau (Werkdocument 49).

Boussofiane, A., S. Martin, and D. Parker. 1997. The Impact on Technical Efficiency of the UK Privatization Programme. *Applied Economics* 29 (3): 297-310.

Byrnes, P. 1991. Estimation of Cost Frontiers in the Presence of Selectivity Bias: Ownership and Efficiency of Water Utilities. In: Rhodes, G.F.(ed.), *Advances in Econometrics, volume 9*. 121-137. London: JAI Press.

Foss, P. 1995. *Economic Approaches to Organisations and Institutions*. Brookfiels U.K. Aldershot.

Helgason, S. 1997. *International Benchmarking Experiences from OECD Countries (Paper Presented at the Conference International Benchmarking Copenhagen 1997)*. Paris: OECD (Puma).

Jorgenson, D.W. 1996. *Investment*. Cambridge: The MIT Press.

Norlander, N., and P. Roos. 1998. Implementing the Malmquist Productivity Index: the Case of the National Corporation of Swedish Fharmacies. In: Färe, R., S. Grosskopf, and R.R. Russell (eds.), *Malmquist Index Numbers: Essays in Honour of Sten*. Boston: Kluwer, 191-215.

Parker, D. 1999. The Performance of BAA Before and After Privatisation: A DEA Study. *Journal of Transport Economics and Policy* 33 (2): 133-145.

Pestieau, P., and H. Tulkens. 1993. Assessing and Explaining the Performance of Public Enterprises. *Finanzarchiv* 50 (3): 293-323.

Pollitt, M.G.1995. *Ownership and Performance in Electric Utilities: The International Evidence on Privatisation and Efficiency*. Oxford: Oxford University Press.

Ricardo, D. *The Works and Correspondence of David Ricardo*. Edited by P. Sraffa 1951-1995. Cambridge: Cambridge University Press.

Ricketts, M. 1994. *The Economics of Bussines Enterprises: An Introduction to Economic Organisations and the Theory of the Firm, 2nd edition*. New York: Simon and Schuster International.

Seiford, L.M. 1999. Data Envelopment Analysis: Twenty Years Out. In: Westerman, G. (ed.), *Data Envelopment Analysis in the Service Sector*. Wiesbaden: Gabler.

Sexton, T., S. Sleeper, and R. Taggart. 1994. Improving Pupil Transportation in North Carolina. *Interfaces* 24 (1): 87-103.

SCRCSSP Steering Committee for the Review of Commonwealth/State Service Provision. 1997. *Data Envelopment Analysis: A Technique for Measuring the Efficiency of Government Service Delivery*. Canberra: AGPS.

SCRCSSP Steering Committee for the Review of Commonwealth/State Service Provision. 1999. *Report on Government Services 1999*. Canberra: Ausinfo.

Smith, A. 1776. *Inquiry into the Nature and Causes of the Wealth of Nations*. Edited by E. Cannan, 1937. New York: Modern Library Inc.

Victoria, Department of Treasury and Finance. 1997. *Output Specification and Performance Measurement Guide*. Melbourne: Department of Treasury and Finance.

Name Index

A

B

C

D

E

F

G

H

N

S

U

V

W

Y

Z

Subject Index

F

H

I

L

M

N

O

P

R

S

T

U

V

W

X

Biographies

Dr. Jos L.T. Blank, the Netherlands

Jos Blank is Head of the program 'Production Public Sector' of the Social and Cultural Planning Office (SCP) in the Netherlands. He graduated in econometrics at the Free University of Amsterdam in1984. He obtained his Ph.D. with a thesis on the effects of school consolidation in the Netherlands. Since 1987 he has worked for the SCP. He coordinates studies on the productivity and efficiency of Dutch public sector services. He has published research on the efficiency of various public services, and has also participated in research on the future demand for public services and the effects of ecotax on the demand for public utilities.

Prof. Bruno De Borger, Belgium

Bruno De Borger is Professor of Economics at the University of Antwerp (UFSIA). He teaches micro-economics and public economics. His recent research focuses on efficiency measurement in the public sector, on the behaviour of public enterprises, and on the correction of externalities in the transport industries.

Dr. Benoit Dervaux, France

Benoît Dervaux is a health economist at the French National Center for Scientific Research (CNRS). He teaches econometrics at the IESEG graduate school of management (Catholic University of Lille). He obtained his Ph.D. in applied economics with a thesis on the economic evaluation of cancer screening in 1992 (Catholic University of Mons, Belgium). Since 1995 he has mainly worked on productivity measurement in the health care industry. His current interest focuses on industry models and industry efficiency measures.

Dr. Evelien Eggink, the Netherlands

Evelien Eggink is Senior Researcher at the program 'Production Public Sector' of the Social and Cultural Planning Office in the Netherlands. She graduated in mathematics at the Catholic University of Nijmegen in 1988. She obtained her Ph.D. with a thesis on labor market economics at the Erasmus University in 1994. Since 1994 she has worked for the Social and Cultural Planning Office. There she studies the productivity and efficiency of Dutch public sector services (nursing homes, hospitals and homes for mentally disabled).

PROF. HAROLD O. FRIED, U.S.A.

Harold Fried is a Professor of Economics at Union College in Schenectady, New York. He received his undergraduate degree in economics from the University of Michigan and his Ph.D. in economics from the University of North Carolina at Chapel Hill. His area of specialization is the measurement of efficiency and the evaluation of performance. He has worked on applications to financial institutions, rural counties, metropolitan areas, hospitals, nursing homes and more. He is particularly interested in collaborating with practitioners to develop models that can be used in the field to improve performance.

PROF. SHAWNA GROSSKOPF, USA

Shawna Grosskopf is Professor of Economics at Oregon State University in Oregon (USA). She received her Ph.D. in 1977 from Syracuse University. Her research interests include public sector economics and performance measurement

PROF. KWABENA GYIMAH-BREMPONG, USA

Kwabena Gyimah-Brempong is Professor of Economics, College of Business Administration, University of South Florida, Tampa, USA. He graduated in Economics (Hons) from the University of Cape Coast, Ghana in 1974 and obtained his Ph.D. with specializations in Urban and Development Economics from Wayne State University in 1981. His research involves estimating production functions for public goods, economics of crime, and Sub-Saharan African development issues. He has published in a wide variety of professional journals. He is currently the Secretary/Treasurer of the African Finance and Economics Association.

PROF. KATHY HAYES, USA

Kathy Hayes is Professor of Economics and Associate Dean of Academic Affairs at Dedman College, Southern Methodist University In Dalls, Texas (USA). She received her PhD in 1980 from Syracuse University. Her research area is state and local public economics.

MR. ANDREW HUGHES, AUSTRALIA

Andrew Hughes is a Senior Economist with the New South Wales Treasury in Australia. He holds a Bachelor degree in Economics (with Honours) from the University of Sydney, Australia and has undertaken graduate work in Accounting and Finance. His primary responsibility is developing methods to enhance and measure public sector performance. He co-developed an analytical framework for integrating productivity measurement and financial performance for commercial enterprises. He also provides advice on overall fiscal, revenue and expenditure strategy to the NSW Government.

Ms Leanne Johnson, Australia

Leanne Johnson is a Principal Analyst at the Bureau of Tourism Research in Australia. Leanne graduated in economics (honours) from the University of Melbourne in 1993. Leanne worked at the Australian Bureau of Statistics from 1994 to 1998. She has published papers on price index theory, tradeables, productivity measurement and the meetings sector. Leanne is currently researching the economic contribution of Australia's tourism sector and developing measures of regional tourism expenditure.

Mr. Steven Kennedy, Australia

Steven Kennedy is Director of the Analytical Services section at the Australian Bureau of Statistics. Steven qualified as a registered general nurse in 1987. He graduated in economics (honours) from the University of Sydney in 1992 and obtained a Masters degree from the Australian National University in 1997. Steven has worked at the Australian Bureau of Statistics since 1993. Steven has published papers on labour market economics, productivity measurement and the health sector. Steven is currently researching the relationship between labour market and health outcomes.

Dr. Kristiaan Kerstens, Belgium

Kristiaan Kerstens is researcher at the Department of Economics and Management of the Catholic University of Lille (LABORES). He graduated in economics and philosophy at the Catholic University of Leuven in 1986. He presented his Ph.D. in economics on some methodological problems in efficiency measurement in 1994 (Catholic University of Brussels, Belgium). In 1997, he joined LABORES to continue his study on mainly public sector performance (urban transit, municipalities, libraries) and methodological issues (efficiency scores, returns to scale measurement).

Prof. J. Douglas Klein, U.S.A.

J. Douglass Klein is Professor of Economics at Union College in Schenectady, New York. He earned his Ph.D. in Industrial Economics in 1975 at the University of Wisconsin, having completed an undergraduate degree in mathematics at Grinnell College. Since joining the Union College faculty in 1979, his research has been in the fields of energy economics and efficiency measurement. He has applied efficiency measurement techniques to the electric power industry in the United States. In addition to examining the efficiency of U.S. metropolitan areas, his is also currently pursuing a study for the U.S. Department of Agriculture on the economic vitality of rural counties in the U.S. mid-west.

DR. HERVÉ LELEU, FRANCE

Hervé Leleu is researcher at the Department of Economics and Management of the Catholic University of Lille. He graduated in statistics and econometrics at the University of Lille III. He defended his Ph.D. in economics at the University of Toulouse in 1997 on the technical efficiency and productivity assessment of French hospitals. At present he pursues his research on this subject and, in particular, works on industry efficiency.

PROF. C.A. KNOX LOVELL, U.S.A.

Knox Lovell holds the Terry Chair in the Department of Economics at the University of Georgia and is Professor in the School of Economics at the University of New South Wales, where he also serves as Director of the Centre for Applied Economic Research. He is Editor-in-Chief of *The Journal of Productivity Analysis*, and his research interests center on the analysis of efficiency and productivity in both the public and private sectors.

PROF. ARNOLD H.Q.M. MERKIES

Arnold Merkies is Emeritus Professor of Mathematical Economics and Econometrics at the Vrije Universiteit, Amsterdam. He studied economics at the University of Amsterdam where he also received his Ph.D. in economics. He worked for Bull-General Electric and in the Central Planning Bureau in the Hague. He has been Visiting Professor in Canada, Thailand and New Zealand. He did research on various subjects, such as model selection, agregation in economics, consumer behavior and efficiency analysis.

DR. LORI L. TAYLOR, USA

Lori Taylor, Ph.D. is a Senior Economist and Policy Advisor at the Federal Reserve Bank of Dallas in Dallas, Texas (USA). She received her Master's and doctoral degrees in economics from the University of Rochester (1987 and 1991, respectively). Her research interests include public finance, regional economics and the economics of education. Ongoing research examines the efficiency of state and local governments and the regional consequences of government policy.

Dr. Frank P. van Tulder, the Netherlands.

Frank van Tulder is a coordinator at the Research Center of the Dutch Ministry of Justice in The Hague. He studied econometrics at the University of Amsterdam. From 1982 till 1998 he did research in the field of public sector economics at the Social and Cultural Planning Bureau. The main focus of his research was on determinants of public services demand and productivity in the public sector. He is particularly interested in law enforcement: the subject of his thesis (1994) was the cost effectiveness of the Dutch law enforcement system.

Prof. William L. Weber, USA

William Weber is currently a professor of economics at Southeast Missouri State University in Cape Girardeau, Mo. USA, where he teaches courses in public policy economics, environmental economics, and financial institutions. He received his Ph.D. from Southern Illinois University at Carbondale in 1986. In addition to his interest in school efficiency and equity issues, he has also examined the consequences of risk-based capital standards on bank efficiency.

Mr. Richard Webster, Australia

Richard Webster is an Assistant Director in the Labour Cost Index and Analysis section of the Australian Bureau of Statistics. Richard graduated in mathematics (honours) from the University of Adelaide in 1991, and in economics from the Flinders University of South Australia in 1995. Richard has worked at the Australian Bureau of Statistics since 1996. Richard has published papers on the impact of data quality on macro-economic models and policy makers and productivity measurement. His current intrest is the measurement of health output and estimating the volume of hidden transactions in the Australian economy.

Prof. Suthathip Yaisawarng, Thailand

Suthathip Yaisawarng is an Associate Professor of Economics at Union College, U.S.A. She holds a Ph.D. in Economics from Southern Illinois University (Carbondale). She specializes in theoretical and empirical applications of efficiency and productivity techniques. She was engaged by the New South Wales (NSW) Treasury in Australia to develop and apply DEA to a number of NSW government agencies. Currently, she is working on issues associated with the impact of a unit's operating environment on its performance and enhancing the credibility of DEA results for use in benchmarking and resource allocation.

Correspondence

Dr. J.L.T. Blank (Chapters 1, 4 and 14)
Social and Cultural Planning Office of the Netherlands
P.O. Box 16164
2500 BD Den Haag
The Netherlands
Phone: 31 (0)70 340 79 35
Fax: 31 (0)70 340 70 44
E-mail: jo@scp.nl

Prof. B. de Borger (Chapter 12)
Universiteit van Antwerpen
UFSIA (SESO)
Prinsstraat 13
2000 Antwerpen
Belgium
Phone: 32 03 220 4057
E-mail: dse.deborger.b@alpha.ufsia.ac.be

Dr. E. Eggink (Chapter 6)
Social and Cultural Planning Office of the Netherlands
P.O. Box 16164
2500 BD Den Haag
The Netherlands
Phone: 31 (0)70 340 79 33
Fax: 31 (0)70 340 70 44
E-mail: ee@scp.nl

Prof. H.O. Fried (Chapter 13)
Union College
Economics
Schenectady NY 12308
United States
Phone: 1 518 388 6368
Fax: 1 518 388 6988
E-mail: friedh@union.edu

Prof. S. Grosskopf (Chapter 7)
Oregon State University
Department of Economics
303 Ballard Exention Hall
Corvallis OR 97331-3612
United States
E-mail: shawna.grosskopf@orst.edu

Prof. K. Gyimah-Brempong (Chapter 9)
Department of Economics
University of South Florida
Tampa, FL 33620-5500
United States
Phone: 1 813 974 6520
Fax: 1 813 974 6510
E-mail: kgyimah@coba.usf.edu

Dr. S. Kennedy (Chapter 3)
Australian Bureau of Statistics
Analytical Services, Methodology Division
P.O. Box 10
Belconnen, ACT 2616
Australia
Phone: 61 06 252 6617
Fax: 61 06 253 1033
E-mail: steven.kennedy@abs.gov.au

Prof. K. Kerstens (Chapter 5)
Université Catholique de Lille
Labores (URA 362)
60 Boulevard Vauban, B.P. 109
F-59016 Lille Cedex
France
Phone: 33 (0)3 20 13 40 80
Fax: 33 (0) 3 20 13 40 70
E-mail: kristiaan.kerstens@flse.fupl.asso.fr

Prof. C.A.K. Lovell (Chapter 2)
School of Economics
University of New South Wales
Sydney, NSW 2052
Australia
Fax: 61 29313 6337
E-mail: k.lovell@unsw.edu.au

Prof. A.H.Q.M. Merkies (Chapter 8)
Zomerlaan 30
2103 AJ Heemstede
The Netherlands
Phone: 31 (0)23 5282578
E-mail: ahqm@worldonline.nl

Dr. F. van Tulder (Chapter 10)
Ministerie van Justitie
WODC (H1403)
P.O. Box 20301
2500 EH Den Haag
The Netherlands
Phone: 31 (0)70 3706631
Fax: 31 (0)70 3707948
E-mail: ftulder@best-dep.minjus.nl

Prof. Suthathip Yaisawarng (Chapter 11)
Department of Economics
Union College, Schenectady NY 12308
United States
Phone: 1 518 3886606
Fax: 1 518 3886988
E-mail: yaisawas@union.edu

www.ingramcontent.com/pod-product-compliance
Lightning Source LLC
LaVergne TN
LVHW021121300926
849160LV00010B/810
* 9 7 8 1 0 3 2 4 3 2 0 2 1 *